Metal Oxide Nanocatalysts for Sustainable Energy Production

Edited by

Abayomi Bamisaye

Sesan Abiodun Aransiola

Nelson Oshogwue Etafo

Mopelola Abidemi Idowu

Naga Raju Maddela

WILEY VCH

Dr. Abayomi Bamisaye
University of Pretoria,
Department of Chemistry,
Pretoria, South Africa

Lead City University,
Department of Chemical Sciences,
Ibadan, Nigeria

Dr. Sesan Abiodun Aransiola
University of South Africa,
Department of Auditing,
Pretoria, South Africa

University of Abuja,
Department of Microbiology,
Abuja, Nigeria

Dr. Nelson Oshogwue Etafo
Universidad Politécnica de Ramos Arizpe,
De Las Americas, 25903,
Ramos Arizpe, Coahuila, Mexico

Universidad Autonoma de Coahuila, Saltillo,
Coahuila, Mexico

Prof. Mopelola Abidemi Idowu
Federal University of Agriculture,
Department of Pure and Applied Chemistry,
Abeokuta, Nigeria

Prof. Naga Raju Maddela
Universidad Técnica de Manabí,
Departamento de Ciencias Biológicas,
Portoviejo, Ecuador

Cover Design: Wiley
Cover Images: ©xrender/Shutterstock, ©vector/stock.adobe.com, ©3d_illustrator/Shutterstock, ©enanuchit/stock.adobe.com, ©Vangly/stock.adobe.com, ©Risat Designer/stock.adobe.com, ©wasanajai/Shutterstock

Typesetting: Lumina Datamatics, Madurai, India

Printing and Binding: CPI Group (UK) Ltd, Croydon, CR0 4YY

Library of Congress Card No.: applied for

British Library Cataloguing-in-Publication Data
A catalogue record for this book is available from the British Library.

Bibliographic information published by the Deutsche Nationalbibliothek
The Deutsche Nationalbibliothek lists this publication in the Deutsche Nationalbibliografie; detailed bibliographic data are available on the Internet at <http://dnb.d-nb.de>.

Print ISBN: 9783527355662
ePDF ISBN: 9783527853885
ePub ISBN: 9783527853878
oBook ISBN: 9783527853892

C9783527355662_270226

Contents

Preface

The world stands at a critical juncture where the intertwined forces of climate change, rapid urbanization, population growth, and escalating energy demands are placing unprecedented pressure on global ecosystems and infrastructure. Record-breaking heatwaves, extreme weather events, and rising atmospheric carbon concentrations are no longer distant projections; they are the lived realities of our time. In this era of acute environmental volatility, the search for sustainable and resilient energy solutions has become not just a scientific pursuit but a moral and strategic necessity.

This book, *Metal Oxide Nanocatalysts for Sustainable Energy Production* is a scholarly response to this defining global challenge. It focuses on one of the most promising classes of materials, metal oxide nanocatalysts, and their evolving role in enabling cleaner, more efficient energy technologies. These nanoscale catalysts offer extraordinary catalytic properties, including high surface area, thermal stability, and tunable redox behavior, making them pivotal in applications such as hydrogen generation, carbon dioxide conversion, fuel cells, and bioenergy systems.

As cities expand and energy infrastructures strain under the weight of growing demand, the need for scalable, low-carbon energy technologies has never been more urgent. This book brings together cutting-edge research from across disciplines to illuminate how metal oxide nanocatalysts can help reimagine our energy future, one that aligns with climate goals, supports sustainable urban development, and builds resilience against global warming.

The entire book shows how metal oxide nanocatalysts and their composites can be used to ameliorate various environmental and energy crises. The chapters were contributed by 74 academicians, scientists, and researchers from 16 different countries, viz. Nigeria, the United Kingdom, India, Kazakhstan, Canada, the United Arab Emirates, the United States of America, Mexico, South Africa, China, the Kingdom of Saudi Arabia, Poland, France, Hungary, Ecuador, and Sweden.

We hope this piece informs, inspires, and catalyzes action in science, policy, and innovation toward a more sustainable and equitable planet.

Abayomi Bamisaye
Sesan Abiodun Aransiola
Nelson Oshogwue Etafo
Mopelola Abidemi Idowu
Naga Raju Maddela

Foreword

The 21st century is witnessing an unprecedented confluence of global challenges—climate change, resource depletion, urban expansion, and the surging demand for clean, affordable, and reliable energy sources. As the impacts of anthropogenic climate disruption intensify, with escalating heatwaves, shifting precipitation patterns, and rising sea levels, the urgency to transition from fossil fuel dependence to sustainable energy systems has never been more pressing. Within this context, the role of catalysis has gained renewed significance. This volume, *Metal Oxide Nanocatalysts for Sustainable Energy Production* offers a timely and insightful contribution to this evolving field. It explores the science, engineering, and practical applications of metal oxide nanocatalysts as enablers of transformative energy technologies capable of ameliorating both environmental and societal challenges.

Metal oxide nanocatalysts possess extraordinary physicochemical properties—high surface-to-volume ratios, redox flexibility, tunable band structures, and thermal stability—that make them ideal for a wide range of sustainable energy applications. These include hydrogen evolution via water splitting, photocatalytic CO_2 reduction, biomass valorization, pollutant degradation, and energy storage systems. By accelerating reaction rates and improving energy efficiency under milder conditions, these materials are essential tools in building decarbonized and climate-resilient energy infrastructures.

This book assembles the collective expertise of leading researchers and practitioners, offering both foundational insights and advanced perspectives. The chapters are grounded in rigorous scientific analysis while also engaging with the broader vision of sustainable development and environmental stewardship. It is particularly commendable for bridging laboratory-scale innovations with real-world challenges, demonstrating how nanocatalysis can translate into scalable solutions for low-carbon energy transitions. I am heartened by the interdisciplinary approach this book adopts—merging material science, nanotechnology, catalysis, and environmental engineering. Such synthesis is critical if we are to move from incremental change to systemic transformation.

I firmly believe that this piece will serve as a valuable reference for scientists, engineers, policymakers, and students committed to building a more sustainable future. It not only deepens our understanding of the science but also underscores the role of innovation in meeting the climate challenge. This is more than a technical compendium—it is a call to action.

Prof. Dr. Enock Olugbenga Oladepo Dare
First Professor of Nanotechnology in Nigeria,
Federal University of Agriculture, Abeokuta
Nigeria

Acknowledgment

We extend our sincere gratitude to all chapter contributors for their insightful input and unwavering commitment throughout the development of this volume. Their timely responses and enthusiasm greatly facilitated the completion of this work within the projected timeline. We deeply value their spirit of collaboration and scholarly dedication. Our appreciation also goes to the anonymous reviewers whose critical and constructive feedback significantly enhanced the academic quality and coherence of the book. Their thoughtful comments were instrumental in refining the content.

We are grateful to the Wiley Editorial and Production team for their professional guidance and steadfast support. Their expertise and responsiveness were vital in bringing this project to fruition within such a constrained timeframe. It has been a genuine privilege to work with a team of such reputation and dedication.

Finally, we sincerely appreciate the management and our colleagues at University of Pretoria (South Africa), Lead City University (Nigeria), University of South Africa (South Africa), University of Abuja (Nigeria), Equipo Docente, Universidad Carolina (Mexico), Federal University of Agriculture, Abeokuta (Nigeria), and Universidad Técnica de Manabí, Portoviejo (Ecuador) for their unhampered backing and for the establishment of treasured propositions at the time of the book proposal and final book preparation, whose encouragement and understanding created a conducive environment for the successful completion of this scholarly endeavor.

Abayomi Bamisaye
Sesan Abiodun Aransiola
Nelson Oshogwue Etafo
Mopelola Abidemi Idowu
Naga Raju Maddela

About the Editors

Abayomi Bamisaye received his Bachelor of Science (Hons.), Industrial Chemistry in 2012 at the Department of Chemistry, Ekiti State University, Ado-Ekiti, Nigeria. He also obtained his MSc (2016) and PhD (2021) in Industrial Chemistry with distinction at the Department of Chemistry, Federal University of Agriculture, Abeokuta, where he worked on the Development of Oil-in-Water (O/W) Nanoemulsion systems for drug delivery and Cosmetology purposes. He is a senior lecturer and researcher at the Department of Chemistry, Lead City University, Ibadan, Nigeria. Dr. Bamisaye is a dynamic researcher whose work reflects a rare combination of leadership, analytical rigor, and innovative thinking. His scholarly contributions span a wide range of interdisciplinary domains including nanomaterials synthesis, photocatalysis, biomass valorization, wastewater treatment, energy storage, and drug delivery systems. Currently serving as a Postdoctoral Research Fellow at the University of Pretoria (UP), South Africa, where his research is focused on the synthesis of nano-mesoporous biomass metal-organic frameworks composites for energy storage. His work in this emerging field directly supports the development of sustainable energy technologies through the valorization of biomass waste, addressing critical challenges in environmental and energy sustainability. He is also a recipient of the prestigious Coimbra Postdoctoral Fellowship, underscoring his academic excellence and global recognition. Dr. Bamisaye's dedication to impactful research and academic engagement is evident in his robust publication portfolio, which includes peer-reviewed journal articles, comprehensive review papers, book chapters, and a full-length academic book. His scholarly work cuts across nanotechnology, materials science, environmental engineering, and nanomedicine, contributing valuable insights to both fundamental science and applied research. He is a member of American Chemical Society (ACS) and Chemical Society of Nigeria (CSN). He is also serving as a reviewer of many peer-reviewed local and international journal outlets.

Sesan Abiodun Aransiola is a Lecturer at the Department of Microbiology, University of Abuja, Nigeria. He obtained his BTech, MTech, and PhD in Environmental Microbiology from the Federal University of Technology, Minna, Nigeria. He has demonstrated his research expertise in the production of vermicasts from vermicomposting of organic wastes to assist plants in the remediation of polluted soil with heavy metals. His expertise spans environmental health, phytoremediation, vermicomposting, marine resources, and bioremediation of pollutants in the soil. Aransiola's research is deeply rooted in innovative solutions for soil and water remediation, with significant contributions to microbial biotechnology and waste management. He is an award-winning researcher and has over 100 publications including book chapters, research and review articles of good international repute with a high impact factor. He has co-edited scientific books of global interest, which include *Microbial Biotechnology for Bioenergy, Marine Bioresources: Prospects and Obstacles, Marine Microbial Products: Applications and Opportunities* (Elsevier). *Ecological Interplays in Microbial Enzymology, Prospects for Soil Regeneration and Its Impact on Environmental Protection, Soil Microbiome in Green Technology Sustainability, Marine Bioprospecting for Sustainable Blue-Bioeconomy, Vermitechnology: Economic, Environmental and Agricultural Sustainability, White Pollution: Biodiversity and Hazards in Marine Plastisphere, Wetland Ecosystems: Conservation Strategies, Policy Management and Applications* (Springer). *Microbial Biofilms—Applications and Control, Marine Greens: Environmental, Agricultural, Industrial and Biomedical Applications, Environmental Footprint of Bioplastic Additives, Emerging Contaminants in Food and Food Products, Phytoremediation in Food Safety: Risks and Prospects, Biotoxins in Food: Threats and Benefits* (CRC Press) with many more in the process. He has been a review editor, and as guest editor, he hosted two special issues with Elsevier (Research in Microbiology) and Frontiers on biofilms—community, structure, applications, and mitigation—and effects of microplastics on soil ecosystems, respectively. Also, he rose to the rank of assistant chief scientific officer at National Biotechnology Research and Development Agency, Nigeria, for years, where he was involved in the uses of vinasse (a by-product of ethanol) as bio-fertilizer to reclaim polluted soil for agricultural purposes before joining the university. Dr. Aransiola's academic journey is complemented by robust research (with some grants won), teaching career (with proper mentoring and supervising both undergraduate and postgraduate students), and active involvement in community services, where he leverages his knowledge to foster sustainability. His contributions continue to impact both academia and the broader environmental sciences community. Abiodun is a member of Nigerian Society for Microbiology, Association of Environmental Impact Assessment of Nigeria (Affiliation of International Association for Impact Assessment), and American Society for Microbiology.

Nelson Oshogwue Etafo is a Lecturer at the Carolina Universidad in Saltillo, Coahuila, Mexico, with more than 15 years of experience in teaching different categories of science students. He obtained his doctorate at the esteemed Autonomous University of Coahuila (UadeC), Saltillo, Coahuila, Mexico. His research focused on photoluminescence and photocatalysis of phosphor materials and nanomaterials. He earned his BSc and MSc degrees in chemistry and Industrial chemistry from the Federal University of Agriculture, Abeokuta (FUNAAB), Nigeria. He is a member of many organizations like the Royal Society of Chemistry (RSC), the American Chemical Society (ACS), and the Chemical Society of Nigeria (CSN). As a researcher, he has collaborated with leading institutions and industry partners to develop new materials with unique properties and applications. His research focused on the development of sustainable and eco-friendly materials and their use in healthcare, energy storage and conversion, catalysis, and other fields. He is committed to conducting research that has a positive impact on society and the environment. He has published many articles, book chapters, and a book in high-impact journals and has attended both national and international conferences as well as reviewed many articles and book chapters. He has won awards for his scholarly and teaching skills both in Nigeria and Mexico.

Mopelola A. Idowu is a Professor in the Department of Pure and Applied Chemistry, Federal University of Agriculture, Abeokuta, Nigeria (FUNAAB). She holds a PhD in Chemistry from Rhodes University, South Africa, and was a postdoctoral scholar at the University of Michigan, USA. Earlier, she earned her BSc Honours and MSc in Chemistry from the University of Ibadan, Nigeria. She has participated in many short courses, including International Deans' Course Africa by German Academic Exchange Service, the German Rectors' Conference and the Alexander von Humboldt Foundation. She is a Fellow of the Andrew Mellon Foundation, an Awardee of the Schlumberger Faculty for the Future fellowship, an Awardee of the Africa Scholarship, a grantee of the German/African bilateral agreement Fellowship in Science and CSIC i-COOP, Spanish/African bilateral agreement grant. Professor Idowu's research efforts are toward the design and evaluation of light-absorbing systems with interesting electronic properties; toward the discovery of materials with applications in photoscience, photomedicine, photocatalysis, sensors, environmental remediation, and energy storage. Her collaborative efforts have furthered her work in the area of time-resolved dynamics of photoexcited molecules; photoinduced damage by exogenous sensitizers through formation of adducts with proteins and their components; kinetics and thermodynamics of use of biomass and nanostructured materials for remediation; computational elucidation of molecules for photovoltaic applications; and evaluation of organic compounds and their mechanism of inhibition in corrosion. Professor Idowu's commitment to excellence in research and education is reflected in her extensive publication record and involvement in academic activities. She has authored numerous research papers and book chapters and has mentored many students, majorly and collaboratively, some of which have gained tenure at institutions nationally and internationally. She is actively involved in academic leadership and administrative duties. Her administrative experience spans, but not limited to, Headship of the Department of Chemistry at FUNAAB, Member, University senate, Faculty appointment and promotion committee and service/headship in university committees, which involve a high level of integrity and staff discipline. Professor Idowu's contributions continue to impact both academia and the broader chemistry community. She is a member of the Chemical Society of Nigeria and other international Chemistry bodies, a former Associate Editor of the journal of the Society, and a reviewer of many peer-reviewed International Journals of Chemistry and Materials Science. She has also been actively involved in many international grants' bodies review panels.

Naga Raju Maddela received his MSc (1996–1998) and PhD (2012) in Microbiology from Sri Krishnadevaraya University, Anantapuramu, India. During his doctoral program in the area of Environmental Microbiology, he investigated the effects of industrial effluents/insecticides on soil microorganisms and their biological activities, and he is working as Faculty in Microbiology since 1998, teaching undergraduate and postgraduate students. He worked on Prometeo Investigator (fellowship received from SENESCYT) at Universidad Estatal Amazónica, Ecuador, during 2013–2015, and received "Postdoctoral Fellowship" (2016–2018) from Sun Yat-sen University, China. He also received external funding from "China Postdoctoral Science Foundation" in 2017, internal funding from "Universidad Técnica de Manabí" in 2020. He participated in national/international conferences and presented research data in China, Cuba, Ecuador, India, and Singapore. Currently, he is working as a Full Professor at the Facultad de Ciencias de la Salud, Universidad Técnica de Manabí, Portoviejo, Ecuador. He has been actively publishing scientific articles, books (authored and edited), and chapters since 2007. As of now, he has published 110 articles, 25 books, and 50 book chapters.

AI Use Disclosure Statement

Artificial intelligence tools were primarily used for language editing and summarization. All outputs were reviewed and verified by the author(s). The analysis, interpretation, and conclusions are the sole responsibility of the author(s).

Furthermore, the author(s) confirm that the use of AI tools complies with Wiley's policies on transparency and responsible authorship. Any references, citations, or data incorporated into this book were manually verified and cross-checked against the sources to ensure accuracy and maintain academic integrity.

Part 1

Introduction to Climate Change and Catalysis

1

The Climate Change Crisis: Emerging Causes and Effects

Marell Navarro-Rojas[1], Rodrigo Duarte-Casar[2], Juan Carlos Romero-Benavides[3], Darío Mondavi-Sobby[2], Mauricio Colpari-Pozzo[2], Tatiana Ordóñez-Zambrano[2], and Marlene Rojas-Le-Fort[2]*

[1] *Universidad Politécnica Salesiana, Cuenca, Ecuador*
[2] *Pontificia Universidad Católica del Ecuador Sede Manabí, Portoviejo, Ecuador*
[3] *Universidad Técnica Particular de Loja, Loja, Ecuador*

*Corresponding author: rduarte@pucesm.edu.ec

1.1 Introduction

Climate change has rapidly escalated from a distant environmental issue to a pressing global crisis, with far-reaching consequences for ecosystems, public health, and economic systems. The intensification of global warming, primarily driven by anthropogenic greenhouse gas (GHG) emissions, is linked to changing weather patterns, biodiversity loss, and disruptions to food and water systems [1, 2]. These impacts are already observable, from rising sea levels threatening coastal infrastructure to increased disease burden in vulnerable populations [3, 4]. In addition to environmental degradation, climate change undermines economic resilience by affecting productivity, labor conditions, and the long-term viability of essential sectors such as agriculture and energy [5]. Acknowledging these interconnected threats is crucial to formulating effective mitigation strategies and fostering a transition toward sustainable, climate-resilient development pathways.

Climate change refers to long-term shifts in temperature, precipitation patterns, and other atmospheric conditions on Earth. Although natural factors such as volcanic activity and solar cycles have historically contributed to climate variability, the current trend of rapid global warming is predominantly driven by anthropogenic activities, especially the burning of fossil fuels and deforestation [6, 7]. The Intergovernmental Panel on Climate Change (IPCC) has consistently warned that if GHG emissions are not substantially curtailed, the planet will face increasingly severe and disruptive events [8]. Moreover, the effects of climate

Metal Oxide Nanocatalysts for Sustainable Energy Production, First Edition. Edited by Abayomi Bamisaye et al.
Published 2026 by Wiley-VCH GmbH.

change are intertwined with social, political, and economic systems, making it a multidimensional global challenge beyond merely an environmental one [9, 10].

No longer a distant threat but a lived reality with wide-ranging effects on both natural systems and human society, the consequences of climate change include the intensification of extreme weather events, disruption of ecosystems, and the degradation of essential services such as food supply, water access, and public health [11]. Vulnerable populations, particularly in developing countries, are disproportionately affected due to entrenched social and economic inequalities [12]. Women and marginalized groups often face unique barriers in adapting to climate impacts, as seen in minority and gender-based advocacy challenges within international climate negotiations [13].

Sustainable technologies are crucial for addressing climate change. An emerging promising area for climate mitigation is developing and deploying nano-metal oxide catalysts. These materials have unique physicochemical properties such as high surface area, enhanced catalytic reactivity, and strong selectivity, which make them particularly effective at reducing industrial emissions, catalyzing hydrogen production, and enabling carbon dioxide (CO_2) capture conversion [14, 15]. Unlike traditional catalysts, nano-metal oxides help the development of cleaner energy systems by increasing reaction efficiencies and reducing the environmental impact of processes. Their use supports the goals of green chemistry and nanotechnology, as well as the SDGs, to build more sustainable, circular, and low-carbon economies [16, 17]. This chapter summarizes the current knowledge on the climate crisis, its drivers, implications, the role of energy production in the current climate crisis, and the need for sustainable technologies to confront it.

1.2 An Overview of Climate Change Drivers

The balance of energy entering and leaving the Earth's atmosphere is essential for maintaining global temperature and climate stability. Temperature anomalies and widespread environmental impacts occur when this balance is disturbed, especially by GHG buildup. This section examines underlying concepts such as the greenhouse effect and radiative forcing, the specific roles of primary GHGs, and the difference between natural and human-induced climate drivers.

Climate change arises from both natural processes and human activities. Natural influences include changes in solar energy, volcanic eruptions, and internal Earth system variations such as El Niño–Southern Oscillation (ENSO). While these factors can affect climate over different timescales, they alone cannot account for the rapid warming seen in recent decades [18, 19].

Anthropogenic drivers, rather than natural causes, mainly cause the recent temperature rise. Human activities have released record amounts of carbon and other pollutants into the atmosphere, overwhelming Earth's natural regulatory systems. Climate models that include both natural and human influences provide strong evidence that human activity is the leading cause of current climate change [20].

1.3 Greenhouse Effect and Radiative Forcing

The greenhouse effect is often perceived as a negative process. Still, it is indeed one of the natural processes that enable life on Earth by raising the average temperature from −18 to 15 °C by limiting energy reflection to space. Solar radiation passes through the atmosphere and warms the planet's surface. In response, Earth emits infrared radiation (IR) back into space. GHGs such as CO_2, methane (CH_4), and water vapor absorb some of this outgoing radiation and reemit it, effectively trapping heat in the atmosphere and raising its temperature. Human activities have significantly intensified this vital process [21].

Radiative forcing refers to the change in energy flux in the atmosphere caused by factors that affect the balance between incoming and outgoing radiation [22]. Positive radiative forcing results in warming, while negative forcing causes cooling. Human-made emissions of GHGs contribute to positive radiative forcing, disrupting the energy balance and strengthening the greenhouse effect. The IPCC states that CO_2 is the main driver of this imbalance, followed by CH_4 and nitrous oxide (N_2O), which are much more potent per molecule than CO_2 (Figure 1.1).

CO_2 is the most prevalent long-lived GHG emitted by human activities, primarily through fossil fuel combustion, deforestation, and cement production. It serves as the baseline for comparing the warming potential of other gases. CH_4, although less abundant, has over 25 times the global warming potential (GWP) of CO_2 over 100 years and is released from livestock, landfills, and fossil fuel extraction [23]. N_2O, mainly emitted from agricultural activities, has nearly 300 times the GWP of CO_2. While present in smaller quantities, other synthetic GHGs, such as hydrofluorocarbons (HFCs) are highly potent climate forcers.

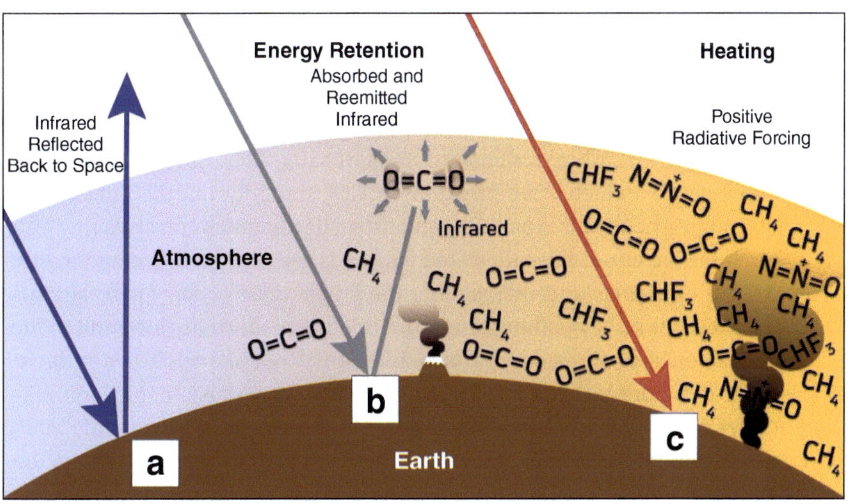

Figure 1.1 Scheme of the greenhouse effect. (a) Solar radiation gets reflected to space. (b) GHGs absorb IR and emit it back to the atmosphere, preventing heat loss. (c) Large concentrations of GHGs result in positive radiative forcing, heating the planet.

Table 1.1 Comparison of key GHGs.

GHG	Primary sources	GWP – 100 years [24]	Relative impact/ abundance
Carbon dioxide (CO_2)	Fossil fuel combustion, deforestation, cement production, and industrialization	1 (baseline)	Most prevalent (75% of GHG emissions); the primary driver of long-term warming
Methane (CH_4)	Livestock, landfills, fossil fuel extraction	28–36	~25 × more potent than CO_2 per molecule; accounts for ~17% of GHG emissions
Nitrous oxide (N_2O)	Agricultural activities (fertilizers), industrial processes	265–298	~300 × more potent than CO_2 per molecule; contributes ~6% of radiative forcing
Hydrofluorocarbons (HFCs)	Synthetic (refrigerants, aerosols, and industrial applications)	12–14 800	Extremely potent but low atmospheric concentration; regulated under Montreal Protocol

Source: Adapted from [25].

The cumulative effect of these gases is an increase in global mean temperatures and a rise in climate variability. This includes altered precipitation patterns, more intense storms, and extended droughts—all linked to the enhanced greenhouse effect. Reducing emissions from all major GHGs is critical for stabilizing the climate. Table 1.1 summarizes the main gases and their contribution to climate change.

1.4 Deforestation

Large-scale deforestation, particularly in tropical regions like the Amazon and Southeast Asia, significantly contributes to climate change [26]. As vital carbon sinks, forests absorb approximately 2.6 billion tons of CO_2 annually, but their destruction through logging, agricultural expansion, and urban development reverses this function. When trees are cut or burned, stored carbon is released, accounting for ~10% of global GHG emissions and intensifying the greenhouse effect. These land-use changes also disrupt local weather patterns, accelerate biodiversity loss, and reduce ecosystem resilience to climate impacts. While initiatives like REDD+ (Reducing Emissions from Deforestation and Forest Degradation) and sustainable land management offer mitigation potential, inconsistent enforcement and economic incentives continue to hinder progress.

The degradation of carbon sinks such as forests, wetlands, and peatlands both amplifies climate change and destabilizes biodiversity hotspots [27]. These ecosystems sequester vast amounts of carbon while supporting millions of species, yet their decline leads to feedback loops: warming temperatures further reduce carbon storage capacity, while biodiversity loss weakens ecosystem services like pollination and soil fertility [28]. Protecting and restoring these habitats is essential for climate mitigation, but competing land-use demands and insufficient conservation funding pose significant challenges to reversing current trends.

Algae, including both macroalgae and microalgae, act as important carbon sinks in marine ecosystems, but their ability to sequester carbon is increasingly threatened by climate change. Rising global temperatures stress algal health, often increasing respiration rates more than photosynthesis, which reduces net carbon uptake [29]. Marine heatwaves worsen this problem by destroying large, efficient macroalgae like kelp and encouraging turf-forming algae with less sequestration potential. Ocean acidification further hampers algal carbon storage, especially for calcifying species, despite some temporary benefits from higher CO_2 levels. These temperature and chemical changes interact with shifts in nutrient availability, where increased water column stratification limits essential nutrients, and eutrophication leads to harmful blooms that release GHGs during decomposition [30].

Additional climate impacts further reduce algal carbon sequestration potential. Changes in light availability due to less ice cover may initially enhance growth but often cause long-term photoinhibition, while increased UV radiation damages cellular structures and decreases productivity [31]. Perhaps most critically, climate change is reshaping algal communities, favoring less efficient carbon sinks like turf algae over canopy-forming species. These interconnected stressors—temperature extremes, acidification, nutrient shifts, light changes, and community reorganization—reveal an apparent decline in algal carbon sink efficiency. This highlights the urgency for targeted conservation strategies to preserve algae's essential role in global carbon cycling amid accelerating climate disruption [30].

1.5 Industrialization

One of the most significant emerging causes of climate change is the rapid industrialization happening in developing countries. As nations pursue economic growth, they often rely on fossil fuels like coal, oil, and natural gas to run factories, transportation, and urban infrastructure, which, consequently, increase industrial activity, resulting in higher GHG emissions, especially CO_2, which traps heat in the atmosphere and speeds up global warming [32]. While developed countries have started shifting to cleaner energy sources, many emerging economies still depend on outdated, carbon-heavy technologies because of cost and access issues, making the climate crisis worse. Urban heat islands (UHIs) exacerbate energy consumption, air and water quality issues, and public health challenges, particularly affecting low-income populations in tropical and subtropical regions. This phenomenon highlights the need for targeted resilience policies and socioeconomic impact mitigation strategies [33].

An effect of industrialization is the increased waste production [34]. Poor waste management practices, particularly in urban areas, are emerging as a growing contributor to climate change [35]. Landfills filled with organic waste produce CH_4, a GHG over 25 times more potent than CO_2 in trapping heat. As global populations grow and consumption increases, so does the volume of waste, leading to higher CH_4 emissions. Additionally, industrial processes, such as oil and gas extraction, release CH_4 through leaks and flaring. While CH_4 remains in the atmosphere for a shorter time than CO_2, its immediate impact on global temperatures is substantial [36]. Addressing these emissions through better waste processing, recycling, and stricter industry regulations is crucial to slowing climate change.

1.6 Significant Implications

1.6.1 Rising Temperatures

One of climate change's most direct and measurable effects is the steady rise in global average temperatures. Since preindustrial times, Earth's surface has already warmed by over 1.1 °C, and without decisive global mitigation efforts, projections indicate that warming could exceed 2 °C by the end of the century [37]. This warming trend is strongly linked to an increased frequency and intensity of extreme weather events, including heatwaves, storms, wildfires, and floods [38].

Heatwaves are becoming more frequent, particularly in urban areas where the heat island effect exacerbates temperature spikes. These events strain public health systems and infrastructure, reduce labor productivity, and disproportionately impact the elderly and socioeconomically disadvantaged populations [39]. Moreover, extreme heat poses significant risks for people with preexisting health conditions such as diabetes, amplifying the burden on healthcare systems.

Temperature shifts affect food production more than water or sunlight [40]. Climate change can also modify the viable cultivation areas of diverse crops [41, 42]. These shifts may exacerbate deforestation due to the expansion of farmland at the expense of forested areas [43]. A dire example is the prolonged drought in the Horn of Africa (2020–2023), where record-high temperatures and failed rainy seasons led to catastrophic crop failures and livestock losses. In Somalia, Ethiopia, and Kenya, millions faced acute food insecurity, with over 20 million people requiring emergency aid [44]. Rising temperatures also increase pest outbreaks, such as desert locust swarms, which devastated East African farmlands in 2020, destroying crops and pastureland.

Warmer ocean temperatures are also contributing to stronger hurricanes and cyclones, which lead to increased devastation in coastal regions through flooding and wind damage [45]. Simultaneously, wildfires have escalated in both size and frequency, destroying ecosystems, homes, and lives while releasing additional GHGs into the atmosphere, thus reinforcing the warming cycle [46].

These climate disruptions are environmental and also humanitarian challenges, disproportionately affecting vulnerable communities and exacerbating health and gender-based inequalities [47].

1.6.2 Sea-level Rise

Sea-level rise is one of the most clear and measurable effects of human-caused climate change, mainly due to warming oceans expanding and land ice melting.

Satellite altimetry data shows that the global average sea level has increased by about 3.7 mm per year since 2006, nearly three times the rate seen for most of the 20th century. This faster rise is especially worrying because more than 600 million people live in low-elevation coastal areas that are vulnerable even to small sea level increases. The processes that cause this include not only melting from Greenland and Antarctic ice sheets but also complex feedback effects like reduced reflectivity from less Arctic sea ice and possible destabilization of glaciers that end in the ocean.

The impacts of sea-level rise show significant regional differences due to factors like ocean circulation patterns, gravitational effects from ice loss, and vertical land movement. For example, many tropical areas experience sea-level rise rates 20–30% higher than the global average because of subsidence and ocean dynamics. Climate models predict that even under moderate emission scenarios (RCP 4.5), global mean sea level could increase by 0.5–0.8 m by 2100, with the possibility of much larger rises if certain ice sheet tipping points are reached. These predictions have serious implications for coastal landforms, including more shoreline erosion, increased nuisance flooding, and the potential for some low-lying island nations and coastal wetlands to become permanently submerged. The long timescales involved in sea-level rise mean these effects will last for centuries, even with strong climate change mitigation efforts.

The world's oceans serve as a crucial buffer against climate change by absorbing a significant portion of atmospheric CO_2. However, this absorption leads to ocean acidification, a chemical process that lowers seawater pH, disrupting marine ecosystems and the biological processes that sustain them [48]. The acidification process impairs the ability of marine organisms, especially calcifying species, to build and maintain their calcium carbonate structures.

Among the most impacted are coral reefs, which are essential to global marine biodiversity. Acidification hampers coral calcification, contributing to bleaching events and the collapse of reef ecosystems [49]. Studies across the Pacific Basin confirm that coral reef systems are increasingly vulnerable to combined stressors from warming and acidification, jeopardizing the stability of these ecosystems [50]. The resulting decline in reef health compromises the structural complexity of reefs and reduces habitat availability for thousands of marine species [51].

These ecological impacts extend to human communities as well. Fisheries that depend on reef-associated species face declining yields, and small-scale fishing economies, particularly in the Pacific Islands, are highly exposed to the consequences of acidification and reef degradation [52]. As coral ecosystems degrade, coastal protection also weakens, increasing vulnerability to erosion and storm surge damage [53].

1.6.3 Extreme Weather Occurrences

The rising frequency and severity of extreme weather events are some of the most visible and destructive signs of climate change. Scientific evidence shows clear trends toward stronger tropical cyclones, longer heatwaves, more intense rainfall, and larger, more destructive wildfires. These changes are caused by fundamental thermodynamic processes. A warmer atmosphere can hold more moisture, resulting in heavier rainfall and flooding, while higher ocean surface temperatures provide extra energy for tropical storms. Attribution studies have shown that many recent extreme events, like the 2021 Pacific Northwest heatwave and Hurricane Harvey's record rainfall in 2017, would have been highly unlikely without human-caused climate change. The combined impact of these events puts pressure on infrastructure, disrupts ecosystems, and challenges communities' ability to adapt worldwide.

The socioeconomic impacts of extreme weather are significant and often disproportionately affect vulnerable populations with limited resources for recovery. Multiyear droughts can destroy agricultural production and lead to water shortages, as observed in recent crises across the Mediterranean and Horn of Africa. Conversely, increasing extreme precipitation events frequently cause devastating flooding, harming homes, transportation systems, and essential facilities. Wildfires, driven by hotter, drier conditions, have become more common in areas from the western United States to Australia, with wide-reaching effects on air quality and biodiversity. These events not only cause immediate physical and economic damages but also pose long-term systemic risks, including population displacement and disruptions to global supply chains. As climate models forecast further intensification of extreme weather under all plausible emission scenarios, the urgency for improved early warning systems, resilient infrastructure, and adaptive land-use planning grows.

1.6.4 Public Health

In recent decades, changing environmental threats have increased the impacts of climate change, creating complex and overlapping risks. The appearance of new ecological hazards like micro- and nano-pollutants and persistent chemical residues, often linked to rapid technological progress, raises new concerns for ecosystems and public health [22]. These human-made stressors worsen the public health crises caused by climate change, which appear through both direct health harms and systemic disruptions to health factors. Extreme heat events, increasing in frequency and severity, disproportionately affect vulnerable groups such as the elderly and those with chronic illnesses. Additionally, stronger storms and wildfires cause trauma, displacement, and strain on medical systems. Meanwhile, changing climate conditions spread vector-borne diseases like malaria and dengue to new areas, putting populations at risk of unfamiliar health threats. These issues demand urgent action to include climate adaptation strategies in public health plans to prevent worsening problems and risks.

The public health effects of climate change go beyond immediate disasters, undermining long-term health stability through environmental degradation and socioeconomic pressures. Falling agricultural yields and decreased nutritional quality of crops—caused by rising CO_2 levels and changing rainfall patterns—threaten food security, leading to malnutrition and related illnesses. At the same time, climate-boosted air pollution from wildfires and industrial sources increases respiratory and heart problems, straining healthcare systems worldwide. Marginalized communities, often least prepared to adapt, face greater risks due to a lack of infrastructure, limited healthcare access, and increased exposure to environmental hazards. As climate change interacts with new pollutants and legacy contaminants, a comprehensive approach is essential—one that includes emissions reduction, equitable health policies, and efforts to address the root causes of environmental problems and injustice.

Climate change is intensifying air pollution by accelerating the formation of ground-level ozone and increasing the frequency and severity of wildfires and dust storms. These environmental shifts significantly deteriorate air quality, with severe public health consequences. For instance, higher temperatures and wildfire smoke have been shown to worsen particulate matter exposure, especially in already polluted regions [54]. This combination creates synergistic risks that magnify respiratory and cardiovascular impacts [55].

Respiratory conditions such as asthma, chronic obstructive pulmonary disease (COPD), and bronchitis are increasingly common, especially among children, the elderly, and vulnerable communities exposed to poor air quality and extreme weather [56]. Furthermore, prolonged exposure to air pollution, especially from wildfire smoke has been linked not only to exacerbation of chronic respiratory illness but also to higher rates of infections, including respiratory pathogens [57].

Cardiovascular diseases, already a leading cause of global mortality, are further amplified by air pollution, which can trigger inflammatory and oxidative stress responses in the cardiovascular system [54]. Simultaneously, wildfire-related $PM_{2.5}$ pollution has been shown to disproportionately impact lower socioeconomic and minority communities, compounding public health disparities [58].

The broader socioeconomic consequences are equally concerning. Increased healthcare costs, reduced labor productivity, and heightened pressure on public health systems are direct outcomes of worsening air quality and climate-driven health burdens [59]. These trends collectively undermine community resilience and economic development on both local and global scales.

1.7 The Role of Energy Production and Use in Accelerating Climate Change

Fossil fuel combustion is the primary driver of anthropogenic carbon emissions, releasing vast quantities of CO_2 into the atmosphere. When coal, oil, and natural gas are burned for energy, the carbon that had been stored underground for millions of years is rapidly oxidized, reacting with oxygen to form CO_2. This process disrupts the natural carbon cycle, where geological sequestration once balanced atmospheric carbon levels. Since the Industrial Revolution, CO_2 concentrations have risen by over 50%, from 280 ppm to over 420 ppm today (https://osf.io/preprints/osf/het6n_v1), with fossil fuels accounting for roughly 75% of global GHG emissions (Figure 1.2). The energy sector—power generation, transportation, and industry—is the dominant contributor, with coal being the most carbon-intensive per unit of energy produced [60].

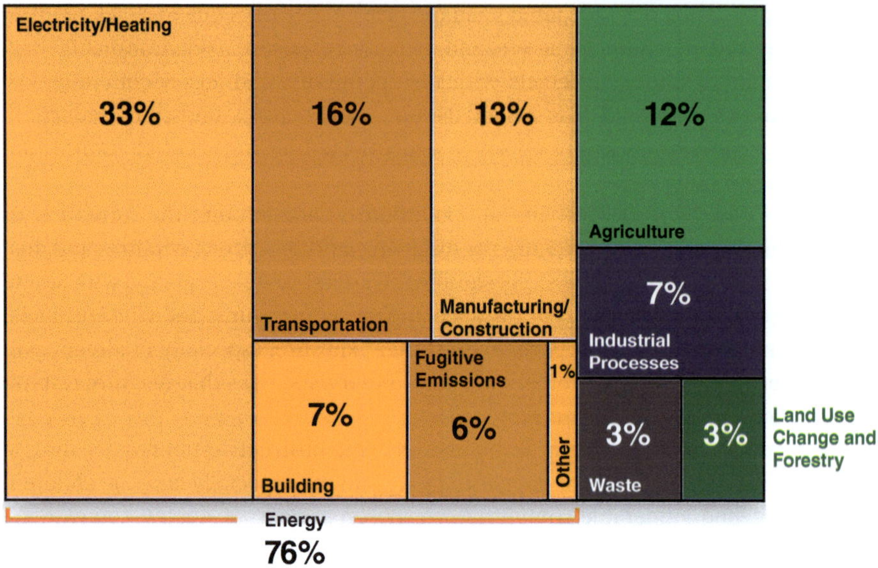

Figure 1.2 Distribution of GHG emissions by sector.
Source: [61] / World Resources Institute / CC BY 4.0.

According to Dutta (2021), burning fossil fuels for the obtention of energy, heating, electricity, transportation, and industry contributes two-thirds of anthropogenic GHGs [61]. The impacts of these emissions are significant, intensifying climate change through the greenhouse effect. Despite progress in renewable energy, fossil fuels still provide about 80% of the world's energy, maintaining reliance on high-emission infrastructure. Mitigation strategies, such as carbon capture and storage (CCS), switching to natural gas (which emits approximately 50% less CO_2 than coal), and policy measures like carbon pricing, are aimed at reducing emissions. However, reaching net-zero targets requires systemic shifts toward renewable energy, greater energy efficiency, and decarbonization across all sectors to counteract the legacy of fossil fuel dependence.

1.8 Economic and Geopolitical Outlook

Global demand for fossil fuels has plateaued in some developed economies but continues to rise in emerging markets, particularly in Asia. China and India, driven by rapid industrialization and urbanization, are the largest consumers of coal. At the same time, the United States and Europe rely more heavily on natural gas due to its lower carbon intensity. Oil remains indispensable for transportation, constituting nearly 90% of the sector's energy supply. The International Energy Agency (IEA) projects that fossil fuel demand may peak by 2030, although this hinges on aggressive policy interventions and the widespread adoption of renewable energy. Fossil fuel markets remain volatile, influenced by geopolitical conflicts, supply chain disruptions, and fluctuating demand. The 2022 energy crisis, exacerbated by Russia's invasion of Ukraine, showed the risks of reliance on hydrocarbon imports. Conversely, oil-exporting nations face economic instability as decarbonization policies threaten long-term demand. Subsidies for fossil fuels, estimated at US$7 trillion annually, distort energy markets and slow the transition to cleaner alternatives [62].

Despite significant progress, current strategies for reducing carbon emissions from burning fossil fuels face significant limitations that hinder their ability to meet global climate goals. Technological solutions, such as CCS, remain energy-intensive, expensive, and limited in scale, with around 50 large-scale projects operating worldwide as of 2024 [63]. While renewable energy use is increasing quickly, it still faces problems with inconsistent output, storage difficulties, and dependence on fossil fuel-based infrastructure for manufacturing and maintaining the power grid. Policy measures, such as carbon pricing and emissions trading schemes, often encounter political opposition, uneven adoption across countries, and prices that are too low to inspire genuine change.

Without strong leadership and societal recognition of necessary changes, vulnerabilities to climate change impacts will increase [64]. Additionally, the ongoing growth of fossil fuel infrastructure, especially in developing countries, locks in future emissions, while market-driven transitions are not keeping pace with the urgent demands of climate science. Without widespread systemic reforms, international cooperation, and accelerated innovation, these issues risk undermining efforts to limit global warming to 1.5 °C, thereby increasing the vulnerability of societies to the growing impacts of climate change. insufficient Nationally Determined Contributions (NDCs) could lead to warming of 3.0–3.5 °C, increasing vulnerability to climate impacts like flooding and drought, particularly in hotspots such as West Africa, India, and North America [65].

1.9 Global Initiatives and Agreements to Reduce Climate Impacts

International climate treaties, such as the Kyoto Protocol and the Paris Agreement, along with their subsequent extensions, establish vital frameworks for global cooperation in reducing GHG emissions. These agreements operate on the principle of common but differentiated responsibilities, recognizing that developed countries (the most significant historical emitters) should take the lead in cutting emissions while supporting developing nations in their shift to sustainable economies. The Paris Agreement introduced NDCs, which permit each country to set its own emission reduction goals while maintaining transparency through regular reporting and review mechanisms. This flexibility has encouraged broader participation, with nearly every country now committed to some form of climate action.

However, despite these advances, significant challenges remain in aligning global efforts with the Agreement's most ambitious goal: limiting global temperature rise to 1.5 °C above preindustrial levels, especially with 2024 as the first year above the 1.5 °C threshold [66]. Current NDCs, even if fully implemented, would still result in warming of about 2.5–2.9 °C by 2100, significantly exceeding the safe threshold needed to avoid catastrophic outcomes [67]. This gap shows a critical paradox: while the 1.5 °C target is often criticized for lacking sufficient ambition, it's also seen as nearly impossible to achieve under current policies [68]. Bridging this gap requires stronger enforcement mechanisms, enhanced financial support for developing nations, and accelerated policy implementation to ensure that commitments are translated into tangible outcomes.

1.10 Obstacles

One of the key weaknesses of current climate agreements is the lack of binding enforcement mechanisms. Unlike the Kyoto Protocol, which imposed legally binding emission cuts on developed nations (albeit with limited success), the Paris Agreement relies on voluntary pledges and peer pressure to drive compliance. While this approach has secured near-universal participation, it has also led to inconsistent progress. Some countries have strengthened their NDCs over time, while others lag or even backtrack on commitments. Strengthening accountability could involve stricter monitoring, independent verification of emissions data, and consequences for noncompliance, such as trade sanctions or exclusion from climate financing programs.

Another major obstacle is the inadequate flow of climate finance to developing countries, which often lack the resources to invest in renewable energy, climate resilience, and low-carbon infrastructure. Developed nations pledged to mobilize US$100 billion annually by 2020 to support mitigation and adaptation efforts in the Global South, yet this target has been repeatedly missed [69]. Without sufficient funding, many vulnerable nations cannot transition away from fossil fuels or build defenses against rising sea levels, droughts, and extreme weather events. Expanding climate finance—through mechanisms like the Green Climate Fund, debt relief for climate-vulnerable nations, and private-sector investments—is essential to ensuring equitable progress.

Even when countries set ambitious NDCs, bureaucratic delays, political resistance, and economic reliance on fossil fuels often slow down implementation. For example, despite global promises to phase out coal, many nations continue to approve new coal-fired power plants due to energy security concerns or lobbying by the fossil fuel industry. A faster transition needs not only stricter regulations but also just transition policies that support workers and communities dependent on high-carbon industries. Retraining programs, green job creation, and social safety nets can help reduce economic disruptions and increase public support for climate action [70].

1.11 Systemic Change

While international agreements are crucial, they must be complemented by systemic changes in energy, agriculture, transportation, and urban planning. Subnational actors—such as cities, corporations, and financial institutions—play an increasingly important role in driving innovation and filling gaps left by national governments. Initiatives like the C40 Cities network and the Science-based Targets initiative (SBTi) demonstrate how local and private-sector leadership can accelerate decarbonization [71].

Ultimately, the success of global climate agreements relies on whether nations can move beyond rhetoric and engage in swift, coordinated action. Enhancing enforcement, boosting financial support, and accelerating policy implementation are essential steps to bridge the gap between current commitments and the missed 1.5 °C target. Without these efforts, the paradox of being both unattainable and insufficient will persist, leaving the world on a perilous path toward irreversible climate damage [72, 73].

1.12 The Importance of Creating Sustainable Technology to Combat Climate Change

The development of more sustainable technology is a vital part of this global effort. Sustainable technologies such as renewable energy systems, energy-efficient infrastructure, and carbon capture solutions provide practical ways to decouple economic growth from environmental harm. By focusing on innovation in clean energy, circular manufacturing, and smart urban planning, societies can reduce their carbon footprints while building long-term resilience. The shift to sustainable technology is an environmental, economic, and social need and opportunity, creating new industries, green jobs, and energy independence for countries worldwide.

However, the widespread adoption of sustainable technology faces challenges, including high initial costs, policy barriers, and uneven access across regions. To overcome these obstacles, governments, corporations, and research institutions must collaborate to increase investments, simplify regulatory frameworks, and promote equitable technology transfer. Emerging innovations like green hydrogen, next-generation batteries, and AI-driven climate modeling have great potential but need coordinated support to reach commercialization. Additionally, public awareness and behavioral changes are essential for creating demand for sustainable solutions. By combining technological advancements with systemic policy reforms and societal engagement, a sustainable future that fights climate change while promoting global equity and prosperity can be achieved.

Catalysis plays a crucial role in reducing GHG emissions by improving the efficiency of chemical reactions across industries. Catalysts decrease energy needs, reduce unwanted byproducts, and support cleaner processes in sectors like petroleum refining, automotive exhaust treatment, and renewable fuel production. Their ability to speed up reactions under gentler conditions makes them essential for reaching net-zero goals and moving toward sustainable industrial practices.

Among emerging solutions, nano-metal oxide catalysts have gained attention for their high surface area, tunable properties, and stability under harsh conditions. Materials such as cerium oxide (CeO_2), titanium dioxide (TiO_2), and zinc oxide (ZnO) show excellent redox activity and selectivity, making them ideal for environmental applications. Their nanostructured form further boosts catalytic performance, offering a promising way to replace traditional, often scarce or toxic, catalytic materials [74].

These advanced catalysts are being deployed in critical climate mitigation strategies, including CO_2 capture and conversion, green hydrogen production via water splitting, and air pollution control through photocatalytic degradation of pollutants. For instance, nano-catalysts can transform captured CO_2 into fuels or plastics, while also improving the efficiency of electrolyzers for hydrogen generation. By integrating these innovations, industries can significantly reduce their carbon footprint while maintaining economic viability.

While renewable energy and advanced technologies promise decarbonization, their production and disposal generate hazardous pollutants. Solar panels, batteries, and electronics release nano-pollutants (e.g., silicon nanoparticles) and heavy metals (e.g., lead, cadmium) into soil and water, while industrial waste from manufacturing often lacks proper recycling infrastructure. These contaminants accumulate in ecosystems, posing risks to human health and wildlife. A circular economy approach, improving material recovery, safer chemical alternatives, and stricter regulations, is needed to minimize trade-offs between technological progress and environmental harm [75].

The rise of nanotechnology and industrial expansion has introduced persistent pollutants, including toxic heavy metals (e.g., mercury, arsenic) and synthetic nanoparticles, into air, water, and food chains [76]. These contaminants resist natural degradation, bioaccumulate in organisms, and cause long-term ecological damage. For example, microplastics and nano waste from electronics infiltrate marine systems, while mining waste leaches heavy metals into groundwater. Monitoring and remediation efforts lag behind pollutant dispersion, underscoring the need for global standards on waste management and more sustainable material design.

Traditional catalytic processes in industries like petroleum refining and chemical manufacturing rely on platinum-group metals (PGMs) and toxic reagents, which are energy-intensive and often generate harmful byproducts (e.g., NO_x, SO_2). While catalysts improve reaction efficiency, their environmental footprint from mining rare metals to emissions during use undermines sustainability goals. Transitioning to bio-based or earth-abundant catalysts (e.g., iron, enzymes) could reduce these impacts, but scalability and cost barriers persist. Green chemistry and its twelve principles are an initiative towards a sustainable chemical industry. Green chemistry focuses on designing chemical products and processes that minimize environmental harm by reducing or eliminating hazardous substances [77]. It promotes principles like atom economy, renewable feedstocks, and energy efficiency to cut waste, lower emissions, and enhance sustainability. By replacing toxic solvents, developing biodegradable materials, and improving catalytic processes, green chemistry aims to make industries from pharmaceuticals to manufacturing cleaner and more resource-efficient. This shift addresses pollution and aligns with circular economy goals, ensuring long-term ecological and economic benefits aligned with UN SDGs [78].

Conventional catalysts, though critical for fuel production and pollution control (e.g., catalytic converters), frequently depend on scarce, expensive materials like palladium and rhodium. Their extraction and processing contribute to significant carbon emissions and ecological damage, while catalyst deactivation over time generates hazardous waste [77]. Innovations in catalyst design, such as single-atom catalysts or photocatalytic systems, could enhance efficiency and reduce reliance on finite resources. Still, industrial adoption remains slow due to technical and economic constraints.

The shift toward sustainable chemical processes demands alternatives to fossil-derived feedstocks and toxic reagents. Electrocatalysis, photocatalysis, and biocatalysis offer promising pathways, using renewable energy or enzymes to drive reactions with minimal waste. For instance, photocatalysts like TiO_2 can break down pollutants under sunlight, while engineered enzymes enable low-temperature industrial processes. However, stability, reaction rates, and large-scale deployment must be addressed to make these alternatives viable for the global industry.

These challenges add to traditional climate stressors and require a broader, interdisciplinary understanding of climate dynamics. Responding effectively and efficiently calls for integrated strategies involving environmental monitoring, land use policy, industrial innovation, and energy system transformation.

1.13 Conclusion

Climate change has quickly shifted from a distant environmental issue to an urgent worldwide crisis, with wide-reaching effects on ecosystems, public health, and economies. The increase in global warming, mainly caused by human-made GHG emissions, is connected to severe weather events, loss of biodiversity, and disruptions to food and water supplies. Tackling these linked threats requires effective mitigation strategies and a move toward climate-resilient, sustainable development paths.

The scientific consensus, as outlined by the IPCC, attributes current warming trends (1.1 °C above preindustrial levels) overwhelmingly to human activities, especially fossil fuel burning and deforestation. Without significant reductions in GHG emissions, the planet will face increasing climate extremes—heatwaves, storms, wildfires, and floods—with cascading socioeconomic impacts. Ocean acidification, caused by CO_2 absorption, is harming marine ecosystems, especially coral reefs that support fisheries and coastal protection. Air pollution, worsened by ground-level ozone and wildfire smoke, raises respiratory and cardiovascular health issues, disproportionately impacting vulnerable populations. Climate-related disruptions to education, infrastructure, and labor productivity further intensify these challenges.

Emerging contributors to climate change include rapid fossil fuel-dependent industrialization in developing economies, large-scale deforestation, and inadequate waste management (a significant source of CH_4 emissions). Despite advances in renewable energy, persistent barriers hinder progress: CCS remains energy-intensive and costly, renewable intermittency challenges grid stability, and political resistance weakens carbon pricing policies. Additionally, producing and disposing of renewable technologies generates hazardous pollutants, such as nanoparticles and heavy metals, creating new environmental trade-offs.

Innovative solutions are critical to overcoming these limitations. Catalysis is pivotal in reducing emissions by optimizing chemical reactions across industries, minimizing energy use and waste. Nano-metal oxide catalysts (e.g., CeO_2, TiO_2, and ZnO) offer enhanced surface reactivity, selectivity, and stability, enabling breakthroughs in CO_2 conversion, green hydrogen production, and air pollution control. Unlike traditional catalysts, these nanomaterials align with green chemistry principles, supporting circular, low-carbon economies.

The future of climate-friendly catalysis lies in earth-abundant materials, AI-driven catalyst design, and solar-fuel systems. However, technological advances must be paired with stronger policy frameworks, including equitable implementation of international agreements like the Paris Agreement, increased funding for clean energy transitions, and stricter regulations on industrial pollutants. By integrating innovation with systemic reforms, catalysis can become a cornerstone of global climate mitigation, helping societies achieve net-zero targets while maintaining economic viability.

References

1 Adeel M, Wang B, Ke J, Mvitu IM. The nonlinear dynamics of CO_2 emissions in Pakistan: a comprehensive analysis of transportation, electricity consumption, and foreign direct investment. *Sustainability*. 2024;17(1):189.

2 Li J, Tao C, Zhang Y. Study on current climate change issues and response strategies. *E3S Web Conf.* 2025;606:03007.

3 Bekele C, Hussaini AS, Fikre A, Duran CP, Mohammed Y, Hungenaw BM, et al. Impact of climate change on human health: exploring challenges and opportunities through a qualitative study in Birhan Field Site, Amhara Region, Ethiopia. *medRxiv*. 2025;2025.04.26.25326478.

4 Ebi K, Boyer C. Building resilience of health systems to climate change in pacific least developed countries: implementation science. *Lancet Planet Health*. 2019;3:s6.

5 Arausi RE, Clark EV, Ikenga FA. Oil production and climate change in the Niger-Delta Region: synergic implication and adaptation. *J Public Adm Finance Law*. 2025;33:78–93.

6 Musa UA, Sanusi ZA, Suleiman H. Causes of climate change: a neglected dimension. *Intellect Discourse*. 2024;32(1):265–90.

7 Rawat A, Kumar D, Khati BS. A review on climate change impacts, models, and its consequences on different sectors: a systematic approach. *J Water Clim Change*. 2023;15(1):104–26.

8 Ekah BJ, Akwagiobe JU, Udo SO, Ewona IO. A review on intergovernmental panel on climate change (IPCC) 1st to 6th report. *Phys Sci Int J*. 2023;27(1):16–24.

9 Bununu YA, Bello A, Ahmed A. Land cover, land use, climate change and food security. *Sustain Earth Rev*. 2023;6(1):16.

10 Usman S, Jayeoba JO, Kundiri A. Climate change at a global concept: impacts and adaptation measures. *Int J Environ Clim Change*. 2024;14(6):445–59.

11 Rakowska J. Searching for a sustainable solution to increasing chemical pollution. *Environ Sci Pollut Res.* 2024;31(28):40903–15.

12 Mohanty S, Subhadra S. Climate justice an integral aspect of environmental anthropology: an Indian perspective. In: *Social Science, Education and Humanities Research.* 2023: p. 149–59.

13 Flavell J. Lessons from the women and gender constituency: interrogating civil society strategies for organising in the UNFCCC. *Int J Polit Cult Soc.* 2023;36(3):385–403.

14 Sah MK, Thakuri BS, Pant J, Gardas RL, Bhattarai A. The multifaceted perspective on the role of green synthesis of nanoparticles in promoting a sustainable green economy. *Sustain Chem.* 2024;5(2):40–59.

15 Thole D, Modibane KD, Mhlaba R, Balogun A, Malgas-Enus R, Botha E, et al. Recent developments of carbon nanotubes-based nanocomposites for hydrogen production, carbon dioxide capture and conversion. *Results Chem.* 2024;7:101496.

16 Fu J, Li P, Lin Y, Du H, Liu H, Zhu W, et al. Fight for carbon neutrality with state-of-the-art negative carbon emission technologies. *Eco-Environ Health.* 2022;1(4):259–79.

17 Johnson FM. Utilizing advancements in chemical sciences for decarbonization: a pathway to sustainable emission and energy reduction. *RSC Sustain.* 2024;2(12):3592–5.

18 McPhaden MJ, Santoso A, Cai W. Introduction to El Niño Southern oscillation in a changing climate. *Geophys Monogr Ser.* 2020;253:1–19.

19 Farooqi ZUR, Sabir M, Qadeer A, Naeem A, Murtaza G, Yousaf H. Understanding the causes of climatic change in the environment. In: Bandh SA, editor. *Climate Change: The Social and Scientific Construct.* Cham: Springer; 2022. p. 37–47.

20 Lyon C, Saupe EE, Smith CJ, Hill DJ, Beckerman AP, Stringer LC, et al. Climate change research and action must look beyond 2100. *Glob Change Biol.* 2022 Jan;28(2):349–61.

21 Smith KA. Greenhouse gas emissions. In: *Encyclopedia of Soils in the Environment* [Internet]. Elsevier; 2005 [cited 2025 Jul 11]. p. 145–53. Available from: https://linkinghub.elsevier.com/retrieve/pii/B0123485304000941

22 Takemura T. Radiative forcing and global warming. In: Akimoto H, Tanimoto H, editors. *Handbook of Air Quality and Climate Change.* Springer: Singapore; 2023. p. 949–66.

23 Boucher O, Friedlingstein P, Collins B, Shine KP. The indirect global warming potential and global temperature change potential due to methane oxidation. *Environ Res Lett.* 2009;4(4):044007.

24 Canadell JG, Monteiro PMS, Costa MH, Cotrim da Cunha L, Cox PM, Eliseev AV, et al. Global carbon and other biogeochemical cycles and feedbacks. In: Masson-Delmotte V, Zhai P, Pirani A, Connors SL, Péan C, Berger S, et al., editors. *Climate Change 2021: The Physical Science Basis. Contribution of Working Group I to the Sixth Assessment Report of the Intergovernmental Panel on Climate Change.* Cambridge and New York: Cambridge University Press; 2021. p. 673–816.

25 Arias PA, Bellouin N, Coppola E, Jones RG, Krinner G, Marotzke J, et al. Technical summary. In: Masson-Delmotte V, Zhai P, Pirani A, Connors SL,

Péan C, Berger S, et al., editors. *Climate Change 2021: The Physical Science Basis. Contribution of Working Group I to the Sixth Assessment Report of the Intergovernmental Panel on Climate Change*. Cambridge and New York: Cambridge University Press; 2021. p. 33–144.

26 Álvarez-García H, Sillero-Medina JA, Pérez-González ME, Ruiz-Sinoga JD. Spatial variations of runoff thresholds associated with changes in land use over the period 1990–2018 in the Mediterranean side of Andalusia. *Investig Geográficas*. 2024;(82):25–42.

27 Raturi A, Raturi A, Singh H. Climate change impacts on wetland ecosystem functioning with special reference to greenhouse gas emissions. In: Singh H, editor. *Forests and Climate Change* [Internet]. Singapore: Springer Nature Singapore; 2024 [cited 2025 Jul 1]. p. 345–64. Available from: https://link.springer.com/10.1007/978-981-97-3905-9_18

28 Abbasi SA, Abbasi N. The likely adverse environmental impacts of renewable energy sources. *Appl Energy*. 2000 Apr;65(1–4):121–44.

29 Tait LW, Schiel DR. Impacts of temperature on primary productivity and respiration in naturally structured macroalgal assemblages. *PLOS One*. 2013 Sep 13;8(9):e74413.

30 Vogel N, Cantin NE, Strahl J, Kaniewska P, Bay L, Wild C, et al. Interactive effects of ocean acidification and warming on coral reef associated epilithic algal communities under past, present-day and future ocean conditions. *Coral Reefs*. 2016 Jun;35(2):715–28.

31 Beardall J, Stojkovic S, Gao K. Interactive effects of nutrient supply and other environmental factors on the sensitivity of marine primary producers to ultraviolet radiation: implications for the impacts of global change. *Aquat Biol*. 2014 Nov;20(22):5–23.

32 Misra AK, Verma M. Impact of industrialization on the dynamics of atmospheric carbon dioxide: a modeling study. *Int J Big Data Min Glob Warm*. 2022 Jun;4(1):2150009.

33 Sadik-Zada ER, Gatto A. Vulnerability to the urban heat islands effect in the Global North and the Global South. In: Khan A, Fiorito F, Niyogi D, editors. *Global Urban Heat Island Mitigation* [Internet]. Elsevier; 2022 [cited 2025 Jul 1]. p. 29–45. Available from: https://linkinghub.elsevier.com/retrieve/pii/B9780323855396000123

34 Shah WUH, Yasmeen R, Sarfraz M, Ivascu L. The repercussions of economic growth, industrialization, foreign direct investment, and technology on municipal solid waste: evidence from OECD economies. *Sustainability*. 2023 Jan 3;15(1):836.

35 Onyanta A. Cities, municipal solid waste management, and climate change: perspectives from the South. *Geogr Compass*. 2016 Dec;10(12):499–513.

36 Lelieveld J, Crutzen PJ, Brühl C. Climate effects of atmospheric methane. *Chemosphere*. 1993 Jan;26 (1–4):739–68.

37 Lu S. The Reality of Climate Change: Evidence, Impacts and Engineering Solutions [Internet]. arXiv. 2024 [cited 2025 Jun 29]. Available from: http://arxiv.org/abs/2410.12412

38 Patz JA, Hatch MJ. Public health and global climate disruption. *Public Health Rev*. 2013 Jun;35(1):1–23.

39 Zilbermint M. Diabetes and climate change. *J Community Hosp Intern Med Perspect*. 2020 Sep 2;10(5):409–12.

40 Ding R, Shi W. Quantitative evaluation of the effects of climate change on cereal yields of Tibet during 1993–2017. *Dili XuebaoActa Geogr Sin*. 2021 Sep 25;76(9):2174–86.

41 González-Jaramillo N, Bailon-Moscoso N, Duarte-Casar R, Romero-Benavides JC. Peach palm (*Bactris gasipaes* Kunth.): ancestral tropical staple with future potential. *Plants*. 2022 Jan;11(22):3134.

42 Lara-Estrada L, Rasche L, Schneider UA. Land in Central America will become less suitable for coffee cultivation under climate change. *Reg Environ Change*. 2021 Sep 1;21(3):88.

43 Magrach A, Ghazoul J. Climate and pest-driven geographic shifts in global coffee production: implications for forest cover, biodiversity and carbon storage. *PLOS One*. 2015 Jul 15;10(7):e0133071.

44 WFP. Regional Drought Response Plan for the Horn of Africa: 2023 | World Food Programme [Internet]. 2023 Jan [cited 2025 Jul 14]. p. 27. Available from: https://www.wfp.org/publications/regional-drought-response-plan-horn-africa-2023

45 Matotek AM, Regodić D, Maljugić B. The influence of climate change challenges on supply chain flexibility. *Entrep Eng Manag Clim Change Eng Chall*. 2025;(Part two):203–10.

46 Chanchal C, Sharma R, Bhushan H, Kumar V, Kumari D, Brahma U, et al. Impact of climate change on agriculture and adaptive strategies: a comprehensive review. *Agric Assoc Text Chem Crit Rev*. 2025;13(1):312–23.

47 Mariam MTB, Nowrozy KJ, Christina GY. Impact of climate change on women's mental health: a narrative review. *OALib*. 2023;10(12):1–15.

48 Craig RK. *Ocean Adaptation* [Internet]. Rochester, NY: Social Science Research Network; 2016 [cited 2025 Jun 29]. Available from: https://papers.ssrn.com/abstract=2600644

49 Hoegh-Guldberg O, Mumby PJ, Hooten AJ, Steneck RS, Greenfield P, Gomez E, et al. Coral reefs under rapid climate change and ocean acidification. *Science*. 2007 Dec 14;318(5857):1737–42.

50 Lebrec M, Stefanski S, Gates R, Acar S, Golbuu Y, Claudel-Rusin A, et al. Ocean acidification impacts in select Pacific Basin coral reef ecosystems. *Reg Stud Mar Sci*. 2019 Apr 1;28:100584.

51 Kennedy EV, Perry CT, Halloran PR, Iglesias-Prieto R, Schönberg CHL, Wisshak M, et al. Avoiding coral reef functional collapse requires local and global action. *Curr Biol*. 2013 May 20;23(10):912–8.

52 Hanich Q, Wabnitz CCC, Ota Y, Amos M, Donato-Hunt C, Hunt A. small-scale fisheries under climate change in the Pacific Islands region. *Mar Policy*. 2018 Feb 1;88:279–84.

53 Keller BD, Gleason DF, McLeod E, Woodley CM, Airamé S, Causey BD, et al. Climate change, coral reef ecosystems, and management options for marine protected areas. *Environ Manage*. 2009 Dec 1;44(6):1069–88.

54 Alahmad B, Khraishah H, Althalji K, Borchert W, Al-Mulla F, Koutrakis P. Connections between air pollution, climate change, and cardiovascular health. *Can J Cardiol.* 2023 Sep 1;39(9):1182–90.

55 Sario MD, Katsouyanni K, Michelozzi P. Climate change, extreme weather events, air pollution and respiratory health in Europe. *Eur Respir J.* 2013 Sep 2;42(3):826–43.

56 Xu J, Su Z, Liu C, Nie Y, Cui L. Climate change, air pollution and chronic respiratory diseases: understanding risk factors and the need for adaptive strategies. *Environ Health Prev Med.* 2025;30:7–7.

57 Mahendran R, Ju K, Yang Z, Gao Y, Huang W, Yu W, et al. Wildfire-related air pollution and infectious diseases: systematic review and meta-analysis. *ACS Environ Au.* 2025 May 21;5(3):253–66.

58 Reid CE, Considine EM, Watson GL, Telesca D, Pfister GG, Jerrett M. Effect modification of the association between fine particulate air pollution during a wildfire event and respiratory health by area-level measures of socio-economic status, race/ethnicity, and smoking prevalence. *Environ Res Health.* 2023 Apr;1(2):025005.

59 Fisk WJ. Review of some effects of climate change on indoor environmental quality and health and associated no-regrets mitigation measures. *Build Environ.* 2015 Apr 1;86:70–80.

60 World Resources Institute. Climate Watch Historical GHG Emissions [Internet]. 2025. Available from: https://www.climatewatchdata.org/ghg-emissions

61 Dutta A. Energy conservation and its impact on climate change. In: Sikdar PK, editor. *Environmental Management: Issues and Concerns in Developing Countries* [Internet]. Cham: Springer International Publishing; 2021 [cited 2025 Jul 14]. p. 139–50. https://doi.org/10.1007/978-3-030-62529-0_8

62 Vernon N, Black S, Parry I. Still not getting energy prices right: a global and country update of fossil fuel subsidies. *IMF Work Pap.* 2021 Sep;2021(236):1.

63 Global CCS Institute. Global Status of Ccs 2024: Collaborating for a Net-Zero Future [Internet]. 2024 [cited 2025 Jul 15]. p. 89. Available from: https://coilink.org/20.500.12592/53g8p7d

64 Matthews HD, Wynes S. Current global efforts are insufficient to limit warming to 1.5 °C. *Science.* 2022 Jun 24;376(6600):1404–9.

65 Warren R, Andrews O, Brown S, Colón-González FJ, Forstenhäusler N, Gernaat DEHJ, et al. Quantifying risks avoided by limiting global warming to 1.5 or 2 °C above pre-industrial levels. *Clim Change.* 2022 Jun 29;172(3):39.

66 Bevacqua E, Schleussner CF, Zscheischler J. A year above 1.5 °C signals that earth is most probably within the 20-year period that will reach the Paris agreement limit. *Nat Clim Change.* 2025 Mar;15(3):262–5.

67 United Nations Environment Programme, Olhoff A, Bataille C, Christensen J, Den Elzen M, Fransen T, et al. Emissions Gap Report 2024: No More Hot Air ... Please! With a Massive Gap between Rhetoric and Reality, Countries Draft New Climate Commitments [Internet]. United Nations Environment Programme; 2024 [cited 2025 Jul 14]. Available from: https://wedocs.unep.org/20.500.11822/46404

68 Bertoldi P. From Kyoto to Paris and Glasgow: overview of international climate agreements and regimes, their limitations, and the role of energy efficiency and sufficiency. *Eceee Summer Study Proceedings*, 2022. p. 391–402. Available from: https://proceedings.eceee.org/papers/proceedings2022/3-284-22_Bertoldi_REV.pdf

69 OECD. Climate Finance Provided and Mobilised by Developed Countries in 2013–2021: Aggregate Trends and Opportunities for Scaling Up Adaptation and Mobilised Private Finance [Internet]. OECD; 2023 [cited 2025 Jul 14]. (Climate Finance and the USD 100 Billion Goal). Available from: https://www.oecd-ilibrary.org/environment/climate-finance-provided-and-mobilised-by-developed-countries-in-2013-2021_e20d2bc7-en

70 Harry SJ, Maltby T, Szulecki K. Contesting just transitions: climate delay and the contradictions of labour environmentalism. *Polit Geogr*. 2024 Jun 1;112:103114.

71 Romito S, Vurro C, Pogutz S. Joining multi-stakeholder initiatives to fight climate change: the environmental impact of corporate participation in the science based targets initiative. *Bus Strategy Environ*. 2024;33(4):2817–31.

72 Scheffer M, Carpenter S, Foley JA, Folke C, Walker B. Catastrophic shifts in ecosystems. *Nature*. 2001 Oct;413(6856):591–6.

73 Solomon S, Plattner GK, Knutti R, Friedlingstein P. Irreversible climate change due to carbon dioxide emissions. *Proc Natl Acad Sci*. 2009 Feb 10;106(6):1704–9.

74 Mohd Raub AA, Bahru R, Mohamed MA, Latif R, Mohammad Haniff MAS, Simarani K, et al. Photocatalytic activity enhancement of nanostructured metal-oxides photocatalyst: a review. *Nanotechnology*. 2024;35(24):242004.

75 Bodar C, Spijker J, Lijzen J, Waaijers-van Der Loop S, Luit R, Heugens E, et al. Risk management of hazardous substances in a circular economy. *J Environ Manage*. 2018 Apr;212:108–14.

76 Kahlon SK, Sharma G, Julka JM, Kumar A, Sharma S, Stadler FJ. Impact of heavy metals and nanoparticles on aquatic biota. *Environ Chem Lett*. 2018 Sep;16(3):919–46.

77 Kate A, Sahu LK, Pandey J, Mishra M, Sharma PK. Green catalysis for chemical transformation: the need for the sustainable development. *Curr Res Green Sustain Chem*. 2022 Jan 1;5:100248.

78 Duarte-Casar R, Romero-Benavides JC, Bailon-Moscoso N, Navarro-Rojas M, Rojas-Le-Fort M. Synergies between green chemistry principles and the sustainable development goals (SDGs) in developing eco-friendly pesticides. In: Babaniyi BR, Babaniyi EE, editors. *The Interplay of Pesticides and Climate Change: Environmental Dynamics and Challenges* [Internet]. Cham: Springer Nature Switzerland; 2025 [cited 2025 Mar 5]. p. 291–312. https://doi.org/10.1007/978-3-031-81669-7_11

2

The Role of Catalysts in Combating Climate Change

Innocent Ojeba Musa[1], Abdulhakeem Idris Abdulhakeem[2], Mustapha Abdulsalam[1], Miracle Uwa Livinus[3], Kolawole Saheed Oluwasola[4], Ummi Aminu Maikano[3], Abraham Esther Joshua[5], and Muhammed Abdulazeez[1]*

[1] *Department of Microbiology, Skyline University Nigeria, Kano, Nigeria*
[2] *Department of Biodiversity and Bioresources, Al-Farabi Kazakh National University, Almaty, Kazakhstan*
[3] *Department of Biochemistry, Skyline University Nigeria, Kano, Nigeria*
[4] *Department of Aquaculture, Istanbul University, Istanbul, Turkiy*
[5] *Department of Biomedicine, Al-Farabi Kazakh National University, Almaty, Kazakhstan*

*Corresponding author: innocentmusa0011@gmail.com

2.1 Introduction

Global warming is undoubtedly one of the most global issues in the modern world developed mainly due to the concentration of greenhouse gases such as carbon dioxide (CO_2), methane (CH_4), and nitrous oxide (N_2O). These emissions include industrial ones, emissions from energy production, agriculture, and deforestation. The consequences are global warming, sea level rise, and extreme weather events affecting ecosystems, human well-being, and economic growth globally [1]. Solving these challenges calls for sound and practical solutions to mitigate emissions of greenhouse gases and reform industrial practices [2]. Catalysis has a major role in this process since it is a switch to more efficient and green chemical processes. A catalyst is a substance that increases the rate of a chemical reaction but remains chemically unaltered after the reaction has occurred it is employed.

In the environmental context, a catalyst is a substance that speeds up the rate of a chemical reaction without being used up or permanently changed. This is achieved by reducing the energy needed for the reaction to occur so that it happens faster and in less energy. Such property makes catalysts relevant in many industries, for instance, in petrochemicals, pharmaceutical, and food industries where they aid in improving the manufacturing processes besides lowering on waste and energy usage [3, 4]. When it comes to climate change mitigation, catalysts are even more

Metal Oxide Nanocatalysts for Sustainable Energy Production, First Edition. Edited by Abayomi Bamisaye et al.
Published 2026 by Wiley-VCH GmbH.

valuable. They enable important activities, including carbon capture and storage (CCS), by which the emission of CO_2 is captured from industrial sources and then buried in the ground to minimize their emission into the atmosphere [5]. Catalysts are also considered to be important to render greenhouse gases useful chemicals and fuels like carbon monoxide (CO) to methanol or CH_4 to synthetic fuels. In particular, they are attached to renewable energy technologies such as hydrogen (H_2) production utilizing water splitting and biofuel production by utilizing biomass [6].

This chapter explains the principles of catalysis, how it exists, and how does it work. It looks at the processes in green chemistry where catalysts form the basis of less pollutive, and more sustainable industrial processes. This chapter also explores the case of catalytic technologies for the minimization of greenhouse gas emissions, the promotion of renewable energy generation, and the resolution of issues regarding catalyst performance, durability, and cost-effectiveness [7]. Including catalytic technologies in climate-related policies and management of industries worldwide will help progress the cause of environmental conservation and extend the globe's lifespan [8].

2.2 Fundamental Concepts of Catalysis

Catalysis is the process through which the rate of a chemical reaction is enhanced by the aid of a substance called a catalyst. Catalysts act as agents and influence this energy in a way that decreases the required reaction energy making the reaction to occur at a faster rate. In most cases, the catalyst does not undergo the shade change and is reusable in the subsequent reactions. This makes catalysis an indispensable activity in most industries and environmental applications [3].

2.2.1 Types of Catalysts

Depending on the physical state, chemical composition, and the kind of reaction they assist; catalysts can be categorized in the following: Homogeneous, heterogeneous, and biocatalysts. Both types have certain features, uses, and functions towards climate change prevention.

2.2.2 Homogeneous Catalysts

Homogeneous catalysts are those which are in the same phase as the reactant, either in a liquid or a gaseous phase. These catalysts work uniformly at the molecular level with the reactants, and a high selectivity and efficiency are observed in the chemical reactions. A good example is in the preparation of biodiesel in which a homogeneous catalytic system based on alkali metals, that is, sodium or potassium hydroxide is employed to convert glycerides into fatty acid methyl esters (FAMEs). From the perspective of climate change, homogeneous catalysis has been applied to carbon capture technologies, for instance, the conversion of CO_2 to useful chemicals [3].

The major disadvantages of homogeneous catalysts are that the catalyst is difficult to separate from the products and the catalyst is easily deactivated. These limitations are being handled with the improvement in the recycling of the catalyst and the introduction of more stable catalyzing systems. Scientists are also looking into the applications of some techniques like ionic liquids and organo-catalysts instead of conventional nonhomogeneous well-known catalysts, which the authors believe will be environmentally friendly [6].

2.2.3 Heterogeneous Catalysts

Heterogeneous catalysts are those which are in a different phase with the reactants; whilst they are in a solid phase while they reactants are in a liquid or a gas phase. These catalysts are also applied broadly in industries due to the ease of their separation from the reaction mixture, hence easy to recover and reuse. The solid catalyst gives a support onto which the reactants can be adsorbed so that are signaled to break and form chemical bonds.

Catalytic cracking used in petroleum reformation is probably one of the most illustrative examples of heterogeneous catalysis. And this process involves the conversion of large hydrocarbons into components like gasoline and diesel with the help of a solid catalyst. In connection with climate change, heterogeneous catalysts are also at the heart of CO_2 utilization processes, during which CO_2 is transformed into ordinary products such as methanol or synthetic fuels. These processes help reduce the amount of CO_2 in the atmosphere and also create valuable alternatives to fossil fuels [8].

2.2.4 Biocatalysts

Biocatalysts are those catalysts, which are formed by living organisms. They are highly specific and powerful possessing the ability to trigger mechanisms at low pressures, temperatures, and physiological pH. Biocatalysts are important in different functions and are used in pharmaceutical and food production. It is expected that their contribution to combating climate change is becoming increasingly important because they can promote reactions that generate biofuels and degrade hazardous substances.

For instance, cellulase, an enzyme, is employed to depolymerize cellulose derived from plant biomass to fermentable sugars, which in turn are used to synthesize bioethanol—an environmentally friendly fuel. Biocatalysis is also involved in the degradation prospects of plastic wastes, especially the synthesis of biodegradable materials. Biocatalysts serve the objectives of the green chemistry principles in bringing about environmentally friendly methods of producing commodities with minimal pollution [9].

2.3 Catalysts in Green Chemistry

Catalysis is one of the most important strategies of green chemistry that enables the processes with low environmental cost and sustainability. In this way, while enhancing the reaction efficiency and minimizing the amount of waste, the use of catalysts constitutes a major innovation in the field of environmentally friendly processes [10]. In this section, it will be an endeavor to understand how indeed catalysts help sustain this process, cut emissions where needed and make reactions smoother.

2.3.1 Role in Sustainable Processes

Catalysts lower the activation energy, thereby minimizing the use of forceful reactants and considerable energy, thus enhancing sustainability [11]. For example, in the generation of large volume chemicals such as ammonia and methanol, the systems guarantee optimum production with low power utilization [12]. Catalytic processes also contribute to the maximizing of atom economy in that most of the atoms on the reactant side of the equation are incorporated into the final product and there will be minimal waste [9].

2.3.2 Reducing Waste and Emissions

Another advantage of the use of catalysts in green chemistry, is the elimination of the formation of byproducts. Catalysts help achieve the maximum conversion of reactants to the products while inhibiting the formation of undesired products, often reducing side product streams [13]. For example, catalytic converters in automobiles drastically reduce toxic emissions of nitrogen oxides (NO_x) and CO, changing them into relatively harmless gases of nitrogen and CO_2 respectively [14]. Moreover, in industrial applications, catalysts have supported the achievement of sustainability by reducing the emission of greenhouse gases [15].

2.3.3 Enhancing Reaction Efficiency

Catalysts reduce the activation energy of the reaction to enhance the rate of the reaction without adding outside energy to the system [16]. This increases the efficiency not only decreases the energy usage but also increases the production rates and lowers the operating expenses [3]. For example, the usage of the catalytic cracking process to convert hydrocarbons in the petroleum industry increases fuel production output and reduces cost as well as environmental impacts [11].

2.4 Catalytic Processes for Greenhouse Gas Reduction

Greenhouse gas emissions are crucially sensitive to catalytic processes; their effective use can help to capture CO_2, convert it and transport. The Figure 2.1 below illustrates the key catalytic processes involved in reducing greenhouse gases. It highlights six main pathways: CO_2 to methanol conversion, Fischer-Tropsch synthesis, electrochemical reduction, H_2 production, photocatalysis, and biofuel production, all of which utilize catalysts to transform or reduce carbon emissions.

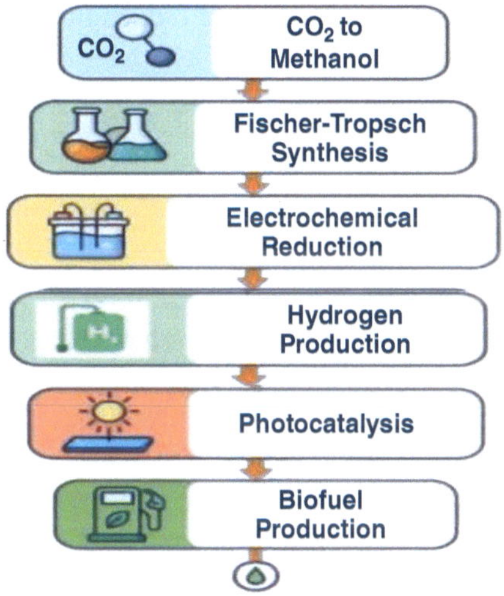

Figure 2.1 Processes for greenhouse gas reduction.

The sections below explain different catalytic technologies that are involved in the mitigation of greenhouse gases such as CO_2, CH_4, and N_2O, which are known to increase world's temperature.

2.4.1 Carbon Capture and Storage (CCS)

CCS is one of the most important technologies for reducing CO_2 emissions of industrial origin. Catalysts are very important during the process of chemisorption and transformation into stable forms for storage of CO_2 [17]. Integrated absorption-desorption processes include amine scrubbing by exploiting proficient catalysts like advanced ones to capture CO_2 from flue gases [18]. In addition, improved solid sorbents and metal-organic frameworks have increased the performance of CO_2 capture and opened the possibility for large-scale application [19].

2.4.2 Catalytic Conversion of CO_2

The conversion of CO_2 to valuable chemical and energetic products is a promising strategy for the mitigation of this greenhouse gas. Transition metal and their complexes are acting as catalysts for the activation of the stable CO_2 molecule and converting it into valuable products, including methanol, formic acid, and hydrocarbons, etc. CO_2 reduction via photocatalytic and electrocatalytic processes is also emerging as a promising method to integrate renewable energy sources for achieving a circular carbon economy [20].

2.4.3 CH_4 and N_2O Reduction

CH_4 and N_2O are greenhouse gases and have global warming potentials more than CO_2. Emerging technologies for lowering these emissions include combustion using a catalytic system and the selective catalytic reduction (SCR) technology [21]. In case of methane, the option of catalytic combustion is provided for its complete oxidation to CO_2 and water, significantly reducing its atmospheric concentration [22]. For N_2O, SCR using vanadium and zeolite-based catalysts is effective in reducing emissions from industrial and agricultural sources [23].

2.4.4 Role in Synthetic Fuels Production

Catalysts are also necessary when using renewable resources to produce synthetic fuels that are a clean replacement for fossil fuels. Fischer-Tropsch synthesis and conversion of CO_2 to methanol depend on highly efficient catalysts, which synthesize syngas (a mixture of CO and H_2 into liquid hydrocarbons and alcohols) [24]. These synthetic fuels can be utilized in existing structures and hence have a significantly less carbon footprint for transportation and other sectors [25].

2.5 Catalysts in Renewable Energy Technologies

Catalysts are a central part of the vast majority of renewable power generation and storage technologies. This part seeks to present catalysts' applications and impacts on H_2 generation, photovoltaic cells, biofuel generation, and electrochemical reduction concerning renewable power systems research [26].

2.5.1 H_2 Production and Fuel Cells

H_2 required for clean energy application is synthesized and used through catalysts, and it is likely to make a profound impact on the reduction of carbon emissions. Some of the industrial processes that H_2 is produced by include steam reforming, water electrolysis, and photocatalytic reactions [27]. During steam reforming, nickel-based catalysts are used to produce H_2 and CO_2 from natural gas where the recent developments concern the minimization of CO_2 emission during the process [28].

In the use of H_2 in fuel cells, platinum-based materials dominate the ORR at the cathode due to their stability and efficiency [28]. There is an active research effort to find the non-precious metal catalysts that could be as effective at a lower cost to promote the economic feasibility of the fuel cell technology [22].

2.5.2 Photocatalysis for Solar Energy Conversion

Photocatalysis exploits the use of light, the catalysts being very vital in converting the sun's light into chemical energy. Among the photocatalysts studied, titanium dioxide (TiO_2) has received the most attention due to its stability, non-toxicity, and photocatalytic efficiency under the UV light [4]. Research to improve its performance under visible light have seen the formation of doped and composite materials [26].

A practical area includes photocatalytic water splitting for H_2 production is another example, where catalysts make water molecules split into H_2 and oxygen using sunlight [29]. This process is a sustainable way of producing H_2 while less dependent on fossil fuels [30].

2.5.3 Catalytic Role in Biofuel Production

They are used in the synthesis of fuels from renewable resources of biomass to improve the reaction rates as well as the selectivity. In biodiesel production through transesterification of vegetable oils and animal fats, alkaline and acid catalysts, such as sodium and sulfuric acids are employed [5]. Recently, solid acids and bases that act as heterogeneous catalysts are preferred over their homogeneous counterparts owing to their features, such as recyclability and eco-friendly nature [31].

Use of the bio-oils obtained from the biomass pyrolysis and hydrothermal liquefaction in the generation of enhanced fuel requires their conversion via catalysts [32]. Catalysts are used in deoxygenation, hydrogenation, and cracking processes to upgrade bio-oils into hydrocarbons that can be used in transport [11].

2.5.4 Catalysts in Electrochemical Reduction

In the electrochemical reduction processes, the catalysts promote the reduction of CO_2 and other molecules into higher-value useable forms of energy, hence provides a renewable method of energy storage and use. These reactions require electrocatalysts, including metal oxides and carbon materials to increase the energy efficiency and selectivity of the reaction process [33]. Catalytic electrochemical CO_2 reduction to fuels such as methane and ethylene is one area of interest, and these processes occur under ambient conditions [34].

Advances in catalyst synthesis, such as single-atom catalysts and nanostructured materials enhance the efficiency of the electrochemical systems placing them on par with traditional energy supplies [35].

2.6 Challenges and Innovations in Catalysis for Climate Change

Several issues are inherent in the creation and utilization of catalytic technologies, such as the need to enhance efficacy, durability, and affordability. On such aspects, effort needs to be made in order for catalysis to make a bigger contribution to the fight against climate change. This section raises and discusses the present problems and developments in catalysis, including efficiency, selectivity, stability, new materials, and costs [21].

2.6.1 Efficiency and Selectivity Improvements

One of the major problems of the catalysis is related to the improvement of the activity and selectivity of catalysts. The effectiveness of any process that uses a catalyst to bring about the desired reactions can be best summarized as follows: catalyst improvement includes single atom catalysts and bimetallic catalyst systems, where improvements in selectivity and activity have been shown [29]. Structured materials and big data learning also assumed significant roles for the prediction and enhancement in catalytic descriptions as well [26].

2.6.2 Catalyst Stability and Longevity

Sintering, coking, or poisoning of the active sites is one of the major problems which affecting the life span and the economy of the catalysts. The new supports and how their surfaces can be treated are more actively being researched to improve stability [26]. For instance, zeolites and metal-organic frameworks as supports allow higher thermal stability and lower tendency towards deactivation [2]. Regenerable and self-healing types of catalysts are likewise on the rise as research to enhance the lifespans of these catalysts reduces operational costs [35].

2.6.3 Emerging Materials and Nanotechnology

Nanotechnology has brought a drastic change in evaluating the catalysts in the field of catalytic processes; the structure and size of catalysts at nanospheres level is possible. These nanocatalysts are characterized by a large surface area and quantum size effects, which results in greater catalytic activity and selectivity [28]. Graphene, carbon nanotubes, metal-organic frameworks, and similar emergent materials are incorporated into the catalytic systems to enhance these systems' performance in multiple aspects of environmental degradation [34].

2.6.4 Overcoming Economic Barriers

The relatively expensive nature of some of the catalytic materials, more so the platinum group metals such as platinum and palladium are economic factor. Efforts to minimize the use of these materials include the use of earth-abundant metal catalysts like those based on iron, nickel, and cobalt [33]. Furthermore, concerning the enhancements of technique, process intensification, and development of catalytic reactors that increase heat and mass transfer efficacy are currently being done with an aim of increasing the cost-efficiency of the catalytic technologies [28].

Policy incentives to overcome financial barriers can also be represented in innovative financing models for the implementation of advanced catalytic systems in important industries that have a major impact on the emission of greenhouse gases [22].

2.7 Industrial Applications of Catalysis in Climate Change Mitigation

Catalysis is of central importance in several industrial processes which are directly related to the emission of greenhouse gases. Since these processes are selective, an improvement of their efficiency has the potential to slow climate change through catalytic technologies. The following part looks at the role of catalysis in today's most critical sectors, including petrochemicals, cement, steel production, and waste management, referring to the emission control [28].

2.7.1 Petrochemical Industry

The petrochemical industry is perhaps one of the most intensive users of catalytic technologies. Catalysts are used in the processing of petroleum, in making transportation fuels, and in the manufacture of useful chemicals. Research advancement in catalysis for cracking, hydrocracking, and reforming has contributed to a decrease in CO_2 emission due to enhanced energy conversion and production of the desired products [36]. Moreover, research and creation of catalytic processes for converting biomass into biofuels provides a renewable alternative to fossil fuels, contributing to the decarbonization of the petrochemical sector [28].

2.7.2 Cement and Steel Production

The industries that give out the largest shares of CO_2 emissions include cement and steel. Efforts are being made to use technology to cut emissions from these sectors by the use of catalytic technologies. In cement production, catalysts may help improve the clinker formation process efficiency and decrease the calcination temperature, leading to less CO_2 emissions [11]. Catalysts are applied in such processes as H_2-based direct reduction of iron, which has the potential to replace carbon-centric methods and therefore reduce the emission of CO_2 by up to two-thirds [35].

2.7.3 Waste Management and Recycling

Catalysis is also well-utilized in waste disposal and recycling, such as in circumstances of pyrolysis and gasification of waste plastics and biomass. The processes upgrade waste materials into potential chemicals and fuels and bring down the use of landfills by minimizing adverse effects of waste management [34]. Moreover, removal of emissions by automobiles, including NO_x and CO using catalytic converters in vehicles will also improve air quality and inhibit climate change [33].

The application of these catalytic technologies of higher efficiency is crucial in these industries to meet the climate goal as well as transform industries towards a cleaner image [22].

2.8 Catalysis in Policy and Regulation

The effectiveness of catalytic technologies depends on policy and legal requirements, at the same time they are shaped by the latter. Catalysts that can help address climate change thus require research, development, and deployment, and policies have an important function in this process. This section analyzes the incorporation of the catalysts in international climate change agreements and the regulatory frameworks of catalytic technologies and the encouragements for catalyst research and facilitation [1].

2.8.1 Catalysts in Global Climate Agreements

International climate treaties, including the Paris Agreement, direct the development of new technologies aimed at reducing greenhouse gases. Catalysts, as one of the key components for reaching the emission reduction goals, are usually discussed in these agreements. For example, the use of catalysts in CCS technologies and renewable energy sources mentioned in the existing global climate policies [37]. That these agreements incorporate the catalytic processes makes them central to international frames of addressing climate change [38].

2.8.2 Regulatory Standards for Catalytic Technologies

Catalytic technologies utilized in industries or with regard to environmental issues are provided with necessary requirements of safety along with efficiency and acceptable emission levels due to the established regulations. The regulation of the application of catalysts is done under various standards charted by bodies like the Environmental Protection Agency (EPA) in the United States or the REACH (Registration, Evaluation, Authorization, and Restriction of Chemicals) regulation in the European Union [39]. These intents of these regulations are to reduce the effects of environmental ramifications, in addition to encouraging the usage of high-performance catalytic systems [1].

2.8.3 Incentives for Catalyst Research and Adoption

Multilateral organizations and governments use diverse motivating factors while encouraging the development and use of catalytic technologies. Some of the incentives are research grants, tax credits, and subsidies to industries that use catalytic techniques to cut emissions. Similarly, the Advanced Research Projects Agency-Energy (ARPA-E) of the U.S. Department of Energy and the Horizon Europe of the European Union fund catalytic research needed to combat climate change. Such incentives are important for the development of the catalyst technology as well as the industrial applications [40].

As the section on policies and regulations shows, catalytic technologies for the mitigation of climate change require supportive policies and regulations that can best be provided by international governments. These policies will enhance the technologies by ensuring they are in line with the international and national policies, thereby facilitating their integration into climate action strategies [41].

2.9 Socioeconomic Impacts of Catalytic Technologies

In addition to the environmental improvement effects, catalytic technologies have profound socioeconomic effects. These effects cover employment generation, efficiency within industrial sectors, awareness, and information dissemination. The successful implementation of catalytic technologies into society can in a way contribute to economic development, enhance sustainability, and promote a greener economy [21].

2.9.1 Job Creation in the Green Economy

Since catalytic technologies are in their developmental stages and deployment, the technological breakthrough is optimistic to open various green economy employment opportunities. Areas such as renewable energy, chemicals, and environmentally friendly industries value the competence needed in making, analyzing and applying catalysts [42]. In line with the emergence of new catalysis alternatives, industries will progress in their improvement for pioneering industries that drive economic development and employment [43].

2.9.2 Cost-effectiveness in Industrial Processes

Catalysts have uses in enhancing the effectiveness of the industrial processes, thus reducing costs greatly. Catalysts increase the reaction rates permitting lower energy consumption, and raw material necessities, for example, in chemical synthesis, production of fuels, and waste disposal [3]. The above efficiencies make catalytic technologies a financially sustainable measure in reducing the operating cost of industries [21].

2.9.3 Public Awareness and Education

Public awareness and education on the function of catalysts in reducing the impacts of climate change are also important to fasten the generation approval on catalytic technologies. Awareness creation by educational programs and outreach activities ensures people get to know that catalysts can be used in obstructing emissions and enhancing sustainability [44]. Awareness campaigns can also create an increased market demand for products and services that incorporate catalytic technologies, thereby stimulating further investment in these solutions by the industry [40].

Catalytic technologies, which can create extensive socioeconomic values, need systemic improvements in environmental, industrial, and community perspectives. These effects amplify why further investment in and backing of catalytic innovation must persist [21].

2.10 Recommendations and Policy Implications

What the subject reveals is that there are many prospects for achieving ambitious goals tied to climate change due to the development of catalytic technologies. However, to realize the potential as highlighted above strategic recommendations and policy implications should be taken into consideration. This section outlines key recommendations for improving catalytic technology adoption and explores the policy frameworks necessary to support their implementation.

2.10.1 Integration with Global Climate Policies

Through full-fledged strategies of global climate policies, catalytic technologies should be incorporated as approaches to mitigating greenhouse gas emissions. Governments should understand that catalysts contribute to the optimization of industrial processes and renewable energy systems, meaning that governments should incorporate catalytic solutions into their national and multilateral climate action plans [43]. The milestones and targets in the use of catalysts should be openly defined and planned, thus working as a stimulant to encourage more investments and development in this age [42].

2.10.2 Promoting Research and Development (R&D)

The method of investment in R&D is important, especially in the improvement of the catalytic technologies. Thus, governments and private sectors should offer funding and incentives to R&D targets relating to efficiency improvement, selectivity, and durability of catalysts [40]. It is deemed that cooperation among academic institutions, industry, and government can enhance the speed of new catalytic material and process discovery and achieve more innovation and technology development [21].

2.10.3 Collaboration Across Sectors

Evaluating implementation of these catalytic technologies will necessitate cooperation from industry and academic sectors, government, and NGOs. Partnerships and networks in this sense enable knowledge exchange, resource consolidation, and synchronize actions towards doing things together [3]. Intersectoral relations can also improve the level of scaling and commercialization of catalytic solutions in order to maximize their effect on as many people as possible [44].

As for the potential paths towards further enhancements of the catalytic technologies, it is proposed to include the climate policies action plans of a range of countries, advance the R&D agenda, and ensure that the catalytic technologies are approached as a tool wherein different branches of industry should collaborate. To fully unlock the potential of catalysts in mitigating climate change and facilitating sustainable development, the following measures are crucial [40].

2.11 Conclusion

Catalytic technologies are crucial in addressing climate change by improving the efficiency of chemical processes, minimizing consumption of energy, and minimizing waste. They play a pivotal role in achieving environmental goals, including reduction in greenhouse gas emissions and providing contributions towards the transition to renewable energy. Irrespective of these advantages, challenges remain in the improvement of catalyst stability, efficiency, and cost-effectiveness.

Continued innovation, particularly in materials science and nanotechnology, is needed. The integration of favorable policies and cross-sectoral cooperation will further enhance their impact. The shift via catalysis to a low-carbon economy is a priority. Its success depends on a continuous combination of technological innovation, strategic partnership, and good policy implementation to ensure long-term sustainability.

References

1 United Nations Environment Programme (UNEP). Emissions Gap Report 2020. 2020. Available from: https://www.unep.org/resources/emissions-gap-report-2020

2 Aransiola SA, Leh-Togi Zobeashia SS, Ikhumetse AA, Musa OI, Abioye OP, Ijah UJJ, et al. Niger Delta mangrove ecosystem: biodiversity, past and present pollution, threat and mitigation. *Reg Stud Mar Sci.* 2024;75:103568. http://doi.org/10.1016/j.rsma.2024.103568

3 Abbas A, Cross M, Duan X, Jeschke S, Konarova M, Huber GW, et al. Catalysis at the intersection of sustainable chemistry and a circular economy. *One Earth.* 2024;7(5):738–41. https://doi.org/10.1016/j.oneear.2024.04.018

4 Musa IO, Samuel JO, Adams M, Nathaniel V, Maude AM, Mahmud ME, et al. Microplastics and nanoplastics in soil. In: *Phytoremediation in Food Safety.* Taylor & Francis Group. CRC Press; 2024. Chapter (8). p. 10. https://doi.org/10.1201/9781032683751-10.

5 Musa IO, Auta HS, Ilyasu US, Aransiola SA, Makun HA, Adabara NU, et al. *Micro- and Nanoplastics in Environment: Degradation, Detection, and Ecological Impact.* Springer Nature; 2023. https://doi.org/1007/s41742-023-00551-9

6 Farooqi AS, Al-Swai BM, Ruslan FH, Zabidi NAM, Saidur R, Syed Muhammad SAF, et al. Catalytic conversion of greenhouse gases (CO_2 and CH_4) to syngas over Ni-based catalyst: effects of Ce-La promoters. *Arab J Chem.* 2020;13(6):5740–9. https://doi.org/10.1016/j.arabjc.2020.04.012

7 Gana M, Isibor PO, Damisa JI, Iseghohi F, Musa IO, Oyewole OA. *Nanoparticles-Based Remediation and Environmental Cleanup.* Springer Nature Switzerland; 2024. Chapter (9). p. 161–82. https://doi.org/10.1007/978-3-03154154-4

8 Ma C, Dai Y, Zhuang G, Zhang H. Catalytic conversion of carbon dioxide into fuel chemicals: progress, challenges, and future directions. *J Environ Chem Eng.* 2025;13(5):117767. https://doi.org/10.1016/j.jece.2025.117767

9 Kaur A, Goyal A, Kumari S, Garg M, Sharma S, Akram M, et al. Green chemistry and catalysis: an emerging sustainable approach in synthetic chemistry. *BIO Web Conf.* 2024;86:01027. https://doi.org/10.1051/bioconf/20248601027

10 Kaur A, Kansal SK. Bi_2WO_6 nanocuboids: an efficient visible light active photocatalyst for the degradation of levofloxacin drug in aqueous phase. *Chem Eng J.* 2016;302:194–203.

11 Akhtar MS, Ali S, Zaman W. Recent advancements in catalysts for petroleum refining. *Catalysts.* 2024;14(12):841. https://doi.org/10.3390/catal14120841

12 Jin Z, Jiang L, He Q. Critical learning from industrial catalysis for nanocatalytic medicine. *Nat Commun.* 2024;15:3857. https://doi.org/10.1038/s41467-024-48319-9

13 Kanwate C, Solunke N, Waddar P, Dengle A, Kamble A. A review on green chemistry and catalysis. *Int J Pharm Sci.* 2025;3(3):3343–52. https://doi.org/10.5281/zenodo.15112605

14 Jena S, Padhy M. The environment and catalytic converter – a review. *Mater Sci Forum.* 2025;7(1):7. https://doi.org/10.33263/Materials71.007

15 Debecker DP, Hii KK(M), Moores A, Rossi LM, Sels B, Allen DT, et al. Shaping effective practices for incorporating sustainability assessment in manuscripts submitted to ACS sustainable chemistry & engineering: catalysis and catalytic processes. *ACS Sustain Chem Eng.* 2021;9(14):4936–40. https://doi.org/10.1021/acssuschemeng.1c02070

16 Wan Isahak WNR, Al-Amiery A. Catalysts driving efficiency and innovation in thermal reactions: a comprehensive review. *Green Technol Sustain.* 2024;2(2):100078. https://doi.org/10.1016/j.grets.2024.100078

17 Emmanuel O, Rozina, Ezeji TC. Advances in carbon dioxide capture and conversion technologies: industrial integration for sustainable chemical production. *Next Sustain.* 2025;6:100108. https://doi.org/10.1016/j.nxsust.2025.100108

18 Noor Y, Almomani F, Qiblawey H. Catalytic CO_2 conversion to C1 value-added products: review on latest catalytic and process developments. *Fuel.* 2023;345:128178. https://doi.org/10.1016/j.fuel.2023.128178

19 Xu K, Zhang Q, Zhou X, Zhu M, Chen H. Recent progress and perspectives on photocathode materials for CO_2 catalytic reduction. *Nanomaterials.* 2023;13(10):1683. https://doi.org/10.3390/nano13101683

20 Qu T, Wei S, Xiong Z, Zhang J, Zhao Y. Progress and prospect of CO_2 photocatalytic reduction to methanol. *Fuel Process Technol.* 2023;251:107933. https://doi.org/10.1016/j.fuproc.2023.107933

21 Jayabal R. Advancements in catalysts, process intensification, and feedstock utilization for sustainable biodiesel production. *Results Eng.* 2024;24:103668. https://doi.org/10.1016/j.rineng.2024.103668

22 Andrade LS, Lima HHB, Silva CTP, Amorim WLN, Poço JGR, López-Castillo A, et al. Metal–organic frameworks as catalysts and biocatalysts for methane oxidation: the current state of the art. *Coord Chem Rev.* 2023;481:215042. https://doi.org/10.1016/j.ccr.2023.215042

23 Elkaee S, Phule AD, Yang JH. Advancements in (SCR) technologies for NOx reduction: a comprehensive review of reducing agents. *Process Saf Environ Prot.* 2024;184:854–80. https://doi.org/10.1016/j.psep.2024.02.046

24 Voeten RLC, Hendriks F, Bezemer GL. Fischer–Tropsch synthesis for the production of sustainable aviation fuel: formation of tertiary amines from ammonia contaminants. *ACS Omega.* 2024;9(29):31974–85. https://doi.org/10.1021/acsomega.4c03734

25 Jung W, Rhim GB, Kim KY, Youn MH, Chun DH, Lee J. Comprehensive analysis of syngas-derived Fischer–Tropsch synthesis using iron-based catalysts with varied acidities. *Chem Eng J.* 2024;484:149408. https://doi.org/10.1016/j.cej.2024.149408

26 Van Pham V, Park S. Catalysis for renewable energy and sustainable development. *Top Catal.* 2024;67:1053–4. https://doi.org/10.1007/s11244-024-02000-z

27 Fujishima A, Honda K. Electrochemical photolysis of water at a semiconductor electrode. *Nature.* 1972;238(5358):37–8. https://doi.org/10.1038/238037a0

28 Kumar N, Aepuru R, Lee S-Y, Park S-J. Advances in catalysts for hydrogen production: a comprehensive review of materials and mechanisms. *Nanomaterials.* 2025;15(4):256. https://doi.org/10.3390/nano15040256

29 Zhou J, Tian Y, Gu H, Jiang B. Photocatalytic hydrogen evolution: recent advances in materials, modifications, and photothermal synergy. *Int J Hydrogen Energy.* 2025;115:113–30. https://doi.org/10.1016/j.ijhydene.2025.03.088

30 Ferreira J, Andrade DI, Fuziki MEK, de Almeida LNB, Colpini LMS, Lenzi GG, et al. Catalytic systems in the reduction of nitrogen oxide emissions in diesel-powered trucks. *Sustainability.* 2022;14(11):6662. https://doi.org/10.3390/su14116662

31 Bhandarkar AD, Deshmukh AV, Gadre AS. Renewable energy sources – a key to protect environment: a review. *IEJRD Int Multidiscip J.* 2021; 6(NCTSRD):6. https://doi.org/10.17605/OSF.IO/YUBQE

32 Huber GW, Iborra S, Corma A. Synthesis of transportation fuels from biomass: chemistry, catalysts, and engineering. *Chem Rev.* 2006;106(9):4044–98. https://doi.org/10.1021/cr068360d

33 Tan M, Huang B, Su L, Jiao X, Feng F, Gao Y, et al. Amorphous nanomaterials: emerging catalysts for electrochemical carbon dioxide reduction. *Adv Energy Mater.* 2024;14:2402424. https://doi.org/10.1002/aenm.202402424

34 Guo S, Li X, Li J, Wei B. Boosting photocatalytic hydrogen production from water by photothermally induced biphase systems. *Nat Commun.* 2021;12:1343. https://doi.org/10.1038/s41467-021-21526-4

35 Molaei MJ. Recent advances in hydrogen production through photocatalytic water splitting: a review. *Fuel.* 2024;365:131159. https://doi.org/10.1016/j.fuel.2024.131159

36 Ashok A, Yuda A, Kumar A. Chapter 6 – photocatalytic water splitting and reduction of CO_2. In: Thomas S, Anas S, Joy J, editors. *Synthesis, Characterization, and Applications of Graphitic Carbon Nitride.* Amsterdam: Elsevier; 2023. p. 111–55. https://doi.org/10.1016/B978-0-12-823038-1.00009-X

37 United Nations Framework Convention on Climate Change (UNFCCC). UN Climate Change Conference Baku - November 2024. 2024. Available from: https://unfccc.int/cop29

38 Rosa L, Gabrielli P. Achieving net-zero emissions in agriculture: a review. *Environ Res Lett.* 2023;18(6):063002. https://doi.org/10.1088/1748-9326/acd5e8

39 European Commission. *The European Green Deal: Striving to be the First Climate-Neutral Continent* [Internet]. Brussels: European Commission;

2019. Available from: https://commission.europa.eu/strategy-and-policy/priorities-2019-2024/european-green-deal_en

40 Li H, Jian S, Baalousha M. Applications of catalytic nanomaterials in energy and environment. *Molecules*. 2023;28(10):4000. https://doi.org/10.3390/molecules28104000

41 International Energy Agency (IEA). *World Energy Outlook 2024*. Paris: IEA; 2024. https://www.iea.org/reports/world-energy-outlook-2024

42 Solecki W, Orlove B, Bolgert A. The role of catalysts in the climate adaptation process. *WIREs Clim Change*. 2025;16:e931. https://doi.org/10.1002/wcc.931

43 Dong H, Deng Q, Li C, Liu N, Zhang W, Hu M, et al. A comprehensive review on renewable power-to-green hydrogen-to-power systems: green hydrogen production, transportation, storage, re-electrification and safety. *Appl Energy*. 2025;390:125821. https://doi.org/10.1016/j.apenergy.2025.125821

44 Kwilinski A, Dacko-Pikiewicz Z, Szczepanska-Woszczyna K, Lyulyov O, Pimonenko T. The role of innovation in the transition to a green economy: a path to sustainable growth. *J Open Innov Technol Mark Complex*. 2025;11(2):100530. https://doi.org/10.1016/j.joitmc.2025.100530

3

Metal Oxide Nanocatalysts in Climate Change Mitigations

Abdulhakeem Idris Abdulhakeem[1], Innocent Ojeba Musa[2], Mustapha Abdulsalam[2], Miracle Uwa Livinus[3], Kolawole Saheed Oluwasola[4], Abraham Esther Joshua[5], and Muhammed Abdulazeez[2]*

[1] *Department of Biodiversity and Bioresources, Al-Farabi Kazakh National University, Almaty, Kazakhstan*
[2] *Department of Microbiology, Skyline University Nigeria, Kano, Nigeria*
[3] *Department of Biochemistry, Skyline University Nigeria, Kano, Nigeria*
[4] *Department of Aquaculture, Istanbul University, Istanbul, Turkiye*
[5] *Department of Biomedicine, Al-Farabi Kazakh National University, Almaty, Kazakhstan*

*Corresponding author: idrisabdulhakeem@yahoo.com

3.1 Introduction

Nanocatalysis refers to the use of nanoscale materials as catalysts to accelerate chemical reactions. The unique properties of nanomaterials, such as high surface area, tunable pore size, and quantum effects, make them highly effective in various catalytic processes [1]. These properties enable enhanced reaction rates, selectivity, and efficiency compared to bulk catalysts, which are crucial for industrial applications and environmental remediation. Nanocatalysis involves the application of catalysts that operate at the nanoscale, typically within the range of 1 to 100 nm [2]. These nanoscale catalysts exhibit unique physical and chemical characteristics that are not commonly found in their bulk counterparts because of the small size, quantum size effects, and the existence of active sites on their surfaces. These characteristics result in enhanced catalytic activities with enlarged rates of reaction, better selectivity towards the products of interest, and reduced activation energies [3].

The essence of nanocatalysts is that they can overcome several of the issues arising in a conventional catalytic process. For instance, their contributed surface activity makes it easier in the utilization of raw materials, hence less emissions and byproducts. This efficiency is most relevant in industrial applications where large flows of those chemicals go through some treatment. Additionally, using nanocatalysts it is possible to design the catalysts for the work at lower temperatures and pressures,

Metal Oxide Nanocatalysts for Sustainable Energy Production, First Edition. Edited by Abayomi Bamisaye et al.
Published 2026 by Wiley-VCH GmbH.

which significantly contributes to the minimal environmental load necessary for chemical production [4].

The nanocatalysts, especially metal oxides, have been the subject of much interest due to the many uses that involve environmental and energy industries. Some of the most frequently used metal oxides are titanium oxide (TiO_2), zinc oxide (ZnO), cerium oxide (CeO_2), and iron oxide (Fe_2O_3). These materials are chemically stable, nontoxic, and high in supply; hence, they can viably be deployed in large quantities [5].

It is worth noting that the role of metal oxide nanocatalysts in combating climate change is broad-based. Catalysts have a central position in the conversion of greenhouse gases such as carbon dioxide (CO_2) or other pollutants into more harmless compounds or useful products. For instance, TiO_2 which has been widely investigated for photocatalytic CO_2 reduction where light energy from natural sunlight is used to propel the conversion process [6]. Similarly, ZnO nanocatalysts are employed in the oxidative degradation of methane (CH_4), a potent greenhouse gas, thereby helping to curb atmospheric CH_4 levels [7].

Additionally, the development of renewable energy technologies depends on metal oxide nanocatalysts. They are applied in such processes as the splitting of water to produce hydrogen and in increasing the efficiency of solar cells. Due to their deposited ability to facilitate reactions at a nanoscale level, new pathways to make cleaner and, therefore, more sustainable energy sources are introduced [1].

3.2 Role of Metal Oxide Nanocatalysts in Reducing Greenhouse Gases

Emission of greenhouse gases has been identified as a factor in climate change; hence, it has to be controlled. Of all these strategies, catalysis employing metal oxide nanomaterial has proven the most effective because of the high conversion rates achievable and the prospect for scale-up. Nanocatalysts as metal oxides help enable reactions that neutralize damaging greenhouse gases or produce desirable chemicals, thus lowering the global warming index [6]. This section explores two significant applications of metal oxide nanocatalysts: CO_2 conversion via the catalytic reduction process as well as the catalysis of the decomposition of CH_4 and other hydrocarbons.

3.2.1 Catalytic Reduction of CO_2

CO_2 minimization is a significant step in combating climate change since the emission is a major influence on global warming. Transition metal oxide nanoparticles, especially TiO_2, have been found to have a potential for the catalytic conversion of CO_2 into value-added products, including methanol, formic acid, and CH_4. This process, sometimes known as photochemical CO_2 reduction, wherein light, in photocatalytic conditions, staggers and transforms CO_2 into renewable fuels while, at the same time, achieving a cut in greenhouse gas emissions [6].

This occurrence is credited to the specific rendition of metal oxide nanocatalysts, for instance, a large surface area, which supports the adsorption of CO_2.

Furthermore, the bandgap energy of such nanocatalysts can be well controlled to enhance the light trapping of the visible light, thereby enhancing the photocatalytic process. For instance, doping TiO_2 with metals like iron or nitrogen can significantly enhance its photocatalytic activity under sunlight [2].

3.2.2 Catalytic Oxidation of CH_4 and Other Hydrocarbons

CH_4 is another greenhouse gas with a much higher global warming potential (GWP) than CO_2 over a 20-year period. One possible solution for the problem of global warming by CH_4 is the catalytic oxidation of CH_4 into CO_2, which is a less dangerous greenhouse gas. This reaction may be carried out with the help of metal oxide nanocatalysts, for instance, CeO_2 and manganese oxide (MnO_2), under reasonably moderate conditions [6].

The oxidation process of CH_4 is also a catalytic one in which CH_4 molecules are adsorbed at the surface of the metal oxide nanocatalysts and oxidized by oxygen to produce CO_2 and water. This reaction is very important in all processes like natural gas processing, especially in emission control systems in automobiles, where CH_4 emissions are minimized. Another development in the nanocatalyst synthesis has been the fabrication of core-shell nanoparticles or the addition of other noble metals, which has enhanced the selectivity and activity of these catalysts in the oxidation of CH_4 [5].

3.3 Metal Oxide Nanocatalysts in Renewable Energy Technologies

The increase in the use of renewable energy sources is undoubtedly necessary for minimizing the use of fossil energy carriers and combating climate change. Metal oxide nanocatalysts form a core part of improving the efficiency of renewable energy technologies, including solar energy conversion and hydrogen generation. It also explores how these nanocatalysts are used to enhance technologies for renewable energy and discusses how nanocatalysts could be advocates of green energy.

3.3.1 Applications in Solar Energy Conversion

Nanoparticles of metal oxides are widely employed in solar energy conversion through photocatalytic water splitting and dye-sensitized solar cells (DSSCs). TiO_2 is among the most investigated nanocatalysts for such uses because of its stability, non-toxicity, and photocatalytic activity. Photolysis of water into hydrogen and oxygen, a clean and renewable source of energy [6].

Additionally, in DSSCs metal oxide nanocatalysts such as ZnO and TiO_2 perform the photoanode roles and increase light-capturing ability in the cells. The mesoporous structure of these catalysts helps in offering a bulky adsorptive bed area for dye accumulation, leading to enhanced efficiency of conversion of solar energy [8].

3.3.2 Hydrogen Production and Storage

Hydrogen is deemed an important part of the new energy systems owing to its superior energy-to-weight ratio and carbon neutrality when combusted. Metal oxide nanocatalysts have a unique position in the generation and stabilization of hydrogen, especially via the splitting and reformulation methods. Thermochemical water splitting cycles employ demand materials such as Fe_2O_3 and CeO_2 that help decompose water into hydrogen and oxygen at elevated temperatures [1].

In addition, metal oxide nanocatalysts are also used in hydrogen storage systems. Consequently, they are good in compact and safe storage of hydrogen due to their capability to reversibly capture and release hydrogen. Nanostructured metal oxides contribute to the increasing rate of hydrogen adsorption and desorption, therefore making the hydrogen storage materials perform and be efficient [2].

3.4 Metal Oxide Nanocatalysts in Pollution Control

Metal oxide nanocatalysts are of importance in pollution control since they provide an efficient solution to numerous pollutants. As shown in Figure 3.1, identification of pollution sources presupposes that industrial effluent, agricultural effluent, and municipal waste contribute essentially to pollution [9]. In case pollution levels reach set standards, notify authorities, and correct measures must be observed. After that, control implementation measures like the use of metal oxide nanocatalysts can potentially lower the number of pollutants. Investigations have shown that such nanocatalysts can remove pollutants in as much as 99% within a very

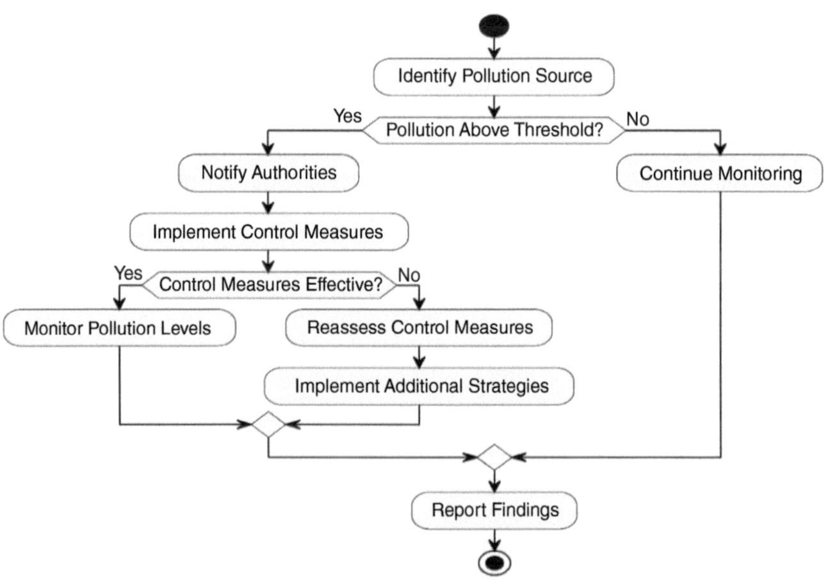

Figure 3.1 Pollution control strategies utilizing metal oxide nanocatalysts.

short time [10]. Subsequent to its usage, one must determine if control measures are effective under evaluation of pollutant levels in the process. In checking over pollution levels, it can be seen that they are continually monitored, which means possible variations can be immediately corrected. If the early methods do not require enhancement, then one may need to return to re-evaluation of control measures and possibly adopt augmenting strategies where they include modifying more advanced oxidation processes with nanocatalysts for increased efficiency [11]. It is important to report the findings on the effectiveness of those interventions to update further pollution-combating actions and improve the methodologies [12].

Nanocatalysts can absorb and degrade pollutants and may transform toxic compounds into less toxic products by catalyzing chemical reactions [13]. Phenomena such as photocatalysis, adsorption, and advanced oxidation reactions (AOR) are integrated with metal oxide nanocatalysts for handling the emergent environmental challenges due to pollution [12].

3.4.1 Catalysts for Air and Water Purification

Some of the metal oxide nanocatalysts have shown higher catalytic efficiency for the purification of both air and water, and they include TiO_2, ZnO, and copper oxide (CuO). These materials' photocatalytic features qualify them to decompose toxic contaminants under varying environmental conditions, giving a green aviary to air and water pollution.

- **Air Purification:** Nanocatalysts of metal oxides are one of the most researched materials because of their effectiveness in decomposing volatile organic compounds (VOCs), nitrogen oxides (NO_x), sulfur compounds, and carbon monoxide (CO) in industrial and urban atmospheres. Of the photocatalysts investigated, most of the knowledge is about TiO_2 nanoparticles, which are particularly effective in the degradation of organic compounds, including VOCs, using UV light. This process entails the production of electrons and holes when the photocatalyst absorbs UV light and uses the resultant electron-hole pairs to reduce the pollutants into harmless products of CO_2 and H_2O [12]. The literature available reveals that both TiO_2 and ZnO are efficient in purifying indoor air, while TiO_2 best degrades formaldehyde, benzene, and other VOCs in indoor environments, particularly in confined areas [9]. In addition, CuO nanocatalysts have been employed to reduce NO_x emissions, which result in the formation of smog as well as acid rain [14].
- **Water Purification:** The pollution of water sources with dangerous organic pollutants and heavy metals is a problem of concern to the entire world. Metal oxide nanocatalysts as TiO_2 has been extensively used for the degradation of dyes, pesticides, pharmaceuticals, and other organic pollutants. The photocatalytic activity of TiO_2 has been examined for the degradation of dangerous substances in the water under the UV or visible light [15]. For instance, the photocatalytic reduction of methylene blue and other toxic dyes has been studied using TiO_2 nanocatalyst, revealing exceptional degradation efficiency of hazardous complex organic compounds [13]. Moreover, transition metal oxides like

Fe_2O_3 and ZnO are synthesized for the elimination of heavy metals including arsenic, lead, and cadmium in water through adsorption and coprecipitation [16]. In addition to effectively enabling the removal of contamination, these materials also contribute towards the optimization of water purification processes to avoid secondary pollution (see Table 3.1).

3.4.2 Degradation of Pollutants and Industrial Waste

Industries emit dangerous substances such as organic compounds, metals, and toxicity in their processes, which has a very negative effect on the environment and people. The metal oxide nanocatalysts has also exhibited high efficiency in the removal of contaminants in the industrial waste effluent containing both organic and inorganic materials.

Table 3.1 Applications of metal oxide nanoparticles in air and water purification.

Nanocatalyst	Application (air purification)	Application (water purification)	References
Titanium dioxide (TiO_2)	Degradation of VOCs (formaldehyde, benzene), organic compounds; photocatalytic oxidation (PCO) to remove air pollutants	Degradation of dyes (methylene blue), pesticides, pharmaceuticals, and organic/inorganic pollutants, heavy metals; killing microorganisms; adsorption; photocatalytic activities	[9, 12, 13, 15]
Zinc oxide (ZnO)	Purification of indoor air; degradation of VOCs	Elimination of heavy metals (arsenic, lead, and cadmium), organic/inorganic pollutants; killing microorganisms; adsorption; photocatalytic activities	[9, 12, 16]
Copper oxide (CuO)	Reduction of NO_x emissions	Degradation of organic pollutants (dyes, pesticides, pharmaceuticals), organic/inorganic pollutants, and heavy metals; killing microorganisms; adsorption; photocatalytic activities	[12, 14]
Cerium oxide (CeO_2)	—	Removal of organic/inorganic pollutants, heavy metals; killing microorganisms; adsorption; photocatalytic activities	[3]
Manganese dioxide (MnO_2)	—	Removal of organic/inorganic pollutants, heavy metals; killing microorganisms; adsorption; photocatalytic activities	[6]
Iron oxide (Fe_2O_3)	—	Elimination of heavy metals (arsenic, lead, and cadmium)	[16]

- **Degradation of Organic Pollutants:** Many of these pollutants include dyes, pesticides, and pharmaceuticals that are found in effluents and are dangerous to the environment and human health. The metal oxide nanocatalysts, especially TiO_2, ZnO, and CuO, have been investigated widely for their photocatalytic activities towards the degradation of organic pollutants. Among all the photocatalysts tested, TiO_2 specifically, demonstrated a high efficiency in the degradation of different organic compounds, including methyl orange, methylene blue, and industrial dyes [12]. These nanocatalysts function in that they can adopt the UV or visible light irradiation to produce the reactive oxygen species (ROS), including the hydroxyl radicals, which oxidize the organic pollutants into nontoxic by-products, such as CO_2 and water [9]. This photocatalytic degradation has applications in the textile, pharmaceutical, and petrochemical industries, where large amounts of organic pollutants are generated (Figure 3.2).
- **Industrial Waste Treatment:** Industrial effluents contain toxic metals and hazardous complex organic compounds that can be prejudicial to the environment and human health. Implementations of metal oxide nanocatalysts have revealed the possibility of cleaning toxic waste, thereby decreasing the degree of toxicity of the effluent. For example, CuO, Fe_2O_3, and Co_3O_4 nanocatalysts have been used to remove polycyclic aromatic hydrocarbons (PAHs), which are characteristic of oil and gas industry emissions [14]. In addition, these materials help to carry out the treatment of industrial gases, including CO, CH_4, and VOCs, by encouraging the oxidation catalytic reactions through which dangerous gases are transformed into less hazardous substances, such as CO_2. The third significant domain of metal oxide nanocatalysts is in the field of water treatment for industrial wastewater containing heavy metals, including cadmium, lead, mercury, and others, through adsorption or photocatalytic reduction [16].

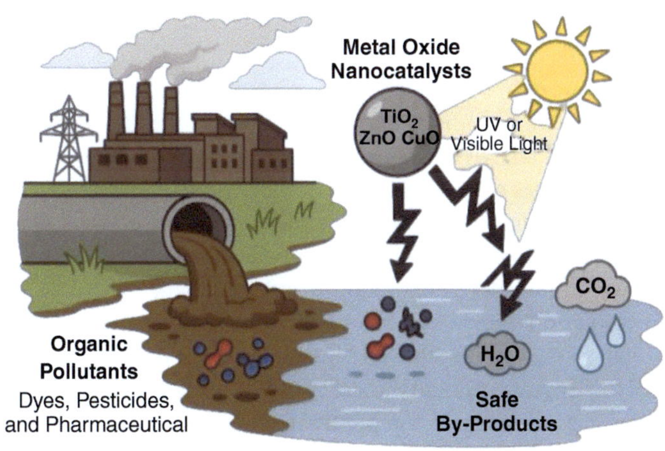

Figure 3.2 Photocatalytic degradation of organic pollutants using metal oxide nanocatalysts.

- **Waste-to-Energy Conversion:** Another application ever being researched on metal oxide nanocatalysts is in waste-to-energy conversion, extending the life span of the catalytic converters. Such materials enable the transformation process of waste materials into important energy products, for instance, hydrogen or CH_4, by use of catalytic methods like reforming or gasification. For instance, palladium (Pd)/TiO_2 nanocatalysts have been described to enhance the process of decomposition of CH_4 into hydrogen, which is a clean energy supply [17]. Affordable and clean power generation is among some of the catalytic processes that the development of renewable energy systems, which is the direction of the world today.

3.5 Mechanisms of Action of Metal Oxide Nanocatalysts

The ways in which metal oxide nanocatalysts can function are essential to bestow catalytic functions onto nanocatalysts. These mechanisms are attributed to the surface chemistry of the nanocatalysts, various reaction steps, and many influential parameters concerning the nanocatalysts' performance. Understanding these mechanisms is important for designing more efficient nanocatalysts and optimizing their applications in various fields, including pollution control, energy production, and chemical synthesis.

3.5.1 Surface Chemistry and Reaction Pathways

The surface properties of metal oxide nanocatalysts are known to be of high importance to the catalytic activity of the latter. It is here that the active sites on the surface of a catalyst that adsorb reactant molecules so that they can transform and then desorb as products are found. The form and properties of these active sites are determined by the crystal structure, surface defects, and the presence of dopants or complementary catalysts [2].

- **Electron-Hole Pair Generation:** It has been demonstrated that metal oxide nanocatalysts active in photocatalytic reactions, like water splitting or pollutant degradation, absorb light energy to create electron-hole pairs. When the adsorbed molecules contact electrons in the conduction band and holes in the valence band, chemical reactions take place, giving rise to the reduction and oxidation processes [9]. The efficiency of the process depends on the rate at which electron-hole pairs are produced and separated.
- **Adsorption and Desorption:** Adsorption of the reactant species on the surface of the catalysts is a central process in the catalytic reactions. Therefore, the availability of these active sites and the strength of this adsorption determine the rate of reaction. For instance, in the photocatalytic degradation process of organic pollutants, the process of how the pollutant molecules contact the metal oxide surface determines the degradation rate of the pollutants [18]. The release of

the products by desorption from the catalyst surface also plays an essential role once the reaction has taken place to prevent catalyst deactivation for continuous operation.

- **Reaction Pathways:** Catalytic reaction mechanisms of metal oxides are closely related to the type and conditions of the metal oxide catalysts involved. Sometimes metal oxide nanocatalysts enable redox reactions, involving electron transfer to or from reactants and products, and thus affecting the chemical type. For example, in the photocatalytic reduction of CO_2, the donation of electrons to CO_2 yields substances such as CH_4 or methanol, whereas the holes in the catalyst oxidize water into oxygen [6].

3.5.2 Factors Affecting Catalytic Efficiency

The characteristics of metal oxide nanocatalysts are determined by the particle size and surface area, which are inherent properties of these materials, as well as by the temperature and concentration of the reacting species, which are the external parameters of the process. In general, awareness of these factors simplifies the process of enhancing catalytic efficacy.

- **Particle Size and Surface Area:** Particle size is one of the most important parameters influencing the performance of nanocatalysts. Particles with a low size have a larger surface area to volume ratio than larger particles; this is because they contain more active sites to which the reactants may adhere. The two characteristics of nanocatalysts that influence their activity are as follows: (a) a comparatively small size of the catalyst and (b) an elevated interaction between the catalyst and reactant molecules [19].
- **Surface Defects and Active Sites:** Lattice defects, namely the surface oxygen vacancies, are found to have significant impacts on the catalytic activities of metal oxide nanocatalysts. These defects can afford more active places for the adsorption of reactants and places for redox reactions, leading to improved catalytic performance [5]. It has been found that the location and density of these defects can be controlled during synthesis for enhancing performance.
- **Co-Catalysts and Dopants:** It has been established that the addition of co-catalysts or dopants to metal oxide-based nanocatalysts considerably enhances their catalytic activity. Occasionally, other additives are employed to facilitate electron transfer, and these are often in the form of noble metals such as platinum, Pd, and so on, to increase the overrate. Likewise, the doping of extra elements like nitrogen, sulfur, or other transition metals in metal oxides can alter the electronic characteristics of the catalyst, which in turn makes them more active towards specific reactions [11].
- **Environmental Conditions:** Temperature, pH, and pressure, which are external factors, significantly affect the efficiency of metal oxide nanocatalysts. For instance, the performance of metal oxide nanocatalysts in the water splitting reactions depends on pH because dissociation of water to form protons is involved in the reduction half reaction [8]. Similarly, higher temperatures may

accelerate the reaction rate but can also lead to catalyst deactivation due to sintering or other forms of degradation [20].
- **Stability and Longevity:** It was ascertained that the long-term stability of metal-oxide nanocatalysts is a prime requisite for their applicability, more especially in giant processes. Catalysts, whose degradation results in reduction of their activity due to sintering, leaching of metal species, or surface poisoning, can only be resolved by developing better synthesis methods and selecting better materials. In all catalytic processes, it is equally important that the catalyst maintain its structural form after being subjected to multiple cycles of usage [5].

3.6 Advancements in Metal Oxide Nanocatalysts Fabrication

A considerable development of synthesis processes along with methods of constructing metal oxide nanocatalysts has been acknowledged, particularly in terms of fabricating new synthesis strategies and tailoring the properties of nanosystems. These improvements are designed to improve the activity and robustness of nanocatalysts for a variety of uses [20].

3.6.1 Synthesis Techniques

The preparation of metal oxide nanocatalysts has progressed through several complex stages, which provide fine-tuning of particle size, morphology, and surface characteristics. These methods are sol-gel processes, hydrothermal methods, chemical vapor deposition (CVD), and flame spray pyrolysis, and each has its merits [21].

- **Sol-gel Synthesis:** Aging of the sol-gel makes this process versatile because it is capable of preparing metal oxide powders or nanoparticles of uniform size at relatively low temperatures. This method includes interconversion between a solution sol phase and a solid gel phase, which provides for better control of the composition and pore size of the catalyst. This is especially significant in creating nanocatalysts with vast surface-to-volume ratios and uniform morphology [22].
- **Hydrothermal Synthesis:** Hydrothermal synthesis employs water at high temperature and pressure to synthesize nanocatalysts characterized by crystallinity and size homogeneity. This technique is scalable and capable of synthesizing nanostructures of different morphologies like nanorods and nanoflowers which promotes the catalytic property [21].
- **CVD:** CVD is a process that is used to put metallic oxide layers on substrates via the chemical reactions of vapor phase reagents. This method provides very good control on the thickness and nature of the layers of the nanocatalyst and therefore is ideal for applications that demand specific structures [23].
- **Flame Spray Pyrolysis:** This involves spraying a precursor solution and exposing it to high-temperature, high-velocity flame, where the formation of metal oxide nanoparticles is virtually instantaneous. Flame spray pyrolysis

has the advantage of being relatively straightforward, providing high production throughput, and achieving high levels of particle purity and narrow size distribution [11].

3.6.2 Nanostructure Design and Modification

The strategy in the preparation and altering of the nanostructures is found to affect the efficiency of metal oxide nanocatalysts. It has been seen that through varying the shape, size, and surface characteristics of the nanoparticles, their catalytic efficiency and stability could be enhanced [24].

- **Nanotubes and Nanowires:** These linear nanostructures have high surface area-to-volume (SA/V) ratios and improved electron transport characteristics. Specifically, the nanotubes and nanowires of metal oxides such as TiO_2 and ZnO are most suitable for photocatalytic and electrocatalytic applications because these kinds of structures provide an opportunity to save charges and reduce recombination [16].
- **Mesoporous Structures:** Mesoporous metal oxides possess large surface areas and pore volumes, which should make them suitable for high catalytic applications, which they are. The structurally well-arranged porous system can also facilitate the diffusion of the mass of the reactants and products through the system, thus improving the reaction rates [5].
- **Doping and Co-Catalysts:** Strategies for regulating the electronic structure of metal-oxide materials include the incorporation of nitrogen, sulfur, or other transition metals into the metal-oxide lattice. For instance, nitrogen-doped TiO_2 has higher photocatalytic performance under visible and infrared light because of the smaller bandgap energy. Additionally, the usage of noble metal co-catalysts such as platinum or Pd deposits on the metal oxides can significantly boost their catalytic efficiency by promoting electron transfer and providing additional active sites [9].

3.7 Challenges and Limitations

Although, the metal oxide nanocatalysts have attracted a lot of attention in many fields of technology, there are still some issues and limitations that affect their applicability and future efficiency of this technology. These are some challenges regarding their stability, reusability, environmental friendliness, and scalability. It may therefore be concluded that comprehending such limitations is vital for enhancing the catalytic effectiveness as well as realistic usability capacity of metal oxide nanocatalysts [24].

3.7.1 Stability and Reusability

One of the major concerns of obtaining metal oxide nanocatalysts is their stability and recyclability over time. Long-term stability, for example, in the form of sintering or aggregation, also limits the surface area of metal oxide nanocatalysts and their activity [2]. These problems are even more obvious in strong reaction conditions like high temperature or corrosive solutions, and where the nanocatalysts could precipitate or have lower catalytic activity.

- **Sintering:** Sintering is the process whereby the particles (nanoparticles) in a given system increase in size while at the same time their overall surface area for catalysis action reduces. This phenomenon is widespread when using metal oxide nanocatalysts and exposure to high temperatures that may degrade the efficiency of the catalyst [19]. Sintering could be reduced by preparing catalysts with high thermal stability or by incorporation of stabilizers that hinder particle sintering.
- **Leaching of Active Sites:** Another problem associated with the recyclability of metal-oxide nanocatalysts concerns the leaching of active sites, mainly in those processes in which the liquid phase participates. The active metal species or components of the metal oxide may also leech into the solution over time, thus reducing the catalytic potential of the material. A possibility for the formulation of this problem can be by enhancing the catalyst's composition and can apply some techniques, such as the utilization of the protective coating [8].
- **Cyclic Stability:** For metal oxide nanocatalysts to be useful in industry, they have to be cyclical stability, that is, their efficiency has to remain glossy subsequent lots of circles. Current research is focused on developing nanocatalysts that can withstand repeated use without significant loss of activity, thus making them more cost-effective for commercial applications [9].

3.7.2 Environmental and Health Impacts

Despite the numerous advantages of metal oxide nanocatalysts for controlling pollution, enhancing crop productivity, generating clean energy, and recycling materials, controversy surrounds the lifetime environmental and health implications of using them on a large scale. Due to their size, nanoparticles have certain characteristics, including a high surface area, reactivity that may have adverse effects on human and environmental health if not controlled [24].

- **Toxicity to Organisms:** Research indicated that metal oxide nanoparticles, including TiO_2 and ZnO can possess toxicity to the aquatic organisms, plants, and microorganisms in the environment [12]. It was established that the level of toxicity of these nanoparticles depends on the size of the particles, their charge, and the nature of the metal oxide. Concerns about their possibility to accumulate in ecosystems and the food chains lead to questions whether they are capable of having negative impacts on both the biodiverse and human populations [24].
- **Leaching of Metals into the Environment:** Another severe issue is that metal ions of metal oxide nanocatalysts can dissolve in the environment. Thus, for instance, in some catalytic reactions or after cleaning up utilized nanocatalysts, metal ions may dissolve into water or soil and pollute it. This could lead to the accumulation of toxic heavy metals in the environment with health risks to humans and hazards to the biophysical environment [18].

- **Biodegradation and End-of-Life Disposal:** The recovery and recycling of metal oxide nanocatalysts after employing them in industrial or environmental applications is another issue. Nanomaterials are not as decomposable as bulk metal oxides and therefore may have high stability in natural surroundings. This might require considering the susceptibility of nanocatalysts to deterioration and the long-term effects of their decomposition: pollution and hazards to wildlife [7]. Thus, emergency solutions with economically feasible and environmentally friendly recycling or biodegradation procedures should be sought.
- **Human Exposure and Occupational Hazards:** There are also concerns about the exposure of humans to nanomaterials during the manufacturing, handling, or use of metal oxide nanocatalysts. Nanoparticle aerosols or dust can be inhaled or come in contact with derm cause respiratory and other health complications [5]. Consequently, safety measures and standards are required in order to safeguard particular workers from probable risks that are related to the manipulation of nanomaterials [25].

3.7.3 Economic and Scalability Issues

Another urgent issue is the further application of metal oxide nanocatalysts in mass production processes. It is proven that the synthesis of high-quality metal oxide nanocatalysts using existing synthesis techniques is very expensive and uncompetitive compared to the cost of large and featureless mass catalysts. One critical challenge in the advancement of metal oxide nanocatalysts is the ability to produce these materials at a commercial scale while preserving their catalytic activity and affordability.

- **Cost of Synthesis:** Some of the common strategies in synthesizing metal oxide nanocatalysts like sol-gel or CVD are costly and need specific instrumentation. This adds to the cost of the catalysts, thus they are relatively expensive compared to the conventional catalysis in fields where price plays a major factor [26]. Further, sustainability and eco-friendly approaches to the synthesis of nanocatalysts are critical factors that may explain the high costs of production by conventional techniques.
- **Process Optimization for Large-Scale Production:** The major challenge lies in replicating these nanocatalysts at a larger scale while maintaining the size, morphology, and catalytic properties of the synthesized metal oxide nanocatalysts. Reproducibility, uniformity, and batch-to-batch consistency are some of the concerns that needs to be solved for industrial applicability. In addition, there arises the problem of how to synthesize the nanocatalysts in a large scale and in a way that does not affect their quality and efficiency [20].

3.8 Recommendations

The use of metal oxide nanocatalysts holds great potential and application in climate change and pollution. Nevertheless, as has been highlighted before in this paper, there are some challenges and limitations that need to be dealt with in order to fully realize their utility. The following recommendations are proposed to further the field of metal oxide nanocatalysts with regard to performance, upscalability, and sustainability.

3.8.1 Emerging Trends in Nanocatalysis

As research on nanocatalysts is advancing the application of nanotechnology in climate change and pollution control, new trends that may improve the function of metal oxide nanocatalysts are also arising. These include:

- **Development of Hybrid and Composite Nanocatalysts:** Another subarea that bears significant potential for future advancement in nanocatalysis is the synthesis of hybrid nanostructured metal oxides either with carbonaceous materials (like graphene or carbon nanotubes) or noble metals. These hybrid materials are favorable due to the synergistic interaction between the two or more materials used to form them so that the activity, selectivity, and stability of the catalyst are improved [6]. For instance, when metal oxides are incorporated with carbon-related materials, the rate of photogenerated charge carriers' separation increases, thus boosting the performances of photocatalytic systems, which include pollutant removal or photocatalytic water splitting.
- **Integration with Renewable Energy Systems:** Renewable energy applications, including solar cells and fuel cells which will have potential benefits when metal oxide nanocatalysts are incorporated. This can help advance energy conversion technologies, which is a futuristic solution since nanocatalysts enable the acceleration of numerous energy conversion processes with the help of metal oxide materials. For instance, metallic oxide photocatalysts are in the experimental stage to promote solar-induced water splitting to prepare hydrogen, a nonpolluting fuel. The capacity to incorporate nanocatalysts into these systems could reform the energy generation process and reduce reliance on oil and coal [9].
- **Advanced Computational Modeling and Design:** The application of computational tools, such as density functional theory (DFT) for the elucidation of the behavior of metal oxide nanocatalysts will remain a key enabler of the fast advancement of more efficient nanomaterials. Since understanding the atomic-level performance of catalyst surfaces and reactants, the researchers can manipulate important properties, including the surface defect and active-site distribution that are responsible for the enhanced catalysis [2].

3.8.2 Potential for Large-Scale Implementation

For metal oxide nanocatalysts to make a significant impact on large-scale industrial processes and climate change mitigation, several aspects must be addressed:

- **Scalability of Synthesis Techniques:** The greatest challenge that has been assessed to hinder vast use of metal oxide nanocatalysts is the difficulties that arise in the scaling up synthesis methods for metal oxide nanocatalysts. Present techniques like sol-gel, hydrothermal, and CVD work well though, they are costly and difficult to upscaling. Pursuing better synthesis techniques that will lower, reasonably meet the need for high-yielding production of metal oxide nanocatalysts without affecting their efficacy for the larger market. Reforms to making sustainable products and solutions include implementing green solvents and numerous low-energy processes [8].

- **Recycling and Reusability:** The requirement of developing novel synthesis approaches that make metal oxide nanocatalysts more economically feasible as well as more environmentally sustainable is to create methods that would enable the recovery and reuse of these metal oxide nanocatalysts. This can be done by enhancing the stabilities of the nanocatalysts under the severe working conditions and optimizing the processes for the reactivation of coking or sintering used catalysts. For instance, the coatings or support structures for the metal oxide nanoparticles that hinder their deactivation could greatly enhance their durability and thus minimize their regular replacement [19].
- **Environmental Impact Assessment and Safety:** As metal oxide nanocatalysts are designed for application at a large scale, performance evaluations of their effects on the environment and human health are also needed. This includes learning about the possible toxic effects of these materials on people and environment and estimate the threats related to their disposal. Measures should be put in place to control risk in the course of the synthesis, storage, and application of these nanomaterials [5]. Specifically, those nanocatalysts having low toxicity and high levels of biodegradation or disposal should be the preferred ones.
- **Cost Reduction and Commercialization:** In order for metal oxide nanocatalysts to be utilized in commercial processes, the cost for its production has to be lowered considerably. This can be done by increasing the yields of synthesis processes, employing low-cost starting materials and enhancing the selectivity of the catalysts so that the amount required is also reduced. To enhance the use of these materials in large-scale climate change mitigation and pollution control, the commercialization of the above materials will require a collaborative effort between academia, industries, and governments [24].

3.8.3 Collaborative Research and Interdisciplinary Approaches

The development of metal oxide nanocatalysts for future technological applications will require interprofessional research. The problem areas should be solved by professionals from materials science, chemistry, environmental science, and engineering. It can initiate a new cycle of research in the area of improving catalytic systems, their potential effects on the environment, and the ultimate establishment of nanocatalyst-based technologies for application in various industries [25].

- **Collaboration with Policy Makers:** Public involvement in the production and application of the nanocatalysts is also crucial in an effort to channel appropriate policies that will control the use of the nanocatalysts in a way that is safer to mankind and the environment. There must be standard guidelines that allow the safe fabrication, application, and elimination of metal oxide nanocatalysts in a way that will assist in climate change prevention as opposed to being detrimental to the process [2].

3.9 Conclusion

Metal oxide nanocatalysts offer great potential in addressing issues of climate change, pollution, and clean energy solutions that affect the environment. They are useful in converting greenhouse gases and degrading the harmful pollutants due to their large surface area, adjustable activity, and high stability. Despite these advantages, issues of stability, reusability, toxicity, and scaling up still exist. Addressing these problems requires the enhancement of synthesis strategies, efficiency, and environmental safety. Advancements in nanotechnology, especially in catalyst design and computational modelling, are expected to enhance the performance of these materials further. Cooperation between researchers, industries, and policymakers will also be significant to ensure their safe and sustainable integration into real-world applications. Further research and innovation may enable metal oxide nanocatalysts to become important in building a cleaner and more sustainable future.

References

1 Shaker LM, Al-Amiery AA, Al-Azzawi WK. Nanomaterials: paving the way for the hydrogen energy frontier. *Discov Nano*. 2024;19(1):3. https://doi.org/10.1186/s11671-023-03949-8

2 McIntyre SM, Garden AL. Computational modelling of nanoparticle catalysis. *Nanoscale*. 2025;17(24):14491–520. https://doi.org/10.1039/D4NR04364B

3 Somwanshi SB, Somvanshi SB, Kharat PB. Nanocatalyst: a brief review on synthesis to applications. *J Phys Conf Ser*. 2020;1644(1):012046. https://doi.org/10.1088/1742-6596/1644/1/012046

4 Ferdous AR, Shah SNA, Shah SS, Aziz MA. Advancements in nanotechnology applications: transforming catalysts, sensors, and coatings in petrochemical industries. *Fuel*. 2024; 371(Part B):132020. https://doi.org/10.1016/j.fuel.2024.132020

5 Lee H, Lee S, An K. Design and application of mesoporous catalysts for liquid-phase furfural hydrogenation. *Molecules*. 2025;30(6):1270. https://doi.org/10.3390/molecules30061270

6 Pascariu P, Gherasim C, Airinei A. Metal oxide nanostructures (MONs) as photocatalysts for ciprofloxacin degradation. *Int J Mol Sci*. 2023;24(11):9564. https://doi.org/10.3390/ijms24119564

7 Zaera F. Role of metal cocatalysts in the photocatalytic production of hydrogen from water revisited. *Energy Fuels*. 2025;39(5):2422–34. https://doi.org/10.1021/acs.energyfuels.4c06100

8 Awogbemi O, Ojo AA, Adeleye SA. Advancements in the application of metal oxide nanocatalysts for sustainable biodiesel production. *Discov Appl Sci.* 2024;6:250. https://doi.org/10.1007/s42452-024-05920-3
9 Ahasan T, Edirisooriya EMNT, Senanayake PS, Xu P, Wang H. Advanced TiO_2-based photocatalytic systems for water splitting: comprehensive review from fundamentals to manufacturing. *Molecules.* 2025;30(5):1127. https://doi.org/10.3390/molecules30051127
10 Venkateshaiah A, Černík M, Padil VVT. Metal oxide nanoparticles for environmental remediation. In: Thomas S, Thomas MS, Pothen LA, editors. *Nanotechnology for Environmental Remediation.* Wiley-VCH GmbH; 2022. Chapter 11. https://doi.org/10.1002/9783527834143.
11 Hübner U, Spahr S, Lutze H, Wieland A, Rüting S, Gernjak W, et al. Advanced oxidation processes for water and wastewater treatment – Guidance for systematic future research. *Heliyon.* 2024;10(9):e30402. https://doi.org/10.1016/j.heliyon.2024.e30402
12 Inamdar AK, Rajenimbalkar RS, Hulsure NR, Kadam AS, Shinde BH, Patole SP, et al. A review on environmental applications of metal oxide nanoparticles through wastewater treatment. *Mater Today Proc.* 2023. https://doi.org/10.1016/j.matpr.2023.05.527
13 Al Miad A, Saikat SP, Alam MK, Hossain MS, Bahadur NM, Ahmed S. Metal oxide-based photocatalysts for the efficient degradation of organic pollutants for a sustainable environment: a review. *Nanoscale Adv.* 2024;6(19):4781–803. https://doi.org/10.1039/D4NA00517A
14 Norouzmehr P, Rabbani AR, Zolfagharpour HR. Effects of metal oxide catalysts on polycyclic aromatic hydrocarbons emissions from large-scale biomass pyrolysis. *Sci Rep.* 2025;15:10903. https://doi.org/10.1038/s41598-025-95847-5
15 Solis Maldonado C, Luna Sánchez RA, Salas AC, Izaguirre Gallegos TL, Ortiz Silos N, Xochihua Juan JL. *Perspective Chapter: Titanium Dioxide as a Photocatalyst for Pharmaceutical Waste Degradation Present in Water.* IntechOpen; 2025. https://doi.org/10.5772/intechopen.1007736
16 Gupta K, Joshi P, Gusain R, Khatri OP. Recent advances in adsorptive removal of heavy metal and metalloid ions by metal oxide-based nanomaterials. *Coord Chem Rev.* 2021;445:214100. https://doi.org/10.1016/j.ccr.2021.214100
17 Zhang J, Fang Y, Zhang Y, Lin Y, Gui Y, Liu L. Photocatalytic degradation of volatile organic compounds over WO_2/SnS_2 nanofibers. *ACS Appl Nano Mater.* 2023;6(23):22301–10. https://doi.org/10.1021/acsanm.3c04496
18 Cheng X, Guan R, Chen Y, Qian Y, Shang Q, Sun Y. Adsorption and photocatalytic degradation process of oxytetracycline using mesoporous Fe-TiO_2 based on high-resolution mass spectrometry. *Chem Eng J.* 2023;460:141618. https://doi.org/10.1016/j.cej.2023.141618
19 Schiopu AG, Iordache DM, Oproescu M, Cursaru LM, Ioța AM. Tailoring the synthesis method of metal oxide nanoparticles for desired properties. *Crystals.* 2024;14(10):899. https://doi.org/10.3390/cryst14100899
20 Afzal S, Ullah S, Shahid M, Najam T, Shaaban IA, Shah SSA, et al. Metal oxide nanoparticles: synthesis and applications in energy, biomedical, and environment sector. *J Chin Chem Soc.* 2025;72(2):96. https://doi.org/10.1002/jccs.202400347

21 Tazim TQ, Kawsar M, Hossain MS, Bahadur NM, Ahmed S. Hydrothermal synthesis of nano-metal oxides for structural modification: a review. *Next Nanotechnol.* 2025;7:100167. https://doi.org/10.1016/j.nxnano.2025.100167

22 Bokov D, Jalil T, Abduladheem A, Chupradit S, Suksatan W, Ansari MJ, et al. Nanomaterial by sol-gel method: synthesis and application. *Adv Mater Sci Eng.* 2021; 2021:5102014. https://doi.org/10.1155/2021/5102014

23 Aragaw TA, Ayalew AA. Application of metal-based nanoparticles for metal removal for treatments of wastewater – a review. In: Ahmad A, Kumar R, Jawaid M, editors. *Emerging Techniques for Treatment of Toxic Metals from Wastewater.* Elsevier; 2023. p. 183–231. https://doi.org/10.1016/B978-0-12-822880-7.00001-7

24 Rathi BS, Ewe LS, Sanjay S, Sujatha S, Yew WK, Baskaran R, et al. Recent trends and advancement in metal oxide nanoparticles for the degradation of dyes: synthesis, mechanism, types and its application. *Nanotoxicology.* 2024;18(3):272–98. https://doi.org/10.1080/17435390.2024.2349304

25 Sahu SK, Palai A, Sahu D. Photocatalytic applications of metal oxide-based nanocomposites for sustainable environmental remediation. *Sustain Chem Environ.* 2024;8:100162. https://doi.org/10.1016/j.scenv.2024.100162

26 Fu J, Jiang X, Han W, Cao Z. Enhancing the cycling stability of transition-metal-oxide-based electrochemical electrode via Pourbaix diagram engineering. *Energy Storage Mater.* 2021;42:252–8. https://doi.org/10.1016/j.ensm.2021.07.037

4

Synthesis and Characterization of Metal Oxide Nanocatalysts for Energy Production

A. O. Esan[1], O. S. Dabo[1], A. Olasupo[2], O. A. Olalere[3], and A. A. Oladunni[4]*

[1] Department of Pure and Applied Chemistry, Ladoke Akintola University of Technology, Ogbomoso, Nigeria
[2] Wonderful Institute for Sustainable Engineering, The University of Kansas, Lawrence, KS, USA
[3] Faculty of Engineering and Science, University of Greenwich, London, UK
[4] Department of Chemical and Petroleum Engineering, Abiola Ajimobi Technical University, Ibadan, Nigeria

*Corresponding author: aoesan@lautech.edu.ng

4.1 Introduction

The increasing global population has further intensified the need for alternative and sustainable energy sources [1]. In the years past, biomass sources such as wood, charcoal, and crop residues served as primary energy sources. However, the industrial revolution in the 19th century led to a transition toward coal, oil, and natural gas as dominant energy carriers. Subsequent technological advancements accelerated the widespread adoption of these fossil fuels. According to recent data, worldwide energy usage has surged by over 33% since 2000, reaching approximately 580 million terajoules in 2023. Projections indicate this figure could escalate to 740 million terajoules by 2040 and 900 million terajoules by 2050 [2–4].

Despite the dominance of fossil fuels, their environmental impact, particularly greenhouse gas emissions, has spurred the search for cleaner, more sustainable energy sources [2, 5]. To address the rising energy demand while ensuring environmental sustainability, researchers have turned to nanocatalysis, specifically metal oxide nanocatalysts [6]. These materials actively enhance the efficiency of various energy technologies, including energy storage and conversion systems. For instance, lithium-ion batteries (LiBs) remain the preferred energy storage solution for portable electronics and electric vehicles. However, conventional anode materials, comprising lithium metal and graphite, exhibit limitations such as shape deformation of Li^+ ions and reduced coulombic efficiency, which restricts storage capacity. "Metal oxide nanocatalysts" have been seen as potential alternatives to address these challenges, improving charge storage and overall battery performance [2, 7, 8, 9].

Beyond LiBs, metal oxide nanocatalysts have demonstrated remarkable applications in supercapacitors, particularly as pseudocapacitors, where they function as redox additives [8]. They also play a vital role in electrocatalysis, particularly in electrochemical water splitting. Conventional catalysts for the "hydrogen evolution reaction" (HER) and "oxygen evolution reaction" (OER), such as platinum, iridium, and ruthenium, are scarce, expensive, and have limited durability. In contrast, metal oxide nanocatalysts offer cost-effective, adaptable, and durable alternatives [10, 11]. Additionally, in biodiesel production, metal oxide nanocatalysts serve as efficient catalysts for transforming free fatty acids and triglycerides to biodiesel, enhancing process efficiency [4].

Various synthesis techniques have been explored for fabricating metal oxide nanocatalysts, each offering unique advantages in tailoring their structural and electronic properties. These methods include chemical vapor deposition, ball milling, sonochemical synthesis, sol-gel processing, green synthesis, and hydrothermal techniques. The choice of synthesis approach significantly influences the catalytic performance of the nanomaterials. For instance, electrical characteristics such as conductance are dependent on the particle sizes of several metal oxide nanocatalysts [2, 7, 12, 13].

The primary objective of this chapter is to explore the synthesis, characterization, and application of metal oxide nanocatalysts for energy production and storage. It will involve an in-depth exploration of the methods for synthesizing and characterizing the "metal oxide nanocatalysts," emphasizing their applications in energy production and storage. Furthermore, the chapter will critically examine the limitations and challenges associated with these methods of synthesis, thereby providing a path to future progress in the field.

4.2 Synthesis Techniques for Metal Oxide Nanocatalysts

There are various methods employed for nanocatalyst synthesis. These methods are generally classified into two namely: physical (top-down) methods which include evaporation and laser ablation, and chemical (bottom-up) approach, such as template-directed synthesis, vapor deposition, sol-gel, hydrothermal/solvothermal, and layer-by-layer assembly, among others. Bottom-up approaches are generally preferred in contrast to top-down techniques due to their efficiency and versatility [13, 14]. There has also been increased attention on the green route of synthesis in recent years. In this method, extract from biological materials (plants and microorganisms) are used for metal oxide nanocatalysts synthesis. These materials contain different bioactive components that are capable of acting as either "reducing" or "capping" agents during synthesis [13].

4.2.1 Sol-Gel Method

The sol-gel method is a well-established technique for synthesizing metal oxide nanocatalysts with tunable characteristics. Generally, it involves the following steps:

- Hydrolysis of the starting material (e.g., metal alkoxides) in solvents which could be water or organic solvents like alcohol. Oxygen is vital for this process and it is supplied either through the water or alcohol. The sol-gel process is classified as "aqueous" when water is the solvent, and "nonaqueous" when an organic solvent is employed. Equation (1) gives an overview of the chemical reactions during hydrolysis [15–18].

$$\text{M—OR} + H_2O \rightarrow \text{M—OH} + \text{R—OH} \quad (1)$$

M represents the metal and R denotes the alkyl group

- During condensation, neighboring molecules react, releasing water or alcohol and forming bonds between metal oxides. This leads to the development of polymer networks that constitute colloids in the liquid phase. Equation (2) reaction is shown below [15, 17].

$$\text{M—OH} + \text{X—O—M} \rightarrow \text{M—O—M} + \text{X—OH} \quad (2)$$

M represents the metal and X represents the hydrogen or alkyl group

- This step involves aging to aid formation of gel. This leads to further structural changes in the gel. Polycondensation and reprecipitation continues further in the solution which consequently leads to reduction in porosity and enhanced thickness between colloid particles [17–19].
- At this stage, the material is dried to remove water and other organic components. Different techniques such as "atmospheric or thermal drying, supercritical drying, and freeze drying" are often employed. Each of these drying methods will lead to changes in the physicochemical features of the nanocatalyst [17–19].
- Heat treatment is the last stage in the sol-gel synthesis technique. The material is calcined to eliminate water molecules and other residual components that could have been trapped within. Pore size and nanocatalyst density are affected by the calcination temperature employed [17].

4.2.2 Hydrothermal/Solvothermal Method

In this method, metal oxide nanocatalysts are synthesized with the aid of a batch reactor at high temperature and pressure creating a controlled environment for reactants to form nanosized crystallites [7, 9, 20]. This method enables the dissolution and recrystallization of materials under conditions that would be otherwise insoluble. This results in a high-purity, uniform nanocatalyst with distinct features and a narrow size distribution [7, 21]. Adjusting parameters like temperature, reaction time, and reactant composition ensures that there is a precise control over morphology and crystallinity [7, 20, 22]. The solvothermal technique shares similarity with hydrothermal. It uses "nonaqueous" solvents instead of water, functioning at temperatures between 100 and 1,000 °C and pressures up to 10 000 bar. Commonly used solvents include diethanolamine, ethanol, and ethylene glycol. Some syntheses require stabilizers to control the reaction [15, 23]. Figure 4.1 illustrates the step-by-step process of hydrothermal synthesis, highlighting the key stages involved in producing nanoparticles or crystals under high-temperature and high-pressure conditions.

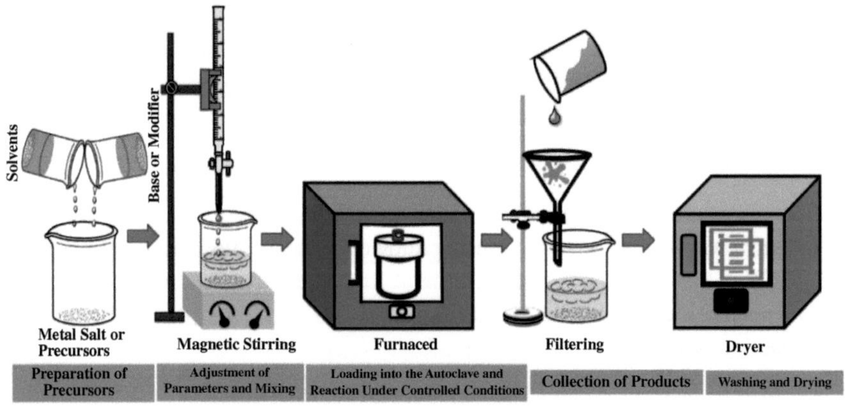

Figure 4.1 Diagrammatic illustration of the hydrothermal "metal oxide nanocatalyst" synthesis.
Source: [22] / Elsevier / CC BY 4.0.

4.2.3 Coprecipitation Method

The coprecipitation technique is a popular adopted approach for synthesizing "metal oxide nanocatalysts," offering advantages in simplicity, cost-effectiveness, and versatility. This technique typically involves dissolving inorganic salts (e.g., nitrates or chlorides) in solvents and water to form a homogeneous solution [24, 25]. The subsequent precipitation of these salts, often in the form of oxalates or hydroxides, is a crucial step in determining the final nanocatalyst properties. There are three vital steps involved in synthesis process:

- Solution preparation: accurate preparation of the precursor solution is essential for achieving desired nanocatalyst properties [24].
- Heat application: the application of heat plays a vital role in deciding the features of the nanocatalyst, including size, shape, and crystallinity [15, 24].
- Precipitation: the use of a precipitation agent (e.g., NaOH, KOH, or $(C_2H_5)_4NOH$ facilitates the formation of nanoparticles [15, 24]. The precipitating agent used influences the size and properties of the synthesized nanocatalyst.

The nanocatalysts' size, shape, and porosity are directly affected by factors, including precursor concentration and solution pH during synthesis. Rapid nucleation occurs once a critical concentration is reached, followed by a growth phase. The final step involves filtering, washing, and calcining the precipitate to obtain a crystalline substance [26, 27].

4.2.4 Chemical Vapor Deposition (CVD) Method

The CVD method involves a gas-phase reaction on a substrate surface, forming a solid material with precise control over composition and architecture [28, 29]. A typical CVD setup includes a precursor delivery system, a CVD reactor chamber, and an exhaust management system, where liquid precursors are often used due to their ability to generate vapor at moderate temperatures, enabling creation of complex materials through precise mixing [30, 31]. This technique allows for tailored material synthesis in both 2D and 3D forms. There are different variations of CVD which include horizontal and vertical chemical vapor deposition (H-CVD and V-CVD) which differ by configuration of the reactor or gas flow directions. Other variations are low-pressure and atmospheric CVD, hot-wall and cold-wall CVD, oxidative CVD, and metal organic CVD [31]. The deposition and quality of nanomaterials are influenced by precursor delivery method, reactor precision, carrier gas properties, substrate temperature, vapor flow rate, and deposition time [32]. Table 4.1 summarizes the synthesis of metal oxide nanocatalysts using various methods, including sol-gel, hydrothermal/solvothermal, coprecipitation, and chemical vapor deposition with relevant references.

4.2.5 Green Synthesis of Metal Oxide Nanocatalysts

Conventional methods for synthesizing nanocatalysts typically require considerable resources, intense energy usage, and environmentally hazardous due to the use of toxic chemicals. These methods typically involve metal precursors, reducing agents, and stabilizing agents, which can be complex and challenging to control. The properties of the synthesized nanocatalysts are controlled by the concentration of the starting materials and the reducing and stabilizing agents' ratio [38, 39]. Given the limitations and environmental concerns associated with traditional methods, researchers are exploring environmentally friendly alternatives. Bio-based synthesis routes offers a cleaner, more sustainable, and advantageous technique for synthesizing nanocatalysts [8, 38].

In the green synthesis of metal oxide nanocatalysts, the active components in plants or microorganisms reduces the metal precursor, then atmospheric oxygen released from the phytochemical breakdown is bound to the metallic species. Afterwards, nanocatalysts are formed through electrostatic attraction and stabilization by phytochemicals to prevent clustering. Phytochemicals serve dual roles as stabilizers and reductants in nanocatalyst synthesis. The quantity of phytochemicals in plant extract influences nanocatalyst size and shape [39].

4.3 Procedures Involved in Green Synthesis of Metal Oxide Nanocatalysts

The green synthesis technique involves several key steps. Initially, plant parts are washed, dried, and ground, then mixed with distilled water (typically 100–250 mL) to create an extract. A common ratio used is 10 g of ground plant parts to 100 mL of water. The mixture is maintained under controlled conditions, with pH levels ranging from 4 to 10, and reaction times varying from 5 minutes to 10 days. After filtering and centrifuging, the extract is cooled to room temperature and stored at approximately 4 °C for future use. To synthesize the nanocatalyst, a precursor solution (e.g., 0.1 M $Zn(CH_3COO)_2 \cdot 2H_2O$ or $CuSO_4 \cdot 5H_2O$) is added to the extract at elevated temperatures (60–100 °C) for a specified duration. A color change usually signals the formation of the nanocatalyst. The resulting nanocatalyst is then dried, calcined, and stored properly [40–42]. Table 4.2 gives an overview of metal nanocatalysts synthesized via green route with relevant references.

Table 4.1 Summary of metal oxide nanocatalysts synthesized via different techniques with their intended applications.

Precursor	Metal oxide	Synthesis technique	Key nanocatalyst properties	Intended applications	References
AlCe–Pt_X intermetallic compound	Pt/Al–CeO_2	Hydrothermal	The AlCe-Pt nanocatalyst featured a rod-like structure with an average crystallite size of 7 nm and a high Pt dispersion, a crystalline CeO_2 phase, and enhanced surface area (83 m^2/g), particularly after H_2/N_2 treatment, indicating its potential for catalytic applications.	CO conversion	Wang et al. [33]
$MnSO_4.H_2O$	Mn_3O_4	Hydrothermal	The Mn_3O_4 nanoparticles formed uniform nanorods with a hausmannite crystalline structure, exhibiting characteristic Mn–O and Mn–O–Mn vibrations. Their average crystallite size is approximately 26 nm. The modified electrode also demonstrated over 88% of its "baseline" performance after 150 cycles.	Electrode for direct urea fuel cell (DUFC)	El-Said et al. [5].
$CoCl_2 \cdot 6H_2O$	Co_3O_4	Coprecipitation	The synthesized Co_3O_4 nanoparticles exhibited a face-centered cubic structure with approximately 50 nm average particle size and undergone morphological transformation from flower-like aggregates to nearly spheroidal particles after calcination.	As catalyst in breaking down ammonium perchlorate thermally, potentially for rocket propulsion or other energy storage systems.	Janjua [34].
A composite of Ni–Mg–Cu–Co–La oxides was created from their nitrate precursors	Ni–Cu/ La_2O_3–MgO	Coprecipitation	The Ni–Cu/La_2O_3–MgO catalyst prepared by coprecipitation exhibited a highly porous structure, smaller crystallite sizes ranging from 25 to 50 nm, and strong metal-support interaction, resulting in enhanced surface area between 96.1–134.1 m^2/g and reducibility. This makes it a promising catalyst for various applications.	Sustainable hydrogen (H_2) production through catalytic steam reforming of glycerol.	Shejale and Yadav [12].

Al-tri-sec-butoxide, zirconyl nitrate solution (ZNS), and $Mg(NO_3)_2$ $6H_2O$	Ni–Co/ Al_2O_3–MgO–ZrO_2	Sol-gel	The MgO-promoted Ni–Co/Al_2O_3–ZrO_2 nanocatalysts showed enhanced surface area between 37–58 m^2/g, uniform dispersion, and smaller particle size (17.1–42.2 nm) with increasing MgO content, making them promising for catalytic applications.	The intended applications include energy-related uses such as hydrogen production, photocatalysis, and photovoltaics	Sajjadi et al. [35].
A composite of $Al_{35}(Mg, Fe, Cu, Ni, Co)_{65}]_3O_4$ was created from their nitrate precursors	(Ni–Co)/ Al_2O_3–MgO–ZrO_2	Sol-gel	Hexanary high-entropy oxides displayed defined crystal structures and metal oxide bonds, demonstrating promising electrocatalytic performance for water oxidation reactions, particularly with non-equiatomic compositions and aluminum-rich formulations.	Electrocatalytic water oxidation reaction	Asim et al. [36].
$Cu(hfa)_2$TMEDA for copper oxides, $Co(dpm)_2$ and $Co(hfa)_2$ TMEDA for cobalt oxides, and Zn(II) bis(ketoiminate) for ZnO-based nanocomposites.	Cu_2O, CuO, Co_3O_4, F-doped Co_3O_4, Ag/ZnO, and CuO/ZnO	Chemical vapor deposition	The nanocatalysts showed diverse properties, including varied morphologies (agglomerates, nanowires, faceted aggregates), compositions (Cu2O, CuO, Co_3O_4, Ag/ZnO, CuO/ZnO), and enhanced photocatalytic activities for hydrogen production.	Sustainable H_2 generation through photocatalytic reforming of aqueous media containing organic oxygenates, utilizing supported oxide-based nanosystems.	Maccato et al. [37].

Table 4.2 Summary of metal oxide nanocatalysts synthesized via green route and their intended applications.

Precursor	Plants used	Metal oxide	Key nanocatalyst properties	Intended applications	References
$Ni(CH_3COO)_2.4H_2O$	Leaves of aloe vera, peels of papaya, and dragon fruit	NiOx	The biosynthesized NiOx nanoparticles demonstrate enhanced electrocatalytic performance, with aloe vera-mediated NiOx exhibiting an overpotential of 413 mV and a Tafel slope of 95 mV dec^{-1}, attributed to its high surface area and conductivity.	Hydrogen production through electrocatalytic water splitting.	Selvanathan et al. [42]
Composite of Ag/ZnO was created from their nitrate precursors	Turmeric root	Ag/ZnO	Ag-doped ZnO nanoparticles exhibit a crystallite size of approximately 45 nm, spherical morphology, and a wurtzite structure, with an absorption peak at 368 nm, demonstrating potential catalytic properties.	Hydrogen production and biodiesel synthesis.	Yadav et al. [43]
$MgNO_3$	*Ziziphus mauritiana* leaves	MgO	The MgO nanoparticles exhibit a band gap of 4.7 eV, an absorption peak at 270 nm, and a crystalline face-centered-cubic structure, with particle sizes ranging from 22 to 30 nm and spongy, spherical morphology.	Catalyst for biodiesel production.	Saman et al. [44]
$Cr(NO_3)_3{\cdot}9H_2O$	Brinjal stalks	Cr_2O_3	The Cr_2O_3 nanocatalyst features a crystallite size of 27 nm, band gap energy of 3.48 eV, and an anionic nature with a zeta potential of −15.5 mV, exhibiting agglomerated morphology and good thermal stability.	Catalyst for biodiesel production.	Bibi et al. [41]

$MnCl_2$ and $CoCl_2 \cdot 6H_2O$	*Rosmarinus officinalis* leaves	Mn_3O_4/CoO	The Mn_3O_4/CoO nanocomposite features a crystalline structure, spherical morphology, and notable optical properties, with a direct bandgap of 2.80 eV and an indirect bandgap of 2.18 eV, showcasing its potential for various applications.	Photocatalytic hydrogen production and CO_2 conversion.	Hasan et al. [45]
Copper sulfate	*Tribulus terrestris* seed	CuO	The biosynthesized CuO nanoparticles exhibit a monoclinic crystalline structure, spherical morphology characterized by an average particle size of 59 nm, and a bandgap of 3.1 eV, showcasing their potential for various applications.	Energy storage applications, specifically in supercapacitors.	Meena et al. [46]
$CuSO_4 \cdot 5H_2O$	*Moringa oleifera* leaves	CuO	The CuO nanoparticles exhibit a nanorod-like morphology with high porosity, a bandgap of 2.5 eV, and notable optical properties, making them suitable for supercapacitor and solar technology applications.	Energy storage applications, specifically in supercapacitors.	Ikhioya et al. [47]

4.4 Challenges in the Synthesis of Metal Oxide Nanocatalysts

The synthesis of "metal oxide nanocatalysts" involves various techniques, each offering unique advantages while also presenting specific limitations. Understanding these challenges is critical to optimizing the performance, scalability, and sustainability of nanocatalysts for energy applications. Table 4.3 shows a comparative overview of synthesis methods for metal oxide nanocatalysts, detailing each method's advantages and disadvantages and their corresponding drawbacks.

Table 4.3 Overview of the benefits and limitations of various methods employed for nanocatalyst synthesis.

Synthesis method	Advantages	Disadvantages	Applications	References
Coprecipitation	Small particle size, uniform distribution, simplicity, and cost-effectiveness.	Low crystallinity which can affect catalytic performance, risk of contamination leading to poor reproducibility and purity of nanocatalysts.	General catalysis and environmental remediation.	Negrescu et al. [23]
Hydrothermal	Controlled morphology, high purity, reduced aggregation, precise structural control, high crystallinity, and suitable for scaling.	Time-consuming, requires high temperature and pressure, and complex post-processing.	Solar energy and photocatalysis.	Negrescu et al. [23], Sharma et al. [51]
Sol-gel	High purity, uniform distribution size, tunable morphology, versatile and effective, offers control over particle features.	High solvent usage, long reaction time, limited precursor, and requires heat treatment.	Photocatalysis, sensors etc.	Zhang et al. [16], Negrescu et al. [23]
Chemical vapor deposition	High precision, versatility, and scalable for industrial applications.	Requires precise control of multiple parameters, high temperature, equipment-intensive, and potentially high operational costs.	Semiconductor fabrication, nanocatalysts for energy, and environmental applications.	Vallejos et al. [30], Sun et al. [31]

4.4.1 Challenges of Green Synthesis of Metal Oxide Nanocatalysts

Green synthesis offers a promising, environmentally friendly approach to traditional methods for fabricating metal oxide nanocatalysts. While this approach reduces environmental hazards and eliminates the use of toxic reagents, it is not without limitations. Two major routes in green synthesis, microorganism-mediated and plant extract-mediated present distinct challenges that hinder their optimization and large-scale application.

- **Microorganism-Mediated Synthesis:** The use of microorganisms such as bacteria, fungi, and algae in nanocatalyst synthesis offers a biologically sustainable route. However, one of the primary drawbacks is the inherently slow reaction kinetics, which complicates the control of nanoparticle nucleation and growth. This results in inconsistent size distribution and poor uniformity of the synthesized nanocatalysts. Additionally, the complexity of handling live cultures demands trained personnel, which increases production costs and introduces variability in outcomes [10, 38].
- **Plant Extract-Mediated Synthesis:** Although plant-mediated synthesis offers a simpler and more accessible method, it suffers from a lack of standardization. Currently, there are no established stoichiometric protocols for determining the quantity of plant material, solvent (typically distilled water), or extract volume required for consistent synthesis. This trial-and-error approach is largely due to the variable concentration of phytochemicals in different plant parts and species, which significantly affects the synthesis process and properties of the resulting nanocatalysts. The absence of standardized procedures hampers reproducibility and scalability, making it difficult to integrate this method into industrial applications [48].

4.4.2 Challenges of Commercialization of Metal Oxide Nanocatalysts

The global nanotechnology industry has been growing rapidly in recent years, and based on several reports, the market capitalization is expected to grow into several billions of dollars in the coming decades. Consequently, scaling nanocatalysts produced in the laboratory to large-scale production is viewed as a profitable venture [49]. However, there are still several hurdles that need to be overcome before full scale commercialization of these materials. Prominent among these challenges is the "valley of death" which typically refers to the distance between favorable laboratory results and acquiring the needed resources to take the materials to the market. There are other challenges, such as policy/regulatory, for instance, the time taken for patent approval could range between 1 and 3 years. Unavailability of necessary infrastructure for large-scale production, environmental concern, public reception, and the absence of evaluation standards are other common challenges to the commercialization of these materials [49, 50].

4.5 Characterization of Metal Oxide Nanocatalysts

The effective characterization of "metal oxide nanocatalysts" is crucial for understanding their structural, morphological, and compositional properties, which directly influence their performance in energy-related applications. Various advanced analytical techniques offer high-resolution insights that aid in optimizing synthesis methods and tuning material features for specific usage [52, 53].

X-Ray Diffraction (XRD) is a fundamental analytical tool for understanding the crystal structure, phase composition, lattice parameters, and atomic arrangement of nanocatalysts. It enables researchers to correlate crystal phases with energy storage and conversion efficiencies, thereby informing the design of more effective materials [10, 23, 53].

High-Resolution Transmission Electron Microscopy (HRTEM) and Scanning Tunneling Microscopy (STM) provide atomic-level visualization of surface features and lattice fringes. These techniques are instrumental in analyzing surface morphology and defects, enabling precise control over synthesis parameters to enhance catalytic activity and stability [23].

Scanning Electron Microscopy (SEM) delivers high-resolution, 3D images of the nanocatalyst surface by utilizing a focused electron beam. This allows for detailed assessment of particle shape, size distribution, and surface texture [23, 52, 54].

Transmission Electron Microscopy (TEM), on the other hand, penetrates ultrathin samples with an electron beam to reveal internal structures, including grain boundaries, porosity, and crystalline defects, which are critical to the functionality of nanocatalysts in electrochemical applications [23, 52, 54].

Energy-Dispersive X-Ray Spectroscopy (EDX or EDS) is often integrated with SEM or TEM to elucidate the chemical composition and distribution pattern of elements within the sample. This compositional analysis helps verify the presence of desired elements and detect potential contaminants or dopants [10, 23].

Other Characterization Techniques such as thermal analysis of metal oxide nanocatalysts can be performed using various instruments, with "thermogravimetric–differential thermal analysis" (TG/DTA) being the commonly utilized method to assess thermal stability and performance. Additionally, particle size analysis (PSA) helps determine the mean size of the synthesized nanocomposites. The "Brunauer–Emmett–Teller" (BET) technique is also useful for examining the surface area and porosity of nanomaterials [54]. Together, these techniques offer a comprehensive toolkit for researchers to fine-tune the synthesis and functionality of metal oxide nanocatalysts, facilitating their application in energy storage systems, fuel cells, photovoltaics, and beyond.

4.6 Conclusion and Future Perspectives

The synthesis and characterization of metal oxide nanocatalysts have been regarded as pivotal areas of research in addressing the global energy crisis and advancing the transition toward environmentally friendly energy technologies. Metal oxide nanocatalysts possess remarkable physicochemical features such as extended surface area, tunable morphology, and redox versatility that make them highly effective in energy production, conversion, and storage applications. From enhancing the efficiency of LiBs and supercapacitors to facilitating key reactions in solar cells, fuel cells, and water splitting, these nanomaterials continue to redefine the frontiers of modern energy systems.

Despite the significant progress made, challenges persist. Issues such as the reproducibility and scalability of synthesis methods, environmental and economic sustainability, and the need for standardized protocols in green synthesis must be addressed. Moreover, the integration of metal oxide nanocatalysts into real-world energy technologies requires continued collaboration among chemists, material scientists, and engineers. Future research should prioritize the development of cost-effective, environmentally benign synthesis methods, alongside advanced characterization tools that can provide deeper insights into structure-performance relationships. In conclusion, metal oxide nanocatalysts represent a transformative solution to current and future energy challenges. As the global community continues to confront the twin demands of energy security and environmental sustainability, these materials are poised to play an active role in shaping a cleaner, more resilient energy future.

References

1 Zhou ZQ, Wang HM, Yang LB, Ma C, Wang JT, Qiao WM, et al. A review of the use of metal oxide/carbon composite materials to inhibit the shuttle effect in lithium-sulfur batteries. *New Carbon Mater*. 2024 Apr;39(2):201–20.

2 Eisa T, Abdelkareem MA, Jadhav DA, Mohamed HO, Sayed ET, Olabi AG, et al. Critical review on the synthesis, characterization, and application of highly efficient metal chalcogenide catalysts for fuel cells. *Prog Energy and Combust Sci*. 2023 Jan;94:101044.

3 Awogbemi O, Ojo AA, Adeleye SA. Advancements in the application of metal oxide nanocatalysts for sustainable biodiesel production. *Discov Appl Sci*. 2024 May 7;6(5):250.

4 Bekhradinassab E, Haghighi M, Shabani M. A review on acidic metal oxide-based materials towards heterogeneous catalytic biodiesel production via esterification process. *Fuel*. 2025 Jan;379:132986.

5 El-Said WA, Alsulmi A, Alshitari W. Hydrothermal synthesis of Mn_3O_4 nanorods modified indium tin oxide electrode as an efficient nanocatalyst towards direct urea electrooxidation. Barakat NAM, editor. *PLoS One*. 2022 Aug 4;17(8):e0272586.

6 Sharma A, Sharma S, Yadav S, Arora B, Dutta S, Dixit R, et al. Covalently functionalized graphene oxide metal complexes: versatile nanocatalysts for organic transformations. *Mater Sci Eng B*. 2024 Dec;310:117671.

7 Boyjoo Y, Wang M, Pareek VK, Liu J, Jaroniec M. Synthesis and applications of porous non-silica metal oxide submicrospheres. *Chem Soc Rev*. 2016;45(21):6013–47.

8 Aigbe UO, Osibote OA. Green synthesis of metal oxide nanoparticles, and their various applications. *J Hazard Mater Advances*. 2024 Feb;13:100401.
9 Xu W, Bai G, Pan E, Li D, Zhang J, Xu S. Advances, optical and electronic applications of functional materials based on rare earth sulfide semiconductors. *Mater Des*. 2024 Feb;238:112698.
10 Danish MSS. Exploring metal oxides for the hydrogen evolution reaction (HER) in the field of nanotechnology. *RSC Sustainability*. 2023;1(9):2180–96.
11 Jamadar AS, Sutar R, Patil S, Khandekar R, Yadav JB. Progress in metal oxide-based electrocatalysts for sustainable water splitting. *MRE*. 2024 Aug;4(3):100283.
12 Shejale AD, Yadav GD. Ni–Cu and Ni–Co supported on La–Mg based metal oxides prepared by coprecipitation and impregnation for superior hydrogen production via steam reforming of glycerol. *Ind Eng Chem Res*. 2018 Apr 11;57(14):4785–97.
13 Chavali MS, Nikolova MP. Metal oxide nanoparticles and their applications in nanotechnology. *SN Appl Sci*. 2019 Jun;1(6):607.
14 Qiao SZ, Liu J, Max Lu GQ. Synthetic chemistry of nanomaterials. In: Xu R, Xu, Y, editors. *Modern Inorganic Synthetic Chemistry*. Elsevier; 2011 [cited 2025 May 24]. p. 479–506. Available from: https://linkinghub.elsevier.com/retrieve/pii/B9780444535993100216
15 Stankic S, Suman S, Haque F, Vidic J. Pure and multi metal oxide nanoparticles: synthesis, antibacterial and cytotoxic properties. *J Nanobiotechnol*. 2016 Dec;14(1):73.
16 Zhang F, Wang X, Liu H, Liu C, Wan Y, Long Y, et al. Recent advances and applications of semiconductor photocatalytic technology. *Appl Sci*. 2019 Jun 18;9(12):2489.
17 Parashar M, Shukla VK, Singh R. Metal oxides nanoparticles via sol–gel method: a review on synthesis, characterization and applications. *J Mater Sci: Mater Electron*. 2020 Mar;31(5):3729–49.
18 Ben Arbia M, Helal H, Comini E. Recent advances in low-dimensional metal oxides via sol-gel method for gas detection. *Nanomaterials*. 2024 Feb 14;14(4):359.
19 Bokov D, Turki Jalil A, Chupradit S, Suksatan W, Javed Ansari M, Shewael IH, et al. Nanomaterial by sol-gel method: synthesis and application. Wang Z, editor. *Adv Mater Sci Eng*. 2021 Jan;2021(1):5102014.
20 Kigozi M, Ezealigo BN, Onwualu AP, Dzade NY. Hydrothermal synthesis of metal oxide composite cathode materials for high energy application. In: Ezema FI, Lokhande CD, Jose R, editors. *Chemically Deposited Nanocrystalline Metal Oxide Thin Films*. Cham: Springer International Publishing; 2021 [cited 2025 May 24]. p. 489–508. Available from: https://link.springer.com/10.1007/978-3-030-68462-4_19
21 Yang Q, Lu Z, Liu J, Lei X, Chang Z, Luo L, et al. Metal oxide and hydroxide nanoarrays: hydrothermal synthesis and applications as supercapacitors and nanocatalysts. *Prog Nat Sci Mater International*. 2013 Aug;23(4):351–66.
22 Tazim TQ, Kawsar Md, Sahadat Hossain Md, Bahadur NM, Ahmed S. Hydrothermal synthesis of nano-metal oxides for structural modification: a review. *Next Nanotechnology*. 2025;7:100167.
23 Negrescu AM, Killian MS, Raghu SNV, Schmuki P, Mazare A, Cimpean A. Metal oxide nanoparticles: review of synthesis, characterization and biological effects. *JFB*. 2022 Dec 5;13(4):274.

24 Yaqoob AA, Ahmad A, Ibrahim MNM, Karri RR, Rashid M, Ahamd Z. Synthesis of metal oxide–based nanocomposites for energy storage application. In: Koduru JR, Karri R, Mubarak NM, Bandala ER, editors. *Sustainable Nanotechnology for Environmental Remediation*. Elsevier; 2022 [cited 2025 May 24]. p. 611–35. Available from: https://linkinghub.elsevier.com/retrieve/pii/B978012824547700028X

25 AbdulRasheed T, Afotey B, Anang DA. Recent advances in methods for synthesizing composite nanocatalysts and their applications: a literature review. *Int J Thin Fil Sci Tec*. 2024 Jan 1;13(1):27–36.

26 Khalid H, Chaudhry AA. Basics of hydroxyapatite—structure, synthesis, properties, and clinical applications. In: Khan S, Chaudhry AA, editors. *Handbook of Ionic Substituted Hydroxyapatites*. Elsevier; 2020 [cited 2025 May 24]. p. 85–115. Available from: https://linkinghub.elsevier.com/retrieve/pii/B9780081028346000045

27 Varanda LC, Souza CGS de, Perecin CJ, Moraes DA de, Queiróz DF de, Neves HR, et al. Inorganic and organic–inorganic composite nanoparticles with potential biomedical applications: synthesis challenges for enhanced performance. In: Holban AM, Grumezescu AM, editors. *Materials for Biomedical Engineering*. Elsevier; 2019 [cited 2025 May 24]. p. 47–99. Available from: https://linkinghub.elsevier.com/retrieve/pii/B9780128184318000040

28 Bandas C, Orha C, Nicolaescu M, Morariu (Popescu) MI, Lăzău C. 2D and 3D nanostructured metal oxide composites as promising materials for electrochemical energy storage techniques: synthesis methods and properties. *IJMS*. 2024 Nov 21;25(23):12521.

29 Jasim AM, He X, White TA, Xing Y. Nano-layer deposition of metal oxides via a condensed water film. *Commun Mater*. 2020 Mar 2;1(1):9.

30 Vallejos S, Di Maggio F, Shujah T, Blackman C. Chemical vapour deposition of gas sensitive metal oxides. *Chemosensors*. 2016 Mar 1;4(1):4.

31 Sun L, Yuan G, Gao L, Yang J, Chhowalla M, Gharahcheshmeh MH, et al. Chemical vapour deposition. *Nat Rev Methods Primers*. 2021 Jan 14;1(1):5.

32 Piszczek P, Radtke A. Silver nanoparticles fabricated using chemical vapor deposition and atomic layer deposition techniques: properties, applications and perspectives: review. In: Seehra MS, Bristow AD, editors. *Noble and Precious Metals - Properties, Nanoscale Effects and Applications*. InTech; 2018 [cited 2025 May 24]. Available from: http://www.intechopen.com/books/noble-and-precious-metals-properties-nanoscale-effects-and-applications/silver-nanoparticles-fabricated-using-chemical-vapor-deposition-and-atomic-layer-deposition-techniqu

33 Wang SF, Umesh NM, Kameoka S, Sriram B. Promotional effects of heat treatments on Pt–CeO_2 nanocatalyst derived via hydrothermal leaching for CO oxidation. *ACS Appl Nano Mater*. 2024 Jun 28;7(12):14372–9.

34 Janjua MRSA. Synthesis of Co_3O_4 nano aggregates by co-precipitation method and its catalytic and fuel additive applications. *Open Chemi*. 2019 Oct 16;17(1):865–73.

35 Sajjadi SM, Haghighi M, Rahmani F. Sol–gel synthesis of Ni–Co/Al_2O_3–MgO–ZrO_2 nanocatalyst used in hydrogen production via reforming of CH_4/CO_2 greenhouse gases. *J Nat Gas Sci Eng*. 2015 Jan;22:9–21.

36 Asim M, Hussain A, Khan S, Arshad J, Butt TM, Hana A, et al. Sol-gel synthesized high entropy metal oxides as high-performance catalysts for electrochemical water oxidation. *Molecules*. 2022 Sep 13;27(18):5951.
37 Maccato C, Barreca D, Carraro G, Gasparotto A, Gombac V, Fornasiero P. Photoassisted H_2 production by metal oxide nanomaterials fabricated through CVD-based approaches. *Surf Coat Tech*. 2013 Sep;230:219–27.
38 Osman AI, Zhang Y, Farghali M, Rashwan AK, Eltaweil AS, Abd El-Monaem EM, et al. Synthesis of green nanoparticles for energy, biomedical, environmental, agricultural, and food applications: a review. *Environ Chem Lett*. 2024 Apr;22(2):841–87.
39 Deka K, Nongbet RD, Das K, Saikia P, Kaur S, Talukder A, et al. Understanding the mechanism underlying the green synthesis of metallic nanoparticles using plant extract(s) with special reference to silver, gold, copper and zinc oxide nanoparticles. *Hybrid Adv*. 2025 Jun;9:100399.
40 Disha SA, Sahadat Hossain Md, Habib MdL, Ahmed S. Green synthesis of nano-sized metal oxides (Ag_2O, CuO, ZnO, MgO, CaO, and TiO_2) using plant extract for a sustainable environment. *Nano Select*. 2025 Feb 21;6:e70000.
41 Bibi H, Ahmad M, Osman AI, Alsahli AA, Munir M, Al-Muhtaseb AH, et al. Green synthesis of highly active and recyclable chromium oxide nanocatalyst for biodiesel production from novel nonedible oil seeds. *GCB Bioenergy*. 2024 May; 16(5):e13140.
42 Selvanathan V, Shahinuzzaman M, Selvanathan S, Sarkar DK, Algethami N, Alkhammash HI, et al. Phytochemical-assisted green synthesis of nickel oxide nanoparticles for application as electrocatalysts in oxygen evolution reaction. *Catalysts*. 2021;11(12):1523.
43 Yadav LSR, Pratibha S, Manjunath K, Shivanna M, Ramakrishnappa T, Dhananjaya N, et al. Green synthesis of Ag ZnO nanoparticles: structural analysis, hydrogen generation, formylation and biodiesel applications. *J Sci Adv Mater Dev*. 2019 Sep;4(3):425–31.
44 Saman S, Balouch A, Talpur FN, Memon AA, Mousavi BM, Verpoort F. Green synthesis of MgO nanocatalyst by using *Ziziphus mauritiana* leaves and seeds for biodiesel production. *Appl Organom Chem*. 2021 May;35(5):e6199.
45 Hasan GG, Laouini SE, Osman AI, Bouafia A, Althamthami M, Meneceur S, et al. Green synthesis of Mn_3O_4@CoO nanocomposites using *Rosmarinus officinalis* L. extract for enhanced photocatalytic hydrogen production and CO_2 conversion. *J Environ Chem Eng*. 2024 Oct;12(5):113911.
46 Meena J, Kumaraguru N, Sami veerappa N, Shin P, Tatsugi J, Kumar AS, et al. Copper oxide nanoparticles fabricated by green chemistry using *Tribulus terrestris* seed natural extract-photocatalyst and green electrodes for energy storage device. *Sci Rep*. 2023 Dec 15;13(1):22499.
47 Ikhioya IL, Onoh EU, Nkele AC, Abor BC, Obitte BCN, Maaza M, et al. The green synthesis of copper oxide nanoparticles using the moringa oleifera plant and its subsequent characterization for use in energy storage applications. *East Eur J Phys*. 2023 Mar 2;(1):162–72.

48 Peralta-Videa JR, Huang Y, Parsons JG, Zhao L, Lopez-Moreno L, Hernandez-Viezcas JA, et al. Plant-based green synthesis of metallic nanoparticles: scientific curiosity or a realistic alternative to chemical synthesis? *Nanotechnol Environ Eng.* 2016 Dec;1(1):4.

49 Rambaran T, Schirhagl R. Nanotechnology from lab to industry – a look at current trends. *Nanoscale Adv.* 2022;4:3664.

50 Aithal PS, Aithal S. Nanotechnology innovations and commercialization-opportunities, challenges & reasons for delay. *I.J. Engineering and Manufacturing.* 2016 Nov;6:15–25.

51 Sharma RK, Bandichhor R, Mishra V, Sharma S, Yadav S, Mehta S, et al. Advanced metal oxide-based nanocatalysts for the oxidative synthesis of fine chemicals. *Mater Adv.* 2023;4(8):1795–830.

52 Vijayaram S, Razafindralambo H, Sun YZ, Vasantharaj S, Ghafarifarsani H, Hoseinifar SH, et al. Applications of green synthesized metal nanoparticles – a review. *Biol Trace Elem Res.* 2024 Jan;202(1):360–86.

53 Yoon Y, Truong PL, Lee D, Ko SH. Metal-oxide nanomaterials synthesis and applications in flexible and wearable sensors. *ACS Nanosci Au.* 2022 Apr 20;2(2):64–92.

54 Jamjoum HAA, Umar K, Adnan R, Razali MR, Mohamad Ibrahim MN. Synthesis, characterization, and photocatalytic activities of graphene oxide/metal oxides nanocomposites: a review. *Front Chem.* 2021 Sep 21;9:752276.

Part 2

Metal Oxide Nanocatalysts: Mechanisms and Reactions

5

Surface Chemistry of Metal Oxide Nanocatalysts

Nelson Oshogwue Etafo[1,2,3], Abayomi Bamisaye[4,5], Nouf Alharbi[6], Sreedeep Dey[7], Olubusayo Funmilola Semire[8,9], Jose Refugio Parga Torres[2], Mopelola Abidemi Idowu[10], and Henrietta W. Langmi[4]*

[1] Universidad Carolina, Saltillo, Coahuila, Mexico
[2] Department of Metallurgy and Materials Science, Institute Technology of Saltillo, Saltillo, Coahuila, Mexico
[3] Facultad de Ciencias Químicas, Universidad Autónoma de Coahuila, Saltillo, Coahuila, Mexico
[4] Department of Chemistry, University of Pretoria, Pretoria, South Africa
[5] Department of Chemistry, Faculty of Natural and Applied Sciences, Lead City University, Ibadan, Nigeria
[6] Physics Division, Department of Physical Sciences, College of Science, Jazan University, Jazan, Saudi Arabia
[7] Department of Chemical Engineering, University of Calcutta, Kolkata, West Bengal, India
[8] Department of Chemistry, Faculty of Physical Earth Science, University of Lagos, Lagos, Nigeria
[9] Department of Chemistry, Universite Grenoble Alpes, Grenoble, France
[10] Department of Chemistry, College of Physical Science, Federal University of Agriculture, Abeokuta, Nigeria

*Corresponding author: etafonelson@yahoo.com

5.1 Introduction

The rise of nanocatalysis has marked a turning point in chemistry, resulting in notable progress in developing metal oxide-based materials that are both highly efficient and selective, with properties that can be precisely adjusted at the nanoscale. Studies have shown that these specially designed nanostructures show great promise in enhancing important industrial oxidation reactions by improving activity, selectivity, and product output [1]. This is due to their very reactive surfaces. It might rearrange itself, change its oxidation state, or even separate certain elements [2]. All of this can greatly affect how well the catalyst works. For example, in materials like cerium oxide (CeO_2) or titanium dioxide (TiO_2), missing oxygen atoms help oxygen move more easily, which boosts oxidation reactions [3]. These vacancies also help capture molecules like carbon monoxide (CO), nitrogen monoxide (NO), or water (H_2O), and they can donate electrons to other molecules, influencing how charges move across the surface [4, 5].

Metal oxide nanomaterials have been made in many different shapes and sizes. For instance, when the particles are shaped like tiny spheres, they are known as zero-dimensional (0D) nanomaterials. One-dimensional (1D) types include long and

Metal Oxide Nanocatalysts for Sustainable Energy Production, First Edition. Edited by Abayomi Bamisaye et al.
Published 2026 by Wiley-VCH GmbH.

thin structures such as nanotubes, nanorods, nanowires, and nanobelts. Platforms like nanosheets and nanofilms belong to the two-dimensional (2D) category. More complex three-dimensional (3D) designs, such as core-shell structures and layered nanostructures are considered advanced and efficient materials [6, 7].

These nanostructured materials are vital in a wide range of catalytic applications, including energy production [2, 8], pollution control [9, 10], and chemical manufacturing [7]. Their high performance mainly comes from their special surface properties, which control how they interact with reacting molecules at the nanoscale. Because of their large surface area compared to their volume, they provide many active spots where reactions can happen. These active sites often include oxygen vacancies (OVs) (missing oxygen atoms), metal ions with different charge states, and hydroxyl (−OH) groups. These features are essential for driving oxidation-reduction (redox) reactions and for helping molecules stick to the surface [11].

Another important factor is how metal nanoparticles interact with the oxide surfaces they sit on, known as strong metal-support interaction, changing the electronic behavior of both the metal and the support. It can make the catalyst more stable, help it resist clumping (sintering), and even change how it behaves during a reaction [12]. Additionally, surface features like hydroxyl groups and acidic or basic sites can further improve the catalyst's performance in reactions involving acids, bases, or light (photocatalysis) [13–15].

Furthermore, studies have shown that the size and shape (morphology) of nanomaterials strongly influence their physical and chemical behavior [16]. Different crystal surfaces (facets) have unique atomic arrangements and electronic structures, which play a key role in improving the catalytic performance of metal oxide nanomaterials [17]. This provides more reactive surfaces or active sites, which helps control how well the catalyst works and what products are formed. Therefore, the concept of shape- and surface-specific nanocatalysis is believed to connect the fields of surface science and materials science. It provides new strategies to control how fast reactions happen, how selective they are, and which products are mainly formed.

It is worth noting that among the many types of catalysts, metal oxide nanocatalysts are particularly important and have been widely used in solid-phase (heterogeneous) catalysis. One key application is the hydrogenation of carbon dioxide (CO_2) to produce useful single-carbon (C1) chemicals [18], which is considered a promising method for achieving a carbon-neutral energy future. Studies suggest that the efficiency of metal oxide nanocatalysts in CO_2 hydrogenation often results from the combined (synergistic) effects at the interface between the metal and oxide components [19]. These interfaces are influenced by factors like the type of metal and oxide used, and how strongly they interact. Scientists often attach functional groups as shown in Figure 5.1, either physically or chemically to metal oxide surfaces using structured supports like ferrites, nano-silica, or alumina. Sharma et al. [7] and Ningthoujam et al. [20] to produce metal oxide nanomaterials suitable for various applications, thereby enhancing is applicability across various fields in material science.

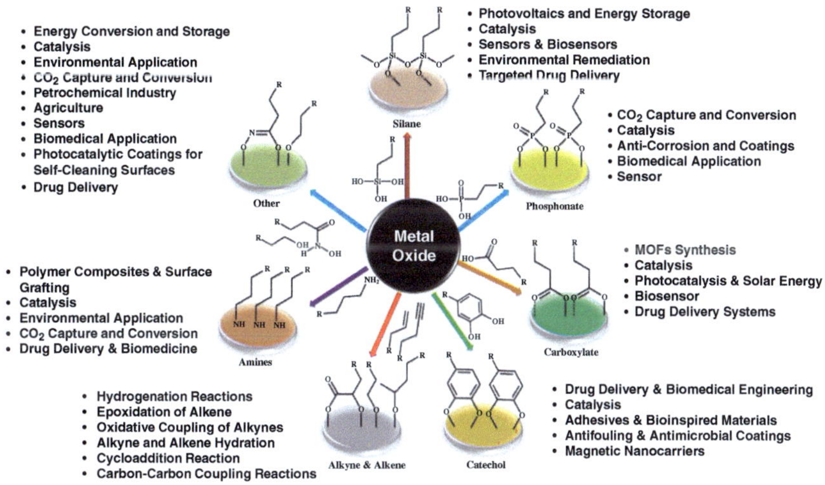

Figure 5.1 Modified surface of metal oxide nanocatalysts and their applications.

5.2 Processes Involved in Surface Chemistry

5.2.1 Surface Interactions of Metal Oxide Nanocatalysts

We must first understand what adsorption is, how thermodynamics works, and how kinetics influence adsorption, to comprehend the catalytic activity of metal oxide nanocatalysts. Surface chemistry involves several reactions, including adsorption, absorption, desorption, and occlusion. These are some of the procedures we will explore in this section.

- **Absorption and Adsorption:** Adsorption occurs at the surface of metal oxide nanocatalysts, whereas absorption occurs in the mass of nanocatalysts. Adsorption is the process by which metal oxide nanocatalyst molecules from one medium aggregate on the surface of another. However, in absorption, when a nanocatalyst material meet a medium, the nanocatalysts of that medium are evenly dispersed throughout the substance. Absorption is known as a bulk phenomenon and adsorption as a surface phenomenon [21].
- **Sorption:** In some circumstances, nanocatalysts' absorption and adsorption occur concurrently. This process is known as sorption [22].
- **Desorption:** Desorption is the opposite of adsorption. This technique eliminates adsorbate nanocatalyst molecules from the adsorbent's surface. Desorption occurs primarily at high temperatures and low pressure. At high temperatures, nanocatalyst molecules accumulate energy, which destabilizes them. This energy causes the adsorbed molecules to detach from the surface. By lowering the pressure, the molecules might begin to move away [23].
- **Occlusion:** Occlusion is the deposition of gas molecules on the surface of a solid. The absorption of gas onto a solid nanocatalyst is known as occlusion. For example, in the hydrogenation of oil, catalysts such as nickel or platinum are utilized to stabilize the hydrogen molecules and ensure proper hydrogenation. Hydrogen

gas molecules are absorbed on the surfaces of nickel or platinum nanocatalyst particles [24].

- **Physisorption and Chemisorption:** The bonding in physisorption is simply a van der Waals contact, whereas in chemisorption, it is a covalent bond. The energy involved in both circumstances results in a negative ΔG for the process. The ΔG for the procedure should be less than 60 kJ/mol [25]. As chemisorption involves the formation and breaking of covalent bonds, it is substantial, with values larger than 60 kJ/mol. Chemisorption is a chemical reaction breaking bonds in adsorbate nanocatalyst molecules, but physisorption remains constant [26]. As chemisorption is a chemical reaction; there will be an activation energy. But as physisorption, a physical reaction; there will be no activation energy. With physisorption, nanocatalyst molecules adhere to the surface, one at a time, before adding another layer, producing multiple layers. Chemisorption is a chemical reaction with the surface, so once covered; it can no longer react, resulting in only one layer [27]. We have represented these differences between physisorption and chemisorption using potential energy diagrams in Figure 5.2.

5.2.2 Kinetics Involved in Processes Involved in Surface Chemistry

Adsorption rates of metal oxide nanocatalysts are directly proportional to surface area activity. Contaminants are generated due to oxidation, reduction, or other chemical reactions. After many days of storage, contaminants accumulate on the adsorbent's surface. Impurities accumulate on the adsorbent. By eliminating these contaminants from the adsorbent, we can boost the metal oxide nanocatalysts' surface activity. The rate of adsorption rises as contaminants are removed from the adsorbent molecule. Adsorption activation refers to the process of boosting

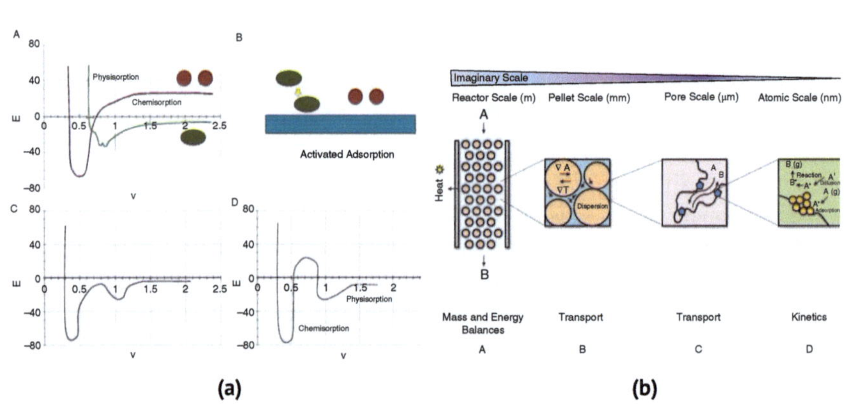

Figure 5.2 (a) Potential energy diagrams showing differences between physisorption and chemisorption. (b) Diagrammatic representation of catalyst design considerations and scales, highlighting turnover frequency (TOF), turnover number (TON), robustness, recovery, and regeneration, with visual emphasis on mass and energy balances, transport phenomena, and kinetics across reactor, pellet, pore, and atomic scales.

adsorption rates. Nickel, platinum, and palladium are among the metals we use as adsorbents. These metals are mostly used to absorb gases. For example, during the hydrogenation of oil, hydrogen gas is absorbed by these adsorbent metals. The surface area of these metals rises as they are broken down into microscopic particles known as nanocatalysts. Larger pieces of metal have fewer adsorption sites. However, when broken down into smaller nanocatalysts, the surface areas roughly double [28]. This indicates that the number of adsorption sites increases, as does the activity of the surface area. The rate of adsorption is increased by two processes: activation of the adsorbent and expanding the adsorbent's surface area [29]. Figure 5.2b illustrates catalyst design considerations and scales. Iruretagoyena Ferrer D elaborated, "Desorption is the physical process by which adsorbed atoms or molecules are released from a surface into the surrounding vacuum or fluid. In other words, adsorption (*physisorption*) phenomena are reversible, meaning that with the addition of a stimulus, such as thermal energy, the gas species or molecule will desorb from the surface. The long-range bonds seen in physical absorption can be broken by applying energy to the adsorbed molecule. The net effect could be to reduce the overall rate of adsorption or eliminate the adsorbed species from the surface" [30]. Rate kinetics can be used to describe desorption, which has been seen in a variety of material systems [31]. In thermal desorption, the heat source provides the energy required for the water molecules to break the hydrogen bonds that hold them together [32]. When an atom is physically or chemically adsorbed to the surface of metal oxide nanocatalysts, it can be freed by adding energy via photon absorption [33]. Photon-stimulated desorption (PSD) happens when energetic photons strike the vacuum chamber walls, resulting in electron emission, known as photoelectrons [34]. Ultrasonic desorption is based on energy absorption by the absorbate, resulting in localized heating. As a result, metal oxide nanocatalyst molecules are released with the addition of thermal energy. Water vapor is the most typical species to vibrate away from metal oxide nanocatalysts' surfaces using this method. The ultrasonic power is transmitted to the system via impedance matching barium titanate crystals to the metal and the thickness of the structure. Ultrasound-stimulated desorption is a feasible method for cleaning metal vacuum systems while they are under vacuum [35]. EID yield is defined as "the number of molecules that desorb from the chamber wall for every electron that strikes the surface." When high-energy electrons bombard the chamber's surface, they can excite the electrons of molecules that are weakly attached to it, potentially breaking bonds and allowing the atoms to desorb as ions or neutrals [36]. The beam-induced desorption (BID) depends on many parameters, including material type, surface roughness, cleanliness of the surface, temperature, integrated particle dose, and heterogeneous catalysts. Usually, solids that accelerate reactions between fluid-phase reactants are easily separated/recovered from products. Examples include metal nanoparticles dispersed on a support, metal oxides/sulfides like TiO_2 or MoS_2, metal atoms grafted to amorphous silica/alumina or SiO_2, zeolites having solid acid sites and shape-selective pores. The metal oxide nanocatalysts must be active and selective for hours or perhaps years. During a lifetime, each site of metal oxide nanocatalysts may facilitate thousands or millions of reactions. They must be able to resist several contaminants. The catalysts must have certain properties determining their functionalities. For example, (1) high TOF (turnover frequency) determines the rate, high TON (turnover number) determines the lifetime, and robustness to harsh conditions also influences the process. (2) Easy recovery/regeneration or inexpensiveness increases the utility of catalysts [37]. Adsorption reactions of metal oxide nanocatalysts can be presented in equation (1) as,

$$A(g) + * \rightarrow A^* \tag{1}$$

where reactants must adsorb onto the catalyst's active sites. Physisorption is a weak adsorption that does not significantly alter the reactant. Metal oxide nanocatalysts frequently necessitate chemisorption, which occurs when new bonds develop between the reactant and catalyst. Chemisorption weakens bonds inside the reactant, hence we refer to a chemisorbed species as "activated" during the chemisorption step [38]. Adsorption is typically, but not always, rapid. When rapid, quasi-equilibrium occurs in the adsorption stage. Adsorbates may need to travel over the surface in order to interact with one another. The diffusion processes might be slow. Unimolecular surface reactions can be presented in equation (2) as,

$$A^* \rightarrow B^* + C(g) \tag{2}$$

where, a surface intermediate may isomerize or disintegrate, producing new surface intermediates and/or gas-phase species. Eley-Rideal surface reactions can be presented in equation (3) as,

$$A(g) + B^* \rightarrow \text{Products} \tag{3}$$

where, surface intermediates can directly react with gas-phase species. An example of such reaction is,

$$CO(g) + O^* \rightarrow CO_2(g) \tag{4}$$

Langmuir-Hinshelwood surface reactions are the reactions between two adsorbed surface intermediates of metal oxide nanocatalysts. However, if products cannot desorb, they will clog all sites and contaminate the metal oxide nanocatalysts [39]. An automated thermal desorber is used for analyzing volatile organic chemicals. There are several odor compounds or pollutants in indoor air quality that must be regulated due to their negative impacts on human beings. It is difficult to identify volatile substances since they occur in trace amounts in solid, liquid, and gaseous samples. The thermal desorption approach overcomes this challenge. Thermal desorption is one of the sample preparation in metal oxide nanocatalyst procedures used to concentrate compounds into a sorbent tube and a focusing trap via absorption-desorption, allowing for the detection of volatile compounds at trace levels with high sensitivity. Essentially, volatile organic molecules are selectively collected from the sample, transferred to the separation area, and then determined by MS [40].

5.3 The Effects of OVs and Defect Locations on Catalytic Activity Enhancement

OVs and defect locations determine the catalytic characteristics of metal oxide nanocatalysts. These structural properties significantly interfere with surface reactivity, electrical structure, and redox behavior. Therefore, they are regarded as important properties improving catalytic activity in a variety of processes. Metal oxide crystal lattices contain OVs. These vacancies are essentially flaws induced by the absence of oxygen atoms. They behave like active sites. These active sites affect catalytic performance in a variety of pathways. (1) Electronic properties: OVs result in localized energy levels in the bandgap. As a result, electron density increases at the surface, improving electron transport to adsorbed molecules and lowering reaction activation energy. Not only that, but these vacancies function as oxygen stores. They promote effective oxidation and reduction cycles. These oxidation-reduction reactions (ORRs) have been seen in the context of CO oxidation and water splitting. (2) Surface reactivity: adsorption is higher in oxygen-deficient locations. Activated reactants, such as O_2 and CO, cause faster catalytic turnover [41]. We can determine catalytic activity based on the distribution pattern and type of faults. (1) Surface vacancies are more active than bulk vacancies, which affects long-term electrical characteristics. (2) Isolated vacancies have more catalytic activity than aggregated defects. It has the potential to significantly alter the structure's stability. Specific crystallographic orientations, such as {110} or {111} planes, exhibit unique catalytic properties due to differences in local bonding and electronic density [42]. OVs play an important role in catalytic processes. During CO oxidation, vacancies act as binding sites for O_2, allowing it to be transformed into reactive intermediates. Vacancies improve photocatalytic processes by improving charge separation and electron transport, which leads to higher efficiency. Vacancies can regenerate during the catalytic cycle, ensuring long-term catalytic activity [43]. To boost catalytic performance, sometimes OVs are intentionally produced using a variety of methods: (1) Doping with Nb or Mo is an effective method to stabilize OVs and boost their activity. (2) Vacancies can also be created by controlled annealing in decreasing atmospheres. Plasma and irradiation techniques precisely control vacancy density and location [43]. OVs often interact with other structural defects, such as metal interstitials or lattice distortions. For example, in CeO_2-based catalysts, the combination of OVs and dopants enhances redox activity for CO_2 hydrogenation [44]. Understanding and optimizing OVs calls for specialist tools. High-resolution microscopy techniques, such as TEM and STEM, can detect faults at the atomic scale. XPS and Raman spectroscopy can disclose the electronic states and chemical environments of vacancies. Simulations with Density Functional Theory (DFT) can predict vacancy behavior and guide catalyst design [45].

5.4 Surface Modification Techniques to Improve Catalytic Performance

Surface modification techniques play a crucial role in enhancing the catalytic performance of metal oxide nanocatalysts. These modifications, according to Wu et al. and Chen et al., aim to improve various aspects of the catalyst, including active site accessibility, providing more surface for catalytic reactions to occur, tuning the electronic structure of the catalyst to optimize adsorption and reaction rates, stability, which prevents agglomeration, leaching, and deactivation of the nanocatalysts, and selectivity to obtain the desired products [46, 47].

Some common surface modification techniques used for metal oxide nanocatalysts:

5.5 Metal Deposition

These include the following:

5.5.1 Loading with Noble Metals (e.g., Au, Pt, Pd)

Depositing small amounts of noble metals onto the metal oxide support can significantly enhance catalytic activity for various reactions like oxidation, reduction, and CO oxidation. The noble metal nanoparticles act as active sites, and the metal oxide support can influence their dispersion and electronic properties. Zhang et al. [48] reviewed that high CO concentrations (COCs) pose significant health and environmental risks. Catalytic oxidation using noble metal catalysts is highly effective for CO removal (COC). This paper reviews the COC reaction mechanism, catalytic performance, and optimization of typical noble metal catalysts based on existing research. Key conclusions include: COC mechanisms vary: Langmuir-Hinshelwood prevails in dry conditions for most noble metals, while single-atom catalysts and supported catalysts with high oxygen storage capacity follow Termolecular Eley-Rideal and Mars-van Krevelen mechanisms, respectively. Water content can either enhance or hinder COC.

Pt and Au-based catalysts exhibit superior performance among typical noble metals (Pt, Pd, Au, Ru, Ag, and Rh). Catalyst preparation, morphology, size, support, and reaction conditions (including water content) significantly impact CO conversion. They further discovered that COC activity can be enhanced by dopant introduction/alloy formation, size/dispersion control, surface valence state manipulation, and optimized synthesis/reaction conditions to improve CO adsorption.

5.5.2 Formation of Bimetallic or Multimetallic Nanoparticles

Combining different metals can create synergistic effects, leading to improved activity, selectivity, and stability compared to single-metal catalysts. Zou et al. reported that alloying at the subnanoscale maximizes these synergistic effects, with Pt-Zr combinations showing the greatest improvement in hydrogen evolution reactions [49]. Israr et al. reported that multimetallic catalysts demonstrate increased activity, selectivity, and stability across various electrochemical reactions, including CO_2 reduction, hydrogen evolution, and oxygen evolution [50].

5.6 Surface Coating and Functionalization

5.6.1 Coating with Another Metal Oxide

Applying a thin layer of a different metal oxide can modify the surface properties, enhance stability, or introduce new active sites. Khoury et al. reviewed that coating with oxides can improve thermal stability, prevent sintering, and introduce new active sites [51]. Zhou et al. demonstrated that submonolayer WOx with tunable coverage preferentially decorates well-coordinated Pt terrace sites as a stable ligand. By combining experimental kinetics with probe molecules, in situ spectroscopies, and first-principles modeling, we show that the WOx coverage on Pt modifies the metal-to-acid site balance while retaining the acid strength intact and results in optimal reactivity for metal-acid catalyzed reactions at a specific metal, size, and support-dependent WOx coverage. The oxide can also alter the reactant adsorption mode, reversing selectivity and pathways from terrace- to step-dominated, as evidenced in furfural decarbonylation and hydrogenation. The insights open avenues for improving metal/metal oxide catalysts beyond the specific system. Therefore, submonolayer oxide coatings, such as WOx on Pt, can modify metal-to-acid site balance and alter reactant adsorption modes, leading to improved selectivity and reactivity [52]. Siebenhofer et al. investigated using in-situ impedance spectroscopy during pulsed laser deposition (i-PLD) under clean conditions. They examined how ultrathin decoration layers affect the oxygen exchange kinetics of pristine mixed conducting oxide thin films. The study reveals that basic decorations with a lower surface work function significantly accelerate oxygen exchange on various material surfaces. Ultrathin oxide decoration layers can tailor the surface dipole and work function of mixed conducting oxides, with the ionic potential of surface cations serving as a descriptor for these effects [53].

5.6.2 Organic Functionalization

Grafting organic molecules (e.g., amines, thiols, polymers) onto the surface can alter the hydrophobicity/hydrophilicity, prevent agglomeration, or introduce specific functionalities for targeted reactions. Silane coupling agents are commonly used for this purpose. This approach, according to Leonova et al. [54], can alter hydrophobicity/hydrophilicity and introduce specific functionalities for targeted reactions. Trimethylsilyl groups grafted onto aluminosilicate catalysts have been shown to increase hydrophobicity and enhance catalytic performance in epoxide ring opening and ethanol dehydration reactions.

5.6.3 Carbon Coating or Doping

Introducing carbon materials like graphene, carbon nanotubes, or carbon dots can enhance electron transport, increase surface area, and improve catalytic activity, especially in electrocatalysis. Bai et al. [55] reported that when incorporated into composites, CQDs can improve conductivity, modify morphology, optimize electronic structure, and provide additional active sites. These carbon materials can regulate the intrinsic structure of metal and metal-free catalysts, affecting dispersibility, utilization efficiency, and coordination environment of active sites.

5.7 Creating Defects and Active Sites

5.7.1 Introducing OV

Controlled reduction treatments can create OVs on the metal oxide surface, which can act as active sites for certain reactions. The reports by Zhuang et al. and Shaw et al. showed that OVs significantly impact the physicochemical properties of catalysts, influencing light absorption and charge separation. The distribution and formation of OVs are strongly dependent on the type of lattice planes and surface morphology. They further stated that the enhanced catalytic performance of metal oxides with OVs has been attributed to the formation of surface oxygen species on reactive surfaces with low OV formation energy [56, 57].

5.7.2 Doping with Heteroatoms

Incorporating small amounts of different metal or nonmetal ions into the metal oxide lattice can modify its electronic and structural properties, leading to enhanced catalytic performance. Al-Naggar et al. reported that doping transition metal oxides with elements like Fe, Co, Mn, Ni, S, N, and P can improve conductivity, optimize reactant adsorption, and increase active sites for water splitting [58]. Similarly, Shrestha et al. reported that metal-ion doping in metal-organic frameworks modifies electronic structure and local coordination, enhancing oxygen evolution reaction activity [59].

5.7.3 Creating High-Energy Facets and Stepped Surfaces

Manipulating the morphology of the nanocrystals to expose specific crystal facets with higher surface energy can increase the number of active sites. Techniques like quenching can induce the formation of disordered stepped surfaces. Li et al. reported that Pd nanocrystals with exposed {100} and {111} facets on g-C_3N_4 exhibited superior photocatalytic hydrogen production compared to those with only {100} facets [60]. Similarly, Zhang et al. demonstrated in Pd/CeO_2 catalysts, CeO_2 nanorods with exposed (110) facets demonstrated the highest benzyl alcohol conversion due to optimal OV concentration and metal-support synergy [61]. While Liu et al. reported for CeO_2/CdS photocatalysts, NR-CeO_2 with exposed (311) facets showed remarkably higher photocatalytic hydrogen evolution activity than NF-CeO_2 with (220) facets, attributed to effective carrier separation and specific surface atomic arrangements [62].

5.8 Support Modification

5.8.1 Using Mesoporous or Hierarchical Structures

Employing supports with well-defined pore structures can enhance mass transport of reactants and products to and from the active sites. Wang et al. [63] highlighted the significance of nanoarchitecture engineering for catalysts, providing insights into optimizing structures with abundant active sites and improved mass transfer. By tailoring pore sizes via confined etching, they fabricated boron- and nitrogen-doped carbon (B,N@C) electrocatalysts with abundant active sites but varying porous structures to demonstrate the importance of mass transfer in the ORR. The ORR performance correlated with reactant diffusion, with the optimized trimodal-porous B,N@C exhibiting enhanced O_2 diffusion and higher activity per hetero-atomic site.

5.8.2 Modifying the Support's Electronic Properties

The interaction between the active metal oxide phase and the support can significantly influence the overall catalytic performance. Lyu et al. [64] quantitatively studied the impact of metal-support interactions on catalytic performance by designing CuxO/ZnO catalysts. They successfully modulated the electronic metal-support interactions between CuxO and ZnO without changing the catalyst structure. The lower work function of ZnO facilitated electron transfer to CuO, promoting the formation of more active Cu species. Experimental and theoretical findings indicate that the resulting electron-rich interface enhances oxygen adsorption and CO oxidation. This strategy offers a novel approach to improve the activity of supported catalysts for diverse applications.

The choice of surface modification technique depends on the specific metal oxide nanocatalysts, the intended reaction, and the desired catalytic properties. Often, a combination of these techniques can be employed to achieve optimal performance. Characterization techniques like TEM, SEM, XRD, XPS, and BET surface area analysis are crucial for understanding the effects of surface modification on the nanocatalyst's structure and properties, which in turn helps in correlating these changes with the observed catalytic activity.

5.9 The Effect of Particle Size and Shape on Catalytic Efficiency

The particle size and shape of metal oxide nanocatalysts are critical factors that significantly influence their catalytic efficiency. These parameters affect several key aspects of the catalyst, including the surface area, the number and type of active sites, the electronic properties, and the mass transport of reactants and products.

5.10 Effect of Particle Size

The particle size has the following effect:

1. **Surface Area and Active Sites:** Sharma et al. and Chen et al. reported that generally, decreasing the particle size of a metal oxide nanocatalyst leads to a significant increase in the surface area to volume ratio. This higher surface area provides more active sites for the reactants to interact with, thus potentially enhancing the overall catalytic activity per unit mass of the catalyst [15, 65, 66]. Smaller nanoparticles often expose a higher proportion of edge and corner sites, which are typically more active than terrace sites due to their lower coordination number and higher surface energy [67].
2. **Quantum Size Effects and Electronic Properties:** According to Sharma et al. and Incio et al., in the nanometer regime, especially below a certain critical size (e.g., <5 nm for some metals), quantum size effects can become significant. These effects can alter the electronic band structure of the metal oxide, which in turn can influence its adsorption properties, activation of reactants, and thus the catalytic activity [67, 68].
3. **Mass Transport:** Smaller nanoparticles can sometimes lead to improved mass transport due to reduced diffusion limitations within the catalyst structure, especially in porous supports [69].
4. **Stability:** Basu and Banik reported that very small nanoparticles can be thermodynamically unstable and prone to agglomeration or sintering at high reaction temperatures, leading to a decrease in surface area and catalytic activity over time. Therefore, an optimal particle size range often exists for maximum catalytic efficiency and stability [70].
5. **Size-Dependent Activity Trends:** The relationship between particle size and catalytic activity is not always straightforward and can exhibit different trends: (1) Positive size effect: smaller nanoparticles show higher activity. The catalytic activity of PNIPAAm-AuNP hybrid microgels exhibits a positive size effect, with smaller 13 nm AuNPs showing higher activity than larger 36 nm AuNPs [71]. (2) Negative size effect: smaller nanoparticles show lower activity. Rao et al. stated that Fe-doped NiO nanoparticles for oxygen evolution showed a negative size effect, with smaller particles exhibiting higher activity due to increased oxidized species density [72]. (3) Volcano-type behavior: an optimal particle size exists that exhibits the highest activity, with both smaller and larger particles being less active. Ge et al. reported that propane oxidation using Pt nanoparticles on CeO_2, a volcano-type behavior was observed, with an optimal size around 4 nm due to the interplay of electronic and geometric factors [73].

5.11 Effect of Particle Shape

The particle shape influences the following:

1. **Exposure of Specific Crystal Facets:** The shape of a metal oxide nanocrystal determines which crystallographic facets are exposed on its surface and their

relative proportions. Different crystal facets have different atomic arrangements and surface energies, leading to variations in their adsorption properties and catalytic activity for specific reactions [74]. Certain facets might have a higher density of specific active sites or exhibit stronger interactions with particular reactants. Shao et al. reported that for Pd-loaded C_3N_4 photocatalysts, different facets exhibited distinct active site distributions, with Pd(100) facets showing more dispersed oxidation sites compared to Pd(111) facets [75]. In iridium oxides, the density and energetics of active sites were found to affect water oxidation activity, with adsorbed oxygen species exhibiting repulsive interactions at higher coverages, according to Liang et al. [76].

2. **Number of Edge and Corner Sites:** Kim et al. and Konsolakis and Lykaki reported that nanocrystals with sharp edges and corners (e.g., cubes, polyhedra) generally have a higher proportion of low coordination number atoms at these sites compared to more spherical nanoparticles. These edge and corner sites can exhibit different catalytic behavior than the atoms on flat surface facets due to their unique electronic and geometric properties [77, 78].
3. **Mass Transport:** Barkaoui et al. reported that the overall shape and morphology of the nanocatalyst can also influence the accessibility of the active sites and the diffusion of reactants and products, especially in structured catalysts or catalytic layers [79]. For instance, anisotropic shapes like nanorods or nanowires might offer different mass transport characteristics compared to spherical nanoparticles.
4. **Shape-Dependent Activity:** The catalytic activity and selectivity can be highly dependent on the shape of the metal oxide nanocatalysts. By controlling the synthesis to obtain specific shapes (e.g., cubes, octahedra, rods, wires), it is possible to tailor catalytic performance for a desired reaction. Sarıbıyık et al. reported that shape-controlled Pt and Pd nanoparticles showed varying selectivity in the hydrodechlorination of trichloroethylene [80]. Similarly, the shape of metal oxides like Mn, Co, Ce, and Cu affects their performance in selective oxidation reactions Burange and Gopinath [81].

In summary, both the size and shape of metal oxide nanocatalysts are crucial parameters that dictate their catalytic efficiency. Optimizing these parameters through controlled synthesis and modification techniques is essential for designing high-performance catalysts for various applications. The ideal size and shape often depend on the specific catalytic reaction and the desired performance characteristics.

5.12 Challenges and Future Directions in the Surface Chemistry of Metal Oxide Nanocatalysts

5.12.1 Challenges

The surface chemistry of metal oxide nanocatalysts faces several critical challenges.

1. **Active Defect Controls:** In dynamic reaction conditions, when their distribution and stability vary, controlling active defects, including OVs, interstitials, and edge sites, remains challenging.

2. **Structural Designs:** Overlapping adsorption, reaction, and desorption processes complicate the establishment of obvious structure-function correlations; while surface analysis has been improved by instruments including XPS, STM, and AFM, real-time tracking of active areas remains limited. Reducing or hydroxylating surface reconstructions during catalysis complicates the design of stable and selective catalysts.
3. **Poor Interfacial Charge Transfer and Electric Fields:** Furthermore, poorly known at the atomic level in heterostructured systems are interfacial charge transfer and electric fields, which restrict control over catalytic routes.
4. **Catalyst Poisoning:** In practical settings, catalyst deactivation resulting from poisoning by pollutants such as sulfur or chlorine remains a key problem. Although they show promise for increased activity and selectivity, multicomponent systems provide complexity that complies poorly with predictive modeling.

5.12.2 Future Directions

Future methodologies should concentrate on operando characterization techniques (e.g., ambient-pressure XPS and synchrotron-based approaches) to investigate surfaces during working circumstances to solve these problems. DFT, molecular dynamics, and machine learning taken together provide a strong path for catalytic behavior prediction. While defect engineering employing ion irradiation or plasma treatment can increase activity by controlled defect generation, tailored surface termination, and exact atomic doping can enable facet-selective reactivity. Finally, entropy-stabilized oxides with multifarious active sites have promise for increasing resistance to deactivation and long-term stability.

5.13 Conclusion

The surface chemistry of metal oxide nanocatalysts is a dynamic and multidisciplinary field that holds immense potential for advancing catalytic technologies. This review has explored key aspects, including the fundamental surface interactions such as adsorption, desorption, and reaction kinetics, which form the basis for understanding catalytic processes. The critical role of OVs and defect locations in enhancing catalytic activity highlights the importance of material engineering at the atomic level. Surface modification techniques offer innovative strategies to optimize catalytic performance. Additionally, the influence of particle size and shape on catalytic efficiency underscores the need for precise control over nanocatalyst design. Despite significant progress, challenges remain, particularly in scalability, stability, and cost-effectiveness. Future directions emphasize the importance of interdisciplinary research, advanced characterization tools, and sustainable practices to address these challenges and unlock the full potential of metal oxide nanocatalysts. Collectively, these efforts will pave the way for more efficient, durable, and environmentally friendly catalytic systems.

References

1 Bamisaye A, Abati SM, Ige AR, Etafo NO, Alli YA, Bamidele MO, et al. Metal-oxide nanocatalysts for spontaneous sequestration of endocrine-disrupting compounds from wastewater. *Chemosphere*. 2024;367:143569.

2 Modupe Abati S, Bamisaye A, Abidemi Adaramaja A, Rapheal Ige A, Adegoke KA, Olurotimi Ogunbiyi E, et al. Biodiesel production from spent vegetable oil with Al_2O_3 and Fe_2O_3-biobased heterogenous nanocatalysts: comparative and optimization studies. *Fuel*. 2024 May 15;364:130847.

3 Liu Z, Guo B, Hong L, Jiang H. Preparation and characterization of cerium oxide doped TiO_2 nanoparticles. *J Phys Chem Solids* [Internet]. 2005 Jan 1 [cited 2025 Jun 7];66(1):161–7. Available from: https://www.sciencedirect.com/science/article/abs/pii/S0022369704002471

4 De Zoysa ASI, Narayana Y, Gaonkar B, Roopashree R. Advancements in catalytic technologies for promoting sustainable healthy environment and green transport. In: *Polymers and Functional Materials for a Cleaner Environment and Human Health* [Internet]. 2025 May 6 [cited 2025 Jun 7]. p. 207–26. Available from: https://www.taylorfrancis.com/chapters/edit/10.1201/9781003639602-8/advancements-catalytic-technologies-promoting-sustainable-healthy-environment-green-transport-agampodi-sandali-inupama-de-zoysa-yashwanth-narayana-bhaskar-gaonkar-roopashree

5 Ogungbesan SO, Etafo NO, Anselm OH, Ejeromedoghene O, Kalulu M, Abdullah M, et al. Transition metal oxide nanohybrid materials: a review of their structures, properties, and applications. *J Mol Struct*. 2025 Aug 15;1337:142209.

6 Makawana JA, Sangani CB, Yao YF, Duan YT, Lv PC, Zhu HL. Recent developments of metal and metal oxide nanocatalysts in organic synthesis. *Mini Rev Med Chem*. 2016 Sep 2;16(16):1303–20.

7 Sharma RK, Bandichhor R, Mishra V, Sharma S, Yadav S, Mehta S, et al. Advanced metal oxide-based nanocatalysts for the oxidative synthesis of fine chemicals. *Mater Adv* [Internet]. 2023 Apr 24 [cited 2025 Jun 7];4(8):1795–830. Available from: https://pubs.rsc.org/en/content/articlehtml/2023/ma/d2ma00977c

8 Ogunbiyi O, Bamisaye A, Abiodun AJ, Owoeye TF, Alli YA, Idowu MA. Biogenic synthesis of zinc oxide nanoparticles for solar cell application and photodegradation of neomycin. *Mater Sci Eng B*. 2025 Sep 1;319:118324.

9 Bamisaye A, Oluboyede OA, Etafo NO, Bamidele MO, Abiola-Kuforiji OT, Idowu MA. Synthesis of zinc oxide nanoparticles using *syzygium malaccense* leaf extract: photocatalytic decomposition of ciprofloxacin and antimicrobial studies. *Nanotechnol Environ Eng* [Internet]. 2025 Sep 30;10(3):54. Available from: https://link.springer.com/10.1007/s41204-025-00459-z

10 Gold VF, Bamisaye A, Adesina MO, Adegoke KA, Ige AR, Adeleke O, et al. Adsorptive uptake of thymol blue from aqueous medium using calcined snail shells: equilibrium, kinetic, thermodynamic, neuro-fuzzy and DFT studies. *J Dispers Sci Technol* [Internet]. 2025;46:1747–63. https://doi.org/10.1080/01932691.2024.2339458

11 Joshi NC, Gururani P, Bhatnagar P, Kumar V, Vlaskin MS. Advances in metal oxide-based nanocatalysts for biodiesel production: a review. *ChemBioEng Rev.* 2023 Jun;10(3):258–71.

12 Liu Y, Zhang L, Niu Y, Zhang B, Li L, Qiao B, et al. Microwave-assisted rapid construction of strong metal-support interactions for sinter-resistant gold nanocatalysts. *Appl Catal A Gen* [Internet]. 2025 Jun 5 [cited 2025 Jun 7];699: 120252. Available from: https://www.sciencedirect.com/science/article/abs/pii/S0926860X2500153X

13 Etafo NO, Bamisaye A, Bamidele MO, Renteria EV, Alli YA, Bakare OC, et al. Beyond the swipe: a review of photocatalytic antimicrobial biocompatible touchscreen technology. *Appl Mater Today.* 2025 Jun;44:102697.

14 Etafo NO, Bamidele MO, Bamisaye A, Alli YA. Revolutionizing photocatalysis: unveiling efficient alternatives to titanium (IV) oxide and zinc oxide for comprehensive environmental remediation. *J Water Process Eng* [Internet]. 2024 May;62:105369. Available from: https://linkinghub.elsevier.com/retrieve/pii/S2214714424006019

15 Nkoh JN, Oderinde O, Etafo NO, Kifle GA, Okeke ES, Ejeromedoghene O, et al. Recent perspective of antibiotics remediation: a review of the principles, mechanisms, and chemistry controlling remediation from aqueous media. *Sci Total Environ.* 2023 Jul;881:163469.

16 Bamisaye A, Adegoke KA, Alli YA, Bamidele MO, Idowu MA, Ogunjinmi OE. Recent advances in nanoemulsion for sustainable development of farm-to-fork systems. *J Clean Prod.* 2023 Dec;429:139226.

17 Liao X, Huang Z, Tian M, Gong J, Ni J, Zhou Q, et al. Facet-dependent electronic modulation of metal-support interactions for enhanced low-temperature benzene oxidation on Pt/Co_3O_4 catalysts. *Chem Eng J.* [Internet]. 2025 Jan 1 [cited 2025 Jun 7];503:158449. Available from: https://www.sciencedirect.com/science/article/abs/pii/S1385894724099406

18 Guo H, Pei LJ, Xie SL, Shen MQ, Yang Y, Liu H. Copper-based nanocatalysts toward chemoselective hydrogenation reaction. *cMat* [Internet]. 2024 Jun 1 [cited 2025 Jun 7];1(1):e9. https://doi.org/10.1002/cmt2.9

19 Alli YA, Bamisaye A, Bamidele MO, Etafo NO, Chkirida S, Lawal A, et al. Transforming waste to wealth: harnessing carbon dioxide for sustainable solutions. *Results Surf Interfaces.* 2024 Oct;17:100321.

20 Ningthoujam R, Singh YD, Babu PJ, Tirkey A, Pradhan S, Sarma M. Nanocatalyst in remediating environmental pollutants. *Chem Phys Impact* [Internet]. 2022 Jun 1 [cited 2025 Jun 7];4:100064. Available from: https://www.sciencedirect.com/science/article/pii/S2667022422000032

21 Soo XYD, Lee JJC, Wu WY, Tao L, Wang C, Zhu Q, et al. Advancements in CO_2 capture by absorption and adsorption: a comprehensive review. *J CO_2 Util.* 2024 Mar;81:102727.

22 Egues S, de Resende NS, Schmal M. Surface characterization of WO_3-TiO_2/Al_2O_3 catalysts and reactivity on selective catalytic reaction of NO by NH_3. *Stud Surf Sci Catal.* 2000;143:933–9.

23 Gunawardene OHP, Gunathilake CA, Vikrant K, Amaraweera SM. Carbon dioxide capture through physical and chemical adsorption using porous carbon materials: a review. *Atmosphere*. 2022 Feb 28;13(3):397.

24 Busca G. From surface science to industrial heterogeneous catalysis. In: Rocca M, Rahman TS, Vattuone L, editors. *Springer Handbook of Surface Science. Springer Handbooks*. Cham: Springer; 2020. p. 1087–1115. https://doi.org/10.1007/978-3-030-46906-1_33

25 Liu L, Luo XB, Ding L, Luo SL. Application of nanotechnology in the removal of heavy metal from water. In: *Nanomaterials for the Removal of Pollutants and Resource Reutilization*. Elsevier; 2019(4):83–147.

26 Groß A. Hot atom chemistry: oxygen at stepped platinum surfaces. *Appl Surf Sci Adv*. 2022 Jun;9:100240.

27 Mudoi MP, Sharma P, Khichi AS. A review of gas adsorption on shale and the influencing factors of CH_4 and CO_2 adsorption. *J Pet Sci Eng*. 2022 Oct;217:110897.

28 Benjelloun M, Miyah Y, Akdemir Evrendilek G, Zerrouq F, Lairini S. Recent advances in adsorption kinetic models: their application to dye types. *Arab J Chem*. 2021 Apr;14(4):103031.

29 Murphy OP, Vashishtha M, Palanisamy P, Kumar KV. A review on the adsorption isotherms and design calculations for the optimization of adsorbent mass and contact time. *ACS Omega*. 2023 May 23;8(20):17407–30.

30 Iruretagoyena Ferrer D. Literature review. Supported Layered Double Hydroxides as CO_2 Adsorbents for Sorption-enhanced H_2 Production Springer Charm 2016. p. 7–43.

31 Knall J, Barnett SA, Sundgren JE, Greene JE. Adsorption and desorption kinetics of In on Si(100). *Surf Sci*. 1989 Mar;209(3):314–34.

32 Mohammed M, Rahman R, Mohammed AM, Adam T, Betar BO, Osman AF, et al. Surface treatment to improve water repellence and compatibility of natural fiber with polymer matrix: Recent advancement. *Polym Test*. 2022 Nov;115:107707.

33 Thornton G, Warburton DR, Owen IW, Richardson CH, Mcgrath R, Mcgovern IT, et al. The use of photon stimulated ion desorption sexafs in the determination of surface structure. *J Phys Colloques*. 1986 Dec; 47(C8):C8-179-C8-183.

34 Bertin M, Fayolle EC, Romanzin C, Poderoso HAM, Michaut X, Philippe L, et al. Indirect ultraviolet photodesorption from $CO:N_2$ binary ices—an efficient grain-gas process. *Astrophys J*. 2013 Dec 2;779(2):120.

35 Arshi Banu PS, Mahbubul IM, Azhar M. Solar-operated vapor absorption cooling system. In: *Renewable Energy Production and Distribution Volume 2*. Elsevier; 2023. p. 63–107.

36 Boeckers H, Rohdenburg M, Swiderek P. Surface science studies on electron-induced reactions of NH_3 and their perspectives for enhancing nanofabrication processes. *Surf Sci*. 2025 Jan;751:122628.

37 Johnson PD. Synchrotron radiation. In 1997. p. 23–43.

38 Weng L, Su L, Xu N, Qian C, Cai Q, Chen R, et al. The preferred adsorption sites and catalytic mechanism of FCC_CoFeGaNiZn multi-principal element alloy for oxygen evolution reaction catalysis based on site preference of constituent atom on sublattice. *Intermetallics*. 2024 Feb;165:108132.

39 Chew WX, Kaizu K, Watabe M, Muniandy SV, Takahashi K, Arjunan SNV. Surface reaction-diffusion kinetics on lattice at the microscopic scale. *Phys Rev E*. 2019 Apr 19;99(4):042411.

40 Duemichen E, Eisentraut P, Celina M, Braun U. Automated thermal extraction-desorption gas chromatography mass spectrometry: a multifunctional tool for comprehensive characterization of polymers and their degradation products. *J Chromatogr A*. 2019 May;1592:133–42.

41 Kaushik S, Xiao X, Xu Q. Design strategies of electrocatalysts for acidic oxygen evolution reaction. *EnergyChem*. 2023 Sep;5(5):100104.

42 Gao S, Zhao S, Tang X, Sun L, Li Q, Yi H. Research on the application of defect engineering in the field of environmental catalysis. *Green Energy Environ*. 2025 Jun;10(6):1187–209.

43 Etim UJ, Zhang C, Zhong Z. Impacts of the catalyst structures on CO_2 activation on catalyst surfaces. *Nanomaterials*. 2021 Nov 30;11(12):3265.

44 Ahn SY, Jang WJ, Shim JO, Jeon BH, Roh HS. CeO_2-based oxygen storage capacity materials in environmental and energy catalysis for carbon neutrality: extended application and key catalytic properties. *Cat Rev*. 2024 Oct 3;66(4):1316–99.

45 Anekwe IMS, Akpasi SO, Enemuo EM, Ashiegbu D, Mustapha SI, Isa YM. Innovations in catalytic understanding: a journey through advanced characterization. *Mater Today Catal*. 2024 Dec;7:100061.

46 Wu D, Han D, Zhou W, Streiff S, Khodakov AY, Ordomsky VV. Surface modification of metallic catalysts for the design of selective processes. *Cat Rev*. 2024 Apr 2;66(2):640–86.

47 Chen S, Zhang H, Yao C, Lou H, Chen M, Lang X, et al. Review of SOFC cathode performance enhancement by surface modifications: recent advances and future directions. *Energy Fuels*. 2023 Mar 2;37(5):3470–87.

48 Zhang J, Shu M, Niu Y, Yi L, Yi H, Zhou Y, et al. Advances in CO catalytic oxidation on typical noble metal catalysts: Mechanism, performance and optimization. *Chem Eng J*. 2024 Sep;495:153523.

49 Zou Q, Akada Y, Kuzume A, Yoshida M, Imaoka T, Yamamoto K. Alloying at a subnanoscale maximizes the synergistic effect on the electrocatalytic hydrogen evolution. *Angew Chem Int Ed Engl*. 2022 Oct 4; 61(40):e202209675.

50 Israr M, Humayun M, Suliman MH, Abdinejad M, Rasheed T, Helal A, et al. Multi-metallic electrocatalysts as emerging class of materials: opportunities and challenges in the synthesis, characterization, and applications. *Cat Rev*. 2024 May 28;67(4):1–61. Available from: https://urldefense.com/v3/__https://research.tudelft.nl/en/publications/multi-metallic-electrocatalysts-as-emerging-class-of-materials-op/*__;Iw!!N11eV2iwtfs!tThC1wMlSYGMR8-9QJQDyzKoQjk6vBgjXGTwbF9T4PddERGB52R0e3-W3YQVIEvXz_DsJ2NJ3zWr91qDmQQ$

51 Khoury C, Montemore MM, Gazit OM. The effect of surface phase oxides on the properties of supported metals and catalysis. In: *The Royal Society of Chemistry*. 2021. p. 153–80.

52 Zhou J, Yang P, Caratzoulas S, Zheng W, Vlachos DG. Engineering active sites of metal/metal oxide catalysts by oxide ligand overlayers. *Angew Chem Int Ed Engl*. 2025 Jan 21;64(4):e202416746.

53 Siebenhofer M, Nenning A, Rameshan C, Blaha P, Fleig J, Kubicek M. Engineering surface dipoles on mixed conducting oxides with ultra-thin oxide decoration layers. *Nat Commun*. 2024 Feb 26;15(1):1730.

54 Leonova L, Moravec Z, Sazama P, Pastvova J, Kobera L, Brus J, et al. Hydrophobicity boosts catalytic activity: the tailoring of aluminosilicates with trimethylsilyl groups**. *Chem Cat Chem*. 2023 Jul 7;15(13):e202300449.

55 Bai M, Fu Z, Liu C, Zhao P, Ma J, Wang W, et al. Advances in understanding the regulation effect of carbon dots on intrinsic activity of metal and metal-free catalysts for electrocatalysis reaction. *Chem Eng J*. 2024 Jun;489:151544.

56 Zhuang G, Chen Y, Zhuang Z, Yu Y, Yu J. Oxygen vacancies in metal oxides: recent progress towards advanced catalyst design. *Sci China Mater*. 2020 Nov 19;63(11):2089–118.

57 Shaw M, Samanta D, Bera S, Mahto MK, Salam Shaik MA, Konar S, et al. Role of surface oxygen vacancies and oxygen species on CuO nanostructured surfaces in model catalytic oxidation and reductions: insight into the structure–activity relationship toward the performance. *Inorg Chem*. 2022 Sep 19;61(37):14568–81.

58 Al-Naggar AH, Shinde NM, Kim JS, Mane RS. Water splitting performance of metal and non-metal-doped transition metal oxide electrocatalysts. *Coord Chem Rev*. 2023 Jan;474:214864.

59 Shrestha NK, Patil SA, Salunke AS, Inamdar AI, Kim H, Im H. Metal-ion doping in metal–organic-frameworks: modulating the electronic structure and local coordination for enhanced oxygen evolution reaction activity. *Dalton Trans*. 2023;52(39):13852–7.

60 Li F, Xu B, You X, Gao G, Xu R, Wang XL, et al. In-situ synthesis of Pd nanocrystals with exposed surface-active facets on g-C_3N_4 for photocatalytic hydrogen generation. *Int J Hydrogen Energy*. 2023 Apr;48(33):12299–308.

61 Zhang L, Chen R, Tu Y, Gong X, Cao X, Xu Q, et al. Revealing the crystal facet effect of ceria in Pd/CeO_2 catalysts toward the selective oxidation of benzyl alcohol. *ACS Catal*. 2023 Feb 17;13(4):2202–13.

62 Liu Z, Zhuang Y, Dong L, Mu H, Li D, Zhang F, et al. Enhancement mechanism of photocatalytic hydrogen production activity of CeO_2 /CdS by morphology regulation. *ACS Appl Energy Mater*. 2023 Jul 24;6(14):7722–36.

63 Wang X, Liu T, Li H, Han C, Su P, Ta N, et al. Balancing mass transfer and active sites to improve electrocatalytic oxygen reduction by B,N codoped C nanoreactors. *Nano Lett*. 2023 Jun 14;23(11):4699–707.

64 Lyu S, Zhang Y, Li Z, Liu X, Tian Z, Liu C, et al. Electronic metal-support interactions between Cu_xO and ZnO for Cu_xO/ZnO catalysts with enhanced CO oxidation activity. *Front Chem*. 2022 May 13;10:912550.

65 Sharma S, Ansari A. Metal and metal oxide sub nano cluster; emerging aspirant for catalytic applications. *Results Chem*. 2023 Jan;5:100982.

66 Chen T, Meng D. A critical review of nanoparticles and nano catalyst. *JCIMS*. 2023 Jan 9;1:12–22.

67 Duan X, Ying L, Li XY, Zhu B, Gao Y. High or low coordination: insight into the active site of Pt nanoparticles toward CO oxidation. *J Phys Chem Lett*. 2023 Nov 9;14(44):9848–54.

68 Inico E, Saetta C, Di Liberto G. Impact of quantum size effects to the band gap of catalytic materials: a computational perspective*. *J Phys Condens Matter.* 2024 Sep 11;36(36):361501.

69 Henrotte O, Kment Š, Naldoni A. Mass transport limitations in plasmonic photocatalysis. *Nano Lett.* 2024 Jul 24;24(29):8851–8.

70 Basu S, Banik BK. Nanoparticles as catalysts: exploring potential applications. *Curr Organocatal.* 2024 Dec;11(4):265–72.

71 Pongsanon P, Kawamura A, Kawasaki H, Miyata T. Effect of gold nanoparticle size on regulated catalytic activity of temperature-responsive polymer-gold nanoparticle hybrid microgels. *Gels.* 2024 May 22;10(6):357.

72 Rao RR, Bucci A, Corby S, Moss B, Liang C, Gopakumar A, et al. Unraveling the role of particle size and nanostructuring on the oxygen evolution activity of Fe-doped NiO. *ACS Catal.* 2024 Aug 2;14(15):11389–99.

73 Ge S, Fan W, Tang X, Cui Y, Wang D, Gong XQ, et al. Revealing the size effect of ceria nanocube-supported platinum nanoparticles in complete propane oxidation. *ACS Catal.* 2024 Feb 16;14(4):2532–44.

74 Zhou D, Zhou Y, Li Y, Shen W. Nanostructured iron oxides for heterogeneous catalysis. *EnergyChem.* 2024 Jul;6(4):100124.

75 Shao DN, Wang XL, Yao YF. Relationship between Pd facets and surface active sites in photocatalysis via NMR molecular probes. *J Phys Chem C.* 2024 Jun 13;128(23):9548–55.

76 Liang C, Rao RR, Svane KL, Hadden JHL, Moss B, Scott SB, et al. Unravelling the effects of active site density and energetics on the water oxidation activity of iridium oxides. *Nat Catal.* 2024 Jun 7;7(7):763–75.

77 Kim S, Kwag J, Machello C, Kang S, Heo J, Reboul CF, et al. Correlating 3D surface atomic structure and catalytic activities of Pt nanocrystals. *Nano Lett.* 2021 Jan 27;21(2):1175–83.

78 Konsolakis M, Lykaki M. Facet-dependent reactivity of ceria nanoparticles exemplified by CeO_2-based transition metal catalysts: a critical review. *Catalysts.* 2021 Mar 31;11(4):452.

79 Barkaoui S, Elboughdiri N, Ghernaout D, Benguerba Y. Well-defined tricobalt tetraoxide's critical morphology effect on the structure–reactivity relationship. *RSC Adv.* 2024;14(30):21745–62.

80 Sarıbıyık OY, Weilach C, Serin S, Rupprechter G. The effect of shape-controlled Pt and Pd nanoparticles on selective catalytic hydrodechlorination of trichloroethylene. *Catalysts.* 2020 Nov 13;10(11):1314.

81 Burange AS, Gopinath CS. Shape-controlled metal oxides for selective catalytic oxidation. In: *Advanced Heterogeneous Catalysts Volume 1: Applications at the Nano-Scale.* ACS Publication; 2020;10(1):291–318.

6

Redox Reactions Catalyzed by Metal Oxide Nanocatalysts

*Monsuru Adewale Adekola[1], Ayodeji Ralpheal Ige[2], Emmanuel Olurotimi Ogunbiyi[3], Olumuyiwa O. Ogunlaja[1,4], and Abayomi Bamisaye[1,5]**

[1] *Department of Chemistry, Faculty of Natural and Applied Sciences, Lead City University, Ibadan, Nigeria*
[2] *Faculty of Civil Engineering and Environmental Sciences, Białystok University of Technology, Białystok, Poland*
[3] *Department of Chemistry, College of Physical Science, Federal University of Agriculture, Abeokuta, Nigeria*
[4] *African Center of Excellence for Water and Environmental Research (ACEWATER), Redeemer's University, Ede, Osun State, Nigeria*
[5] *Department of Chemistry, University of Pretoria, Pretoria, South Africa*

*Corresponding author: abayomibamisaye@gmail.com, abayomi.bamisaye@up.ac.za

6.1 Introduction

Redox reactions or electron transfer processes between chemical species are central to many chemical and biochemical processes. Such reactions are important in many applications, including degradation of pollutants, energy conversion, and catalytic conversion of materials to value-added products. The multiple oxidation states, oxygen vacancies, and high surface area-to-volume ratio allow different metal oxide nanocatalysts (MeONCs) to become a topic of interest as they can perform redox catalysis under mild conditions [1, 2].

Studies have shown that the nanostructured materials, that is, MeONCs like TiO_2, CeO_2, Fe_2O_3, and MnO_2 have great redox properties and are ideally suited to catalyze oxidative and reductive processes for various chemical processes in the field of material science, a typical example is environmental remediation [3, 4] and energy-related applications [5–8]. Morphological, electronic, and structural properties are some of the most important factors that determine their catalytic performance and can be optimized through synthesis techniques, doping, or defect engineering [5, 9]. Redox reactions are foundational chemical processes that control phenomena, as varied as metal corrosion and complex biochemical processes in organisms [10]. These are ultimately associated with the exchange of electrons between chemical species, and a shift of oxidation state [1]. Such processes are always coupled

Metal Oxide Nanocatalysts for Sustainable Energy Production, First Edition. Edited by Abayomi Bamisaye et al.
Published 2026 by Wiley-VCH GmbH.

together, that is, when one species is oxidized, the other is reduced so that there is overall charge conservation in the system [11].

The impetus of redox reactions is based on the variations in electronegativity and electron affinities; species with a more significant electron affinity are inclined to oxidize other species [11]. The knowledge and understanding of electron flow in a chemical system is fundamental in understanding as well as the manipulation of chemical changes across a variety of industries like synthetic chemistry, energy storage, and environmental services [12]. Nevertheless, redox potentials are extremely dependent on factors such as the solvent, electrolyte mixture, and reaction conditions, which, in most cases, make the comparison of data among studies very challenging [13]. Notably, redox processes are often critical in catalysis, whereby some forms of RH obtain a rapidity in the chemical reaction, even though they are not consumed during the reaction [14].

6.2 Role of Nano-Catalysts in Enhancing Redox Processes

Nano-catalysis has become an exciting and useful science, mainly because of the unique morphological characteristics of these metal-based nanomaterials, which constitutively have a much larger surface area to volume ratio when compared to the bulk materials [15, 16]. Among them, MeONCs act as a key to move redox reactions along, by exploiting their peculiar electronic structure, surface chemistry, multi-oxidation states, and optimizing their catalytic efficiency [15].

They are highly active and selective due to their tunable composition, morphology, and particle size, which qualifies them for numerous processes, including redox-driven processes [17]. Although studies have shown that single-component metal or metal oxide catalysts tend to agglomerate or be deactivated when subjected to severe conditions [17].

This necessitated the incorporation of these materials into core-shell nanoreactors, for example, encapsulation of active nanoparticles inside metal-organic frameworks (MOFs) has been demonstrated to provide stability improvement and performance [18]. Also, MeONCs have a great specific surface area, which makes active adsorption sites easily available to reactants and transformations [18]. Similarly, catalytic activity is also enhanced because of the presence of defects, especially oxygen vacancies, to alter the electronic structure and promote the charge transfer [19].

6.3 Importance of Metal Oxides as Catalytic Materials

The electronic and structural properties of metal oxides can be tuned, and this fact plays a vital role in not only catalytic activity but also selectivity in heterogeneous catalysis [2]. They are versatile due to an intense involvement of metal cations and oxygen anions in the catalytic cycles, forming a dynamic surface environment that directs the reaction pathways [20]. Metal oxides have several favorable properties, which primarily include the ability to possess multiple oxidation states and thus be suitable for a variety of reactions [5].

It is possible to improve the catalytic activity much by using the strategic adjustments to the complex, like doping with a foreign element, creating a heterojunction or providing the noble metal that will enhance the charge transfer, surface reactivity, and structural stability [21]. The presence of oxygen vacancies is one of the critical features in enhancing reactivity by acting as active sites of adsorption and activation of reactants, thus enhancing catalytic effectiveness [2]. Metal oxides have a lot of significant benefits when compared to homogeneous catalysts in aspects of recovery, reusability, and environmental friendliness [22]. They are heterogeneous and so can easily be recovered out of reaction mixtures, thus minimizing contamination and with imminent recyclability [23]. Other recent developments have been atomically dispersed noble metals on oxide supports, in which the ligand is the metal oxide, which both stabilizes the metal atoms and increases the catalytic activity [24].

6.4 Fundamentals of Redox Reactions

Redox reactions form the basis of a huge and diverse number of chemical and biological processes, including the rusting of iron, the energy conversion occurring within cells, and many others [12]. As earlier noted, reduction-oxidation processes are directly linked in such a way that oxidation cannot happen without a simultaneous reduction because the free electrons of one compound need to be taken up by a second one [25]. Whether or not a species will gain electrons or can be reduced, there is a scale called a reduction potential that describes its tendency [13]. In thermodynamic terms, the spontaneity of redox reactions is defined by the difference in standard reduction potentials (E^o) [26]. The more positive an E^o value, the more the increased disposition to acquire electrons is characterized, which is why the species can be described as a stronger oxidizing agent [26]. It is the basis of redox chemistry in a wide variety of applications, in synthesis, in energy systems, and in biological regulation [26].

6.4.1 Classification of Metal Oxides

Metal oxides with their unique structural and electronic characteristics have become a popular category of nano-catalysts with high activity and selectivity in a range of chemical reactions [27]. The application of metal oxide nanoparticles is increasing in different areas, such as environmental decontamination, catalysis, and even in the sphere of personal care products, which indicates their widespread use and possibilities of application [28]. The general effect taking place in nanomaterials compared to the bulk substance is a very high surface area to volume density ratio, which, in this case, directly translates into increased catalytic efficiency [29]. The improved catalysis can be explained by the fact that more active sites on the nanoparticle surface are presented, which allows a better interaction with the reactant molecules [30]. The most significant group of metal oxide catalysts are transition metal oxides characterized by partly filled d-orbitals, and thus they can easily undergo redox reactions due to the ease of redox electron transfer processes [31]. It is this natural redox energy that is essential in the catalysis of a wide range of reactions such as oxidation, reduction, and dehydrogenation. Their surface area, electronic structure, and flaws can also significantly alter the catalytic behavior of transition metal oxides and be precisely controlled through controlled synthesis processes, optimizing their activity in specific areas of catalysis.

The other important class of nano-catalysts are the mixed metal oxides, which consist of two or more different metal cations in the same oxide lattice, resulting in synergistic effects that can further improve their catalytic activity more than individual components.

6.4.2 Comparative Evaluation of MeONCs

The redox activity, surface area, band gap, and thermal stability of MeONCs as shown in Table 6.1 vary widely, which determines the suitability of each of them in particular applications of redox. The systematic comparison and contrast of the properties leads to the understanding of how to optimize the selection of catalysts to be used to address issues of environmental, energy, and synthetic chemistry.

6.4.3 Unique Properties at the Nanoscale

Metal oxides are the most important heterogeneous catalysts that have distinct physicochemical properties because of their size-specific and unique characteristics that govern their performance as nano-catalysts [37]. The growth in popularity behind MeONCs is attributed to the fact that their functionality lies in being able to generate

Table 6.1 Performance metrics of common MeONCs.

Metal oxide	Oxidation states	Surface area (m^2/g)	Band gap (eV)	Redox activity	Stability	Applications	References
TiO_2	Ti^{3+}/Ti^{4+}	50–200	3.0–3.2	Moderate	High	Photocatalysis and wastewater treatment	Navidpour et al. [32]
CeO_2	Ce^{3+}/Ce^{4+}	30–150	2.9–3.2	High	Moderate	CO oxidation and energy devices	Singh [33]
Fe_2O_3	Fe^{2+}/Fe^{3+}	20–90	2.0–2.2	High	Moderate	Fenton-like reactions and battery electrodes	Li et al. [34]
MnO_2	Mn^{3+}/Mn^{4+}	100–300	1.3–1.7	High	Moderate	Supercapacitors and OER catalysis	Waheed et al. [35]
CO_3O_4	Co^{2+}/Co^{3+}	60–150	1.6–2.2	Very high	High	OER/HER and fuel cells	Liu et al. [36]

a far-reaching range of chemical reactions, and they have increasingly found utility in fields such as green chemistry and environmental remediation, energy conversion, and chemical synthesis [38]. This nano material's surface area-to-volume ratio aesthetic possesses a great portion of the surface area active sites, allowing an increase in interaction with the reactant molecules, and in turn, the nanomaterial enhances catalytic activity [39]. Metal oxide metal types that exist in the nanoscale exhibit a distorted electronic structure compared to their bulk counterparts, which results in the metal oxide metal type changing redox properties and catalytic behavior [39]. Formation of surface defects, for example, oxygen vacancies and undercoordinated metal atoms on the surface of the nanoparticle metal oxides, is employed as active sites in catalytic reactions [1]. Such defects can trap electrons or holes, which favors the process of charge transfer and enhances reactant activation and adsorption [40]. Moreover, the optics and electronics of nanoscale metal oxides can be altered as a result of the quantum confinement effects; the effect can be used to manipulate the electronic band structure to influence both the optical and electronic properties of nanoscale metal oxides, which subsequently alter the efficiency of the photocatalyst [41]. It is also considered that metal oxide nanoparticles contribute to accelerating catalytic activity because they can stabilize the high-energy intermediates and transition states and thereby reduce the activation energy of catalytic reactions [42].

6.5 Synthesis Methods of MeONCs

The efficiency of MeONCs in redox reactions depends on synthesis parameters that influence crystallinity, surface area, and defect structures. Table 6.2 summarizes the most widely used methods.

Table 6.2 Common synthesis methods for metal oxide nanoparticles.

Method	Key features	Advantages	Limitations	References
Sol-gel	Hydrolysis and condensation of metal alkoxides forming a gel	Fine control of morphology and purity	Requires long drying and calcination	Sharma et al. [27]
Hydrothermal	Crystal growth in a sealed autoclave under high pressure/temperature	Shape-controlled, highly crystalline nanoparticles	Equipment-intensive	Lim [43]
Coprecipitation	Simultaneous precipitation of metal ions as hydroxides or carbonates	Simple, scalable, and cost-effective	Broad particle size distribution	Goyal et al. [44]
Flame spray pyrolysis	Atomized precursor combusted in flame to form oxide nanoparticles	Fast, scalable synthesis of pure nanomaterials	Demands flammable precursors and gas handling	Lee et al. [45]

6.6 Mechanisms of Redox Catalysis by Metal Oxides

6.6.1 Surface Redox Chemistry and Active Sites

The metal oxides redox catalysis assessment is essential to proceed with the development of energy conversion and storage technologies, especially in reactions such as the oxygen reduction reaction and the oxygen evolution reaction (OER) [9]. There is a need to comprehend the dynamics of these reactions at the surface of metal oxides so as to effectively create more efficient catalysts [9]. Metal oxide catalysts exhibit surface redox chemistry, determining their activity and closely corresponding with properties and dynamics of active sites [9]. In contrast, metal semiconducting oxides exhibit the property of carrier density that is capable of being altered over a finite range of energy [46]. The tuning of the electronic states of these materials can be done using defect engineering, doping, or surface modification [47]. Namely, the promising method of heteroatom doping to the improvement of the electrocatalytic performance by altering the electronic structure of the catalyst is proposed, but it may not provide a sufficient number of active sites [48].

Catalysis and energy storage are potential applications of metal oxide resources, as these materials offer large structural flexibility and have all these properties in abundance [49]. The metal oxides have pseudocapacitance based on redox reactions that are mostly on the surface area; therefore, specific surface area is a key concern when it comes to the use of metal oxides in supercapacitors [49]. Nonetheless, the vast majority of metal oxides (except, of course, RuO2) are not conductive, so they could not be used as electrode materials in supercapacitors in their pure form [49].

6.6.2 Oxygen Vacancies and Lattice Oxygen Mobility

The presence of transition metal oxides is beneficial because of the easy alteration of the oxidation state of metal ions, which is vital in redox reactions, thereby making them readily used as catalysts [49]. The molecular mechanism of these reactions remains poorly understood, even though the phenomenon is of biological and renewable energy interest [50]. Nevertheless, oxygen vacancies and lattice oxygen dynamics as shown in Figure 6.1 have become the decisive variables necessary to consider when examining the catalytic efficiency of metal oxides in redox reactions [50]. Placed at interstitial sites of the crystal lattice structure where the oxygen is absent, oxygen vacancies (a point defect) cause a change in both electronic and structural properties that can significantly change the reactivity of the material [50]. The vacancies, which are formed in this manner, may form undercoordinated metal cation active sites useful in the adsorption and activation of reactant

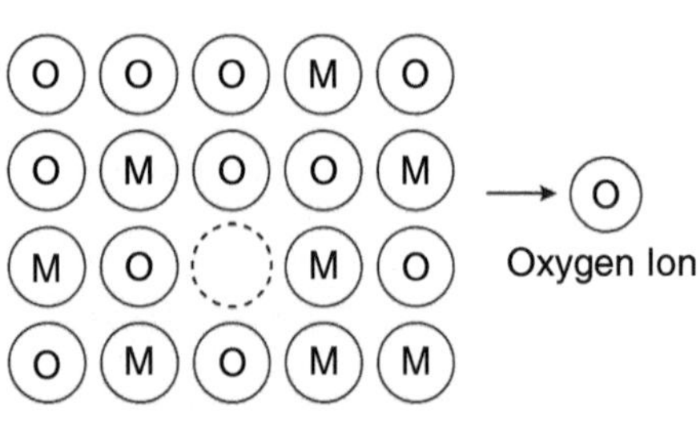

Figure 6.1 Oxygen vacancy formation in metal oxide lattice.

molecules [51]. Many synthesis and post-synthesis techniques enable control of the concentration and distribution of oxygen vacancies, enabling them to tune catalytic activity [51]. Further, the viscosity of lattice oxygen, which is defined as the ease with which the oxygen ions can flow within the oxide lattice, is also an essential factor in redox catalysis because it controls the speed of oxygen interchange between the catalyst and the reactants [51]. This transport is especially notable in oxidation reactions, in which lattice oxygen may easily enter into the oxidation of substrates.

6.6.3 Redox Cycles (Mars-van Krevelen [MvK] Mechanism)

MeONCs are also centered around redox cycles, which play an important role in the catalytic action used in oxidation reactions. The MvK mechanism as shown in Figure 6.2, is one of the most appreciated mechanisms, being a redox mechanism with the involvement of lattice oxygen. In this process, an oxidant (e.g., O_2) does not take part directly in the oxidation of the reactant. Rather, the metal oxide catalyst would contribute lattice oxygen to oxidize the substrate, leaving behind a reduced catalyst (e.g., $MO_x \rightarrow MO_{x-1}$) [52, 53]. Oxygen gas then reoxidizes the catalyst, regenerating the active lattice oxygen, and the cycle is realized.

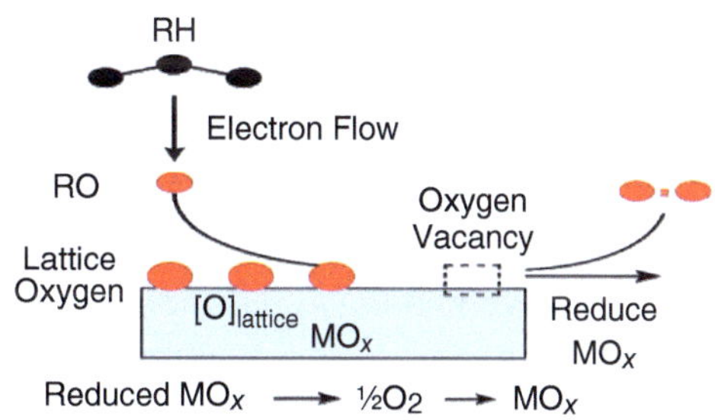

Figure 6.2 Schematic of MvK redox mechanism on a metal oxide surface.

This process can be illustrated as:

1. Substrate oxidation by lattice oxygen: $RH + [O]_{lattice} \rightarrow RO + \text{Reduced } MO_x$
2. Reoxidation of the reduced metal oxide: $\text{Reduced } MO_x + \frac{1}{2}O_2 \rightarrow MO_x$

The MvK mechanism as shown in Figure 6.2 is typical for metal oxides like TiO_2, CeO_2, Fe_2O_3, and MnO_2, and is critical in oxidation catalysis, such as CO oxidation, VOC degradation, and oxidative dehydrogenation.

6.6.4 Interaction with Substrates (Adsorption-Desorption Phenomena)

The process of interaction between substrates and MeONCs is one of the most decisive ones concerning the efficiency of catalysis. Such interaction mostly depends on the adsorption-desorption processes, which determine reactant availability and activation at the surface of the catalyst [51]. In the first stage, which is adsorption, the substrate molecules are attached to active sites on the surface area of the nanocatalyst [54]. It may take place through physisorption (weak van der Waals interactions), but is more usually through chemisorption (strong chemical bonding), which may be required to allow catalytic conversion [51]. Adsorption strength and selectivity are strongly dependent on the surface area, porosity, and surface chemistry of metal oxides, particularly the disposition of oxygen vacancies of defects [54].

After adsorption, the substrate is catalytically transformed, which in many cases includes an electron transfer or a rearrangement of bonds. After that reaction, it is necessary to desorb the products on the surface to release the active sites and enable new substrate molecules to bind [54]. The adsorption and desorption have to be balanced; too strong adsorption will strongly occupy the active sites, and poor adsorption will result in a low level of catalytic activity. Such metal oxides as ZnO, CeO_2, and Fe_2O_3 are well-known because of their tunable surface property, due to which they are highly efficient in terms of enabling substrate interactions related to redox reaction, environmental remediation, and photocatalysis [54].

6.7 Applications of MeONCs in Redox Reactions

6.7.1 Environmental Applications

MeONCs have attracted much attention as a novel material used to resolve the environmental pollution issue, with redox reactions showing great potential in the removal of different pollutants, volatile organic compounds, pesticides, EDCs, dyes, etc. [49, 55, 56]. MeONCs have powerful characteristics due to specific physicochemical properties, especially their large surface area, tunable electronic structure, and redox activity, which enable them to be highly effective in catalyzing oxidation and reduction processes of pollutants [49]. Being small and extremely reactive, nanomaterials can cause extensive ecological impacts and various toxicity levels relative to other larger particles [57]. A high level and the synergy of the interaction of nanomaterials with contaminants contribute to the rapid decrease of the level of the pollutant, caused by the high surface area-to-volume ratio and the numerous reactive sites of nanomaterials [57]. The application of the metal oxide nanoparticles is increasing, and thus causes them to be released into the environment. There has been a great concern about the toxicity of metal oxide nanoparticles to animals and the environment [57]. Furthermore, MeONCs provides an interesting mechanism of causative polemic action to break up the pollutants through the redox mechanisms to convert the toxic materials into harmless ones. Those can be used both in volatile organic compounds and in dyes, covering two significant categories of environmental pollutants [58]. Such nano-catalysts are effective thanks to their special features, such as large surface area, adjustable electronic structure, and redox activity, which enhance an efficient catalytic transformation [58]. Figure 6.3 shows the mechanism of pollutant uptake my MeONCs.

6.7.2 Photocatalytic Water (H_2O) Splitting and Carbon Dioxide (CO_2) Reduction

Photocatalysis as a fast-emerging field has attracted much attention because it may serve as a prospective channel toward global energy needs and environmental

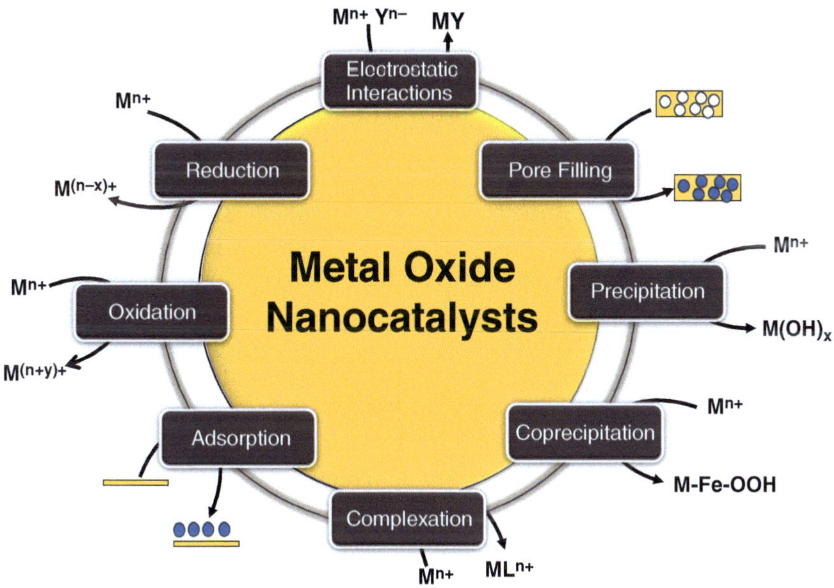

Figure 6.3 Pollutant uptake mechanism of MeONCs.

issues, including redox reactions, including water splitting, and CO_2 reduction [59]. Due to their electronic and structural properties, MeONCs emerge to play central roles in these photocatalytic reactions, and a solution has been provided to utilize solar energy to generate clean fuels and high-value chemicals [59]. The increased catalytic activity of nanoscale metal oxides is mainly explained by their large specific surface area that can offer a large number of active sites where reactants will bind, and further chemical alterations will take place. Also, using quantum confinement effects and raising the surface defect density of nanomaterials, their electronic band structure can be adjusted to maximize light absorption and facilitate efficient charge separation and transport [60].

It is observed that the high-field growth of anodic oxides on metal surfaces offers a diverse path to prepare photocatalyst materials with desired morphologies and compositions through semiconducting thin solid film synthesis [61]. Choice of the right synthetic methodology in the synthesis of metal oxide nanostructures is critical because it directly affects how the material is going to behave in given application with the consideration of operational cost, safety, product yield, ease of use, and capital investment.

6.7.3 Wastewater Treatment (Fenton-Like Reactions)

Generally, metal oxides are used as catalysts in these processes along with hydrogen peroxide to eliminate organic pollutants in wastewater through catalytic oxidation [44]. The beneficial properties of the Fenton reagent, which is used when treating wastewater, are that it is an efficient treatment method since iron is abundant and hydrogen peroxide can be handled safely and easily to remove dangerous substances such as phenols and pesticides [44]. Nevertheless, the classic Fenton process is not devoid of shortcomings, including the generation of sludge comprising iron and the limited range of pH to which it is recommended, which promoted the research and formation of heterogeneous Fenton catalysts consisting of iron with greater efficiency, cost-effectiveness, and the possibility of reuse [62]. The Fenton process has success in the destruction of organic pollutants due to the high oxidation potential of the hydroxyl radical produced in the process [62]. Fenton, electro-Fenton, photo-Fenton, and combinations of them are some of the most promising advanced oxidation processes that have the potential of mineralizing refractory organics in wastewater [62]. Hydroxyl radicals have a nonselective target and can almost destroy any organic pollutant and mineralize into benign molecules of H_2O or CO_2 [63]. This photochemical step of the photo-Fenton reaction can also be improved by the usage of solar radiation, which will result in a complete mineralization of organic compounds within a very short duration [62]. To circumvent costs and increase photo-Fenton performance, several strategies have been put in place, such as the utilization of heterogeneous catalysts and hence chelating agents [64].

6.8 Energy Conversion and Storage

6.8.1 Fuel Cells (e.g., Oxygen Reduction Reactions)

The increasing environmental pressure and the shrinking of reserves of conventional fossil fuels have created an ambitious research and development activity in the field of energy storage and energy conversion, where, in particular, electrocatalysts that can be used in energy storage devices are in demand [65]. Fuel cells convert the chemical energy of a fuel into electricity (Figure 6.4), producing energy using electrochemical reactions, and are thus a promising research area related to sustainable energy [66]. The mass application of fuel cells has, however, been hindered by a number of factors, most notably the high cost and limited availability of effective electrocatalysts, including readily available efficient electrocatalysts, such as platinum, which are typically needed to promote such important reactions as the oxygen reduction reaction [67]. Similarly, the ORR, which takes place at the cathode of the fuel cells, is a slow reaction and greatly constrains the efficiency overall of the device [67]. To cope up with these issues, there has been tremendous focus on the discovery of alternative electrocatalysts constructed using earth-abundant materials, where metal oxides have particularly been given quite a lot of attention owing to the wide redox characteristics they document, high surface area, and the ability of catalytic activity to be increased should such materials be produced within the nano range [67].

MeONCs have been arising as a very good alternative to the conventional platinum catalysts that fuel cells have to apply mainly in the oxygen reduction reaction because they are abundant, cheap, and possess tunable electronics and structure-dependent properties [68].

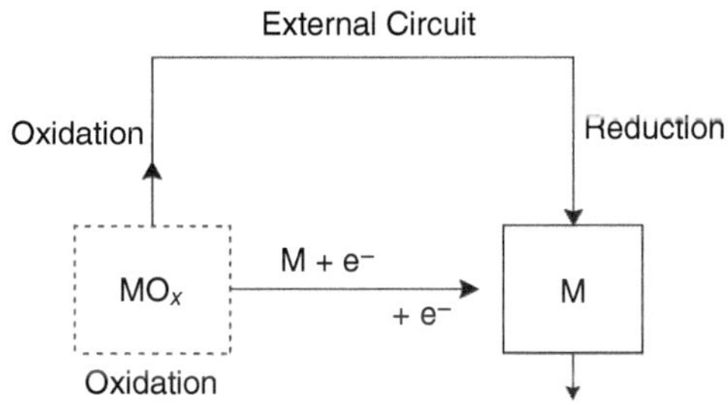

Figure 6.4 Schematic of redox reactions in energy device.

6.8.2 Batteries (Li-Ion, Na-Ion) and Supercapacitors

MeONCs are materials of high density on the surface, redox, and structural flexibility that are important candidates in energy storage devices, including lithium-ion (Li-ion) and sodium-ion (Na-ion) batteries and supercapacitors [69]. The electrodes TiO_2 (Li-ion), Fe_2O_3, MnO_2 (Li-ion), and Co_3O_4 (Na-ion) are metal oxides in Li-ion batteries and Na-ion batteries. These are the materials that are reversibly redox on charge and discharge cycle, where it is possible to store and release energy in them [70]. Metal oxides have high theoretical capacity, but typically have the undesirable expansion during cycling, being handled by nanostructuring and composite formation. Na-ion batteries, being structurally similar to Li-ion systems, benefit from metal oxides with large interlayer spacing to accommodate the larger Na^+ ions [70].

Transition metal oxides such as RuO_2, MnO_2, NiO, and Co_3O_4 have been used to store energy via fast, reversible redox reactions at the electrode/electrolyte interface in supercapacitors, and particularly, the pseudocapacitors. Such materials provide high power density and cycle life. Metal oxide nanostructure properties of electrochemical stability, electrical conductivity, and surface area are key factors to their performance in next-generation energy storage [71].

A sustainable way of generating clean hydrogen fuel is electrochemical water splitting, which consists of two half-reactions: oxidation of water or OER and reduction of water or hydrogen evolution reaction (HER). MeONCs are central in improving efficiency and stability as well as the selectivity of the two processes. The anode part is responsible for the oxidation of the water and is kinetically sluggish because of the complicated 4-electron transfer procedure. High-energy four-electron transfer. Oxides of metals oxides, specifically RuO_2, IrO_2, MnO_2, $NiFe_2O_4$, and Co_3O_4 are extensively researched in view of OER since they have multiple oxidation states and generate surface oxygen products. The OER activity is related to the binding energy of intermediates (e.g., OH, O, OOH), which is well regulated by metal oxides with the use of doping or nanostructuring [72].

Water reduction occurs at the cathode and involves:

$$2H^+ + 2e^- \rightarrow H_2 \text{ (in acidic media)}$$

$$2H_2O + 2e^- \rightarrow H_2 + 2OH^- \text{ (in alkaline media)}$$

Despite the fact that the most efficient HER catalyst is platinum, some transition metal oxides have potential when used together with conductive supports or heterostructures [73].

6.9 Conclusion

MeONCs have proved to be very viable catalysts in a wide range of redox reactions within environmental, energy, and industrial contexts. Their unusual physicochemical properties which include high surface area-to-volume ratio, of surface area-to-volume, tunable electronic energy levels, multiple oxidation states, and numerous oxygen vacancies make them effective in electron transfer, increase in reactant adsorption, and stabilization of reaction intermediates. The various synthesis techniques have enabled the fine-tuning of particle morphology, crystallinity, and density, this also enhances its catalytic activity and stability. Mechanistic understanding has identified the importance of surface redox chemistry in performance, especially the influence of oxygen vacancies, lattice oxygen mobility, and the MvK cycle. Also, the catalytic efficiency, durability, and selectivity have also been enhanced when strategies such as heteroatom doping, defect engineering, and nano-structuring are employed. The scalable and sustainable system of MeONCs provides excellent possibilities of solving the global energy gap requirements, environment cleanup, and the switch to clean energy.

References

1 Yu K, Lou L, Liu S, Zhou W. Asymmetric oxygen vacancies: the intrinsic redox active sites in metal oxide catalysts. *Adv Sci.* 2020;7(2):1901970.

2 Jenkins AH, Medlin JW. Controlling heterogeneous catalysis with organic monolayers on metal oxides. *Acc Chem Res.* 2021;54(21):4080–90.

3 Adaramaja AA, Bamisaye A, Abati SM, Adegoke KA, Adesina MO, Ige AR, et al. Thermally modified nanocrystalline snail shell adsorbent for methylene blue sequestration: equilibrium, kinetic, thermodynamic, artificial intelligence, and DFT studies. *RSC Adv.* 2024;14(18):12703–19.

4 Bamisaye A, Oluboyede OA, Etafo NO, Bamidele MO, Abiola-Kuforiji OT, Idowu MA. Synthesis of zinc oxide nanoparticles using *Syzygium malaccense* leaf extract: photocatalytic decomposition of ciprofloxacin and antimicrobial studies. *Nanotechnol Environ Eng.* 2025 Sep 30;10(3): 54. Available from: https://link.springer.com/10.1007/s41204-025-00459-z

5 Xu S, Liu C, Teng Z, Sun F, Chi J, Liu X, et al. Role and advances in nonmetal doping of electrocatalysts for the hydrogen evolution reaction: a review. *Inorg Chem Front.* 2025;12:5258–88.

6 Ogunbiyi O, Bamisaye A, Abiodun AJ, Owoeye TF, Alli YA, Idowu MA. Biogenic synthesis of Zinc oxide nanoparticles for solar cell application and photodegradation of neomycin. *Mater Sci Eng B.* 2025 Sep 1;319: 118324.

7 Abati SM, Bamisaye A, Abidemi Adaramaja A, Rapheal Ige A, Adegoke KA, Olurotimi Ogunbiyi E, et al. Biodiesel production from spent vegetable oil with Al_2O_3 and Fe_2O_3-biobased heterogenous nanocatalysts: comparative and optimization studies. *Fuel.* 2024 May 15;364: 130847.

8 Abiodun AJ, Alamu GA, Adedokun O, Abati SM, Sanusi YK. Enhancing photovoltaic performance of monolithic dye-sensitized solar cells through green-synthesized CuO–FeO nanocomposite counter electrodes. *Sustain Energy Res.* 2024;11(1):46. https://doi.org/10.1186/s40807-024-00139-7

9 Wang X, Zhong H, Xi S, Lee WSV, Xue J. Understanding of oxygen redox in the oxygen evolution reaction. *Adv Mater.* 2022;34(50):2107956.

10 Mindia Ali A, Waziri I, Garba H. Recent developments in the electron transfer reactions and their kinetic studies. In: *Chemical Kinetics and Catalysis - Perspectives, Developments and Applications.* IntechOpen; 2024; 1–16.

11 Hajalifard Z, Mousazadeh M, Khademi S, Khademi N, Jamadi MH, Sillanpää M. The efficacious of AOP-based processes in concert with electrocoagulation in abatement of CECs from water/wastewater. *NPJ Clean Water.* 2023;6(1):1–25.

12 Huang J, Jones A, Waite TD, Chen Y, Huang X, Rosso KM, et al. Fe(II) redox chemistry in the environment. *Chem Rev.* 2021 Jul 14;121(13): 8161–233. https://doi.org/10.1021/acs.chemrev.0c01286

13 Kwabi DG, Ji Y, Aziz MJ. Electrolyte lifetime in aqueous organic redox flow batteries: a critical review. *Chem Rev.* 2020 Jul 22;120(14): 6467–89. https://doi.org/10.1021/acs.chemrev.9b00599

14 McKee DW. Oxidation protection of carbon materials. In: *Chemistry & Physics of Carbon.* CRC Press; 2021; 173–232.

15 Majumdar D, Bhattacharya SK. Recent developments of methanol electrooxidation using nickel-based nanocatalysts. *ChemistrySelect.* 2022;7(40):e202201807.

16 Bamisaye A, Adegoke KA, Alli YA, Bamidele MO, Idowu MA, Ogunjinmi OE. Recent advances in nanoemulsion for sustainable development of farm-to-fork systems. *J Clean Prod.* 2023 Dec 3;429: 139226.

17 Wang X, Song S, Zhang H. A redox interaction-engaged strategy for multicomponent nanomaterials. *Chem Soc Rev.* 2020;49(3):736–64.

18 Tian Y, Kong L, Zou H, Liu M, Liu G. Boosting hydroxyl radical generation for highly efficient electrooxidation of sulfamethazine via facile doping strategy. *Sustain Horiz.* 2024;11:100103.

19 Zhao S, Yang Y, Bi F, Chen Y, Wu M, Zhang X, et al. Oxygen vacancies in the catalyst: efficient degradation of gaseous pollutants. *Chem Eng J.* 2023;454:140376.

20 Kou Z, Li X, Zhang L, Zang W, Gao X, Wang J. Dynamic surface chemistry of catalysts in oxygen evolution reaction. *Small Sci.* 2021;1(7):2100011.

21 Liu Q, Zhang A, Wang R, Zhang Q, Cui D. A review on metal-and metal oxide-based nanozymes: properties, mechanisms, and applications. *Nanomicro Lett.* 2021;13:1–53.

22 Miceli M, Frontera P, Macario A, Malara A. Recovery/reuse of heterogeneous supported spent catalysts. *Catalysts.* 2021;11(5):591.

23 Islam A, Teo SH, Taufiq-Yap YH, Ng CH, Vo DVN, Ibrahim ML, et al. Step towards the sustainable toxic dyes removal and recycling from aqueous solution—a comprehensive review. *Resour Conserv Recycl.* 2021;175:105849.

24 Sui X, Zhang L, Li J, Doyle-Davis K, Li R, Wang Z, et al. Advanced support materials and interactions for atomically dispersed noble-metal catalysts: from support effects to design strategies. *Adv Energy Mater.* 2022;12(1):2102556.

25 Di Meo S, Venditti P. Evolution of the knowledge of free radicals and other oxidants. *Oxid Med Cell Longev.* 2020;2020(1):9829176.

26 Vogelezang M, Verdonk A. A possible electrochemical route to a thermodynamic redox reaction equilibrium constant in secondary education: an attempt to come from science fiction to science education. *World J Chem Educ.* 2020;8(3):128–40.

27 Sharma RK, Bandichhor R, Mishra V, Sharma S, Yadav S, Mehta S, et al. Advanced metal oxide-based nanocatalysts for the oxidative synthesis of fine chemicals. *Mater Adv.* 2023;4(8):1795–830.

28 Holt D. *Hybrid Metal Oxide Nanostructures for Solar Water Splitting and Inorganic Perovskite Photovoltaics.* University of Sussex; 2023.

29 Asha AB, Narain R. Nanomaterials properties. In: *Polymer Science and Nanotechnology.* Elsevier; 2020 Jan 1; 343–59.

30 Astruc D. Introduction: nanoparticles in catalysis. *Chem Rev.* 2020;120:461–3.

31 Konsolakis M, Lykaki M. Recent advances on the rational design of non-precious metal oxide catalysts exemplified by CuO_x/CeO_2 binary system: implications of size, shape and electronic effects on intrinsic reactivity and metal-support interactions. *Catalysts.* 2020;10(2):160.

32 Navidpour AH, Abbasi S, Li D, Mojiri A, Zhou JL. Investigation of advanced oxidation process in the presence of TiO_2 semiconductor as photocatalyst: property, principle, kinetic analysis, and photocatalytic activity. *Catalysts.* 2023;13(2):232.

33 Singh B. Lanthanide-based catalysts for electrochemical water splitting: unraveling the interplay of structure, properties, and performance. *Chem Commun.* 2025;61:9753–79.

34 Li J, You J, Wang Z, Zhao Y, Xu J, Li X, et al. Application of α-Fe_2O_3-based heterogeneous photo-Fenton catalyst in wastewater treatment: a review of recent advances. *J Environ Chem Eng.* 2022;10(5):108329.

35 Waheed MS, Alsalhi SA, Abdelmohsen SAM, Zaman S, Ahmad S, Aman S, et al. Development of novel cube-embedded MnO_2/ZnO nanocomposite for OER activity and supercapacitor performance evaluation. *JOM.* 2024;76(7):3665–78.

36 Liu Y, Hao L, Ma H, Su M, Zhang Y. Convenient preparation of Ni-Fe layered double hydroxide nanosheets on Co_3O_4 hollow cube for Monitoring the environmental pollutant Nitrobenzene. *Appl Surf Sci.* 2025;691:162682.

37 Jašík J, Fortunelli A, Vajda Š. Exploring the materials space in the smallest particle size range: from heterogeneous catalysis to electrocatalysis and photocatalysis. *Phys Chem Phys.* 2022;24(20):12083–115.

38 Fawy KF, Rodriguez-Ortiz G, Ali A, Jadeja Y, Khan H, Pathak PK, et al. Catalytic exploration metallic and nonmetallic nano-catalysts, properties, role in photoelectrochemistry for sustainable applications. *Rev Inorg Chem.* 2025; 45(4): 783–829.

39 Lakhani P, Bhanderi D, Modi CK. Nanocatalysis: recent progress, mechanistic insights, and diverse applications. *J Nanopart Res.* 2024;26(7):148.

40 Maarisetty D, Baral SS. Defect engineering in photocatalysis: formation, chemistry, optoelectronics, and interface studies. *J Mater Chem A Mater.* 2020;8(36):18560–604.

41 Ko J, Berger R, Lee H, Yoon H, Cho J, Char K. Electronic effects of nano-confinement in functional organic and inorganic materials for optoelectronics. *Chem Soc Rev.* 2021;50(5):3585–628.

42 Du H, Luo H, Jiang M, Yan X, Jiang F, Chen H. A review of activating lattice oxygen of metal oxides for catalytic reactions: reaction mechanisms, modulation strategies of activity and their practical applications. *Appl Catal A Gen.* 2023;664:119348.

43 Lim SN. Synthesis of Li-rich cathode material with spherical shape and high crystallinity by using flame spray pyrolysis. *New Renew Energy.* 2024;20(3):20–7.

44 Goyal R, Singh O, Agrawal A, Samanta C, Sarkar B. Advantages and limitations of catalytic oxidation with hydrogen peroxide: from bulk chemicals to lab scale process. *Catal Rev.* 2022;64(2):229–85.

45 Lee WC, Fang Y, Le H, Hodgson R, Chan HWB, Qian R, et al. Enhanced photoelectrochemical water splitting by a 3D hierarchical sea urchin-like structure: ZnO nanorod arrays on TiO_2 hollow hemisphere. *Nanotechnology.* 2024;35(29):295301.

46 Xia C, Wang H, Kim JK, Wang J. Rational design of metal oxide-based heterostructure for efficient photocatalytic and photoelectrochemical systems. *Adv Funct Mater.* 2021;31(12):2008247.

47 Feng Y, Wu J, Chi Q, Li W, Yu Y, Fei W. Defects and aliovalent doping engineering in electroceramics. *Chem Rev.* 2020;120(3):1710–87.

48 Wang J, Liao T, Wei Z, Sun J, Guo J, Sun Z. Heteroatom-doping of non-noble metal-based catalysts for electrocatalytic hydrogen evolution: an electronic structure tuning strategy. *Small Methods.* 2021;5(4):2000988.

49 Shaheen I, Hussain I, Zahra T, Javed MS, Shah SSA, Khan K, et al. Recent advancements in metal oxides for energy storage materials: design, classification, and electrodes configuration of supercapacitor. *J Energy Storage.* 2023;72:108719.

50 Chen H, Wan K, Zheng F, Zhang Z, Zhang Y, Long D. Mechanism insight into photocatalytic conversion of lignin for valuable chemicals and fuels production: a state-of-the-art review. *Renew Sust Energ.* 2021;147:111217.

51 Yan D, Xia C, Zhang W, Hu Q, He C, Xia BY, et al. Cation defect engineering of transition metal electrocatalysts for oxygen evolution reaction. *Adv Energy Mater.* 2022;12(45):2202317.

52 Fang Y, Wang B, Liu Z, Chen Z, Li M, Xu X. Beyond equilibrated structures: sequential lattice oxygen evolution shapes Mars-van Krevelen catalytic oxidation on β-MnO_2(110). *ACS Catal.* 2025 Jun 6 [cited 2025 Aug 8];15(11): 9940–8. https://doi.org/10.1021/acscatal.5c00169

53 Widmann D, Behm RJ. Dynamic surface composition in a Mars-van Krevelen type reaction: CO oxidation on Au/TiO_2. *J Catal.* 2018 Jan 1 [cited 2025 Aug 8];357: 263–73. Available from: https://www.sciencedirect.com/science/article/pii/S0021951717303950

54 Li F, Han GF, Baek JB. Nanocatalytic materials for energy-related small-molecules conversions: active site design, identification and structure-performance relationship discovery. *Acc Chem Res.* 2021;55(1):110–20.

55 Gold VF, Bamisaye A, Adesina MO, Adegoke KA, Ige AR, Adeleke O, et al. Adsorptive uptake of thymol blue from aqueous medium using calcined snail shells: equilibrium, kinetic, thermodynamic, neuro-fuzzy and DFT studies. *J Dispers Sci Technol.* 2025;46:1747–63. https://doi.org/10.1080/01932691.2024.2339458

56 Bamisaye A, Abati SM, Ige AR, Etafo NO, Alli YA, Bamidele MO, et al. Metal-oxide nanocatalysts for spontaneous sequestration of endocrine-disrupting compounds from wastewater. *Chemosphere.* 2024;367:143569.

57 Saleh TA. Nanomaterials: classification, properties, and environmental toxicities. *Environ Technol Innov.* 2020;20:101067.

58 David E, Niculescu VC. Volatile organic compounds (VOCs) as environmental pollutants: occurrence and mitigation using nanomaterials. *Int J Environ Res Public Health.* 2021;18(24):13147.

59 Bhuyan A, Ahmaruzzaman M. Ternary 3D/2D/3D direct dual Z-scheme MOF-on-MOF-derived α-Fe_2O_3/g-C_3N_4/Fe-MOF photocatalyst for boosted sunlight-driven removal of metronidazole: effect of coexisting ions, mechanistic insights, and water matrices. *Environ Sci Nano.* 2024;11(12):4805–29.

60 Zhang Y, Feng X, Ma D, Li J, Chen Y, Shi L, et al. Confinement effects in photocatalysis: progress and challenges. *J Mater Chem A Mater.* 2025; 13:10431–50.

61 Valadi Palliyalil A, Ullattil SG. TiO_2 mesocrystals: recent progress in synthesis, structure, and photocatalytic applications. *Adv Compos Hybrid Mater.* 2025;8(4):1–46.

62 Ribeiro JP, Nunes MI. Recent trends and developments in fenton processes for industrial wastewater treatment – a critical review. *Environ Res.* 2021;197:110957.

63 Krystynik P. Advanced oxidation processes (AOPs)—utilization of hydroxyl radical and singlet oxygen. In: *Reactive Oxygen Species.* IntechOpen; 2021. p. 336.

64 Ahile UJ, Wuana RA, Itodo AU, Sha'Ato R, Dantas RF. A review on the use of chelating agents as an alternative to promote photo-Fenton at neutral pH: current trends, knowledge gap and future studies. *Sci Total Environ.* 2020;710:134872.

65 Kalair A, Abas N, Saleem MS, Kalair AR, Khan N. Role of energy storage systems in energy transition from fossil fuels to renewables. *Energy Storage.* 2021;3(1):e135.

66 Abdelkareem MA, Elsaid K, Wilberforce T, Kamil M, Sayed ET, Olabi A. Environmental aspects of fuel cells: a review. *Sci Total Environ.* 2021;752:141803.

67 Zaman S, Huang L, Douka AI, Yang H, You B, Xia BY. Oxygen reduction electrocatalysts toward practical fuel cells: progress and perspectives. *Angewandte Chemie*. 2021;133(33):17976–96.

68 Huang L, Zaman S, Tian X, Wang Z, Fang W, Xia BY. Advanced platinum-based oxygen reduction electrocatalysts for fuel cells. *Acc Chem Res*. 2021;54(2):311–22.

69 Fang S, Bresser D, Passerini S. Transition metal oxide anodes for electrochemical energy storage in lithium- and sodium-ion batteries. In: *Transition Metal Oxides for Electrochemical Energy Storage*. Wiley Online Library; 2022;10(1): 55–99.

70 Hariharan S. Nanostructured electrode materials for lithium and sodium battery applications 2013, National University of Singapore. Available from: https://scholarbank.nus.edu.sg/handle/10635/37829.

71 Seok D, Jeong Y, Han K, Yoon DY, Sohn H. Recent progress of electrochemical energy devices: metal oxide–carbon nanocomposites as materials for next-generation chemical storage for renewable energy. *Sustainability*. 2019;11(13):3694.

72 Zhang K, Zou R. Advanced transition metal-based OER electrocatalysts: current status, opportunities, and challenges. *Small*. 2021;17(37):2100129.

73 Das C, Sinha N, Roy P. Transition metal non-oxides as electrocatalysts: advantages and challenges. *Small*. 2022;18(28):2202033.

7

Photocatalysis Using Metal Oxides

Shakirudeen Modupe Abati[1], Abayomi Bamisaye[1,2], Kolawole Kazeem, Ayodeji Ralpheal Ige[3], Oresegun Olakunle Ibrahim[4], Olumuyiwa O. Ogunlaja[2,5], Mopelola Abidemi Idowu[6], and Henrietta W. Langmi[2]*

[1] *Department of Chemistry Faculty of Natural and Applied Sciences, Lead City University, Ibadan, Nigeria*
[2] *Department of Chemistry, University of Pretoria, Pretoria, South Africa*
[3] *Faculty of Civil Engineering and Environmental Sciences, Bialystok University of Technology, Bialystok, Poland*
[4] *Department of Mechanical Engineering, School of Engineering, Zhejiang University, Hangzhou, Zhejiang, China*
[5] *African Center of Excellence for Water and Environmental Research (ACEWATER), Ede, Nigeria*
[6] *Department of Pure and Applied Chemistry, College of Physical Sciences, Federal University of Agriculture, Abeokuta, Nigeria*

*Corresponding author: abayomibamisaye@gmail.com, abayomi.bamisaye@up.ac.za

7.1 Introduction

The escalating global water crisis, driven by unchecked industrialization and human activities, has led to severe ecological damage and significant public health concerns [1, 2]. Traditional water treatment methods often fall short, either failing to achieve complete degradation or generating lethargic secondary waste streams. This inadequacy highlights a critical societal and environmental imperative for innovative and sustainable water purification technologies. Photocatalysis represents a transformative approach in environmental remediation, particularly in addressing the pervasive issue of water pollution [3]. As an advanced oxidation process (AOP), it offers a compelling solution to the degradation of organic contaminants, leveraging the generation of highly reactive oxygen species (ROS) through the synergistic action of light and a catalyst. This method stands apart from conventional chemical treatments due to its inherent sustainability, low energy requirements, and operational simplicity. Its capacity to mineralize organic pollutants into benign end-products, even at trace concentrations, underscores its eco-friendly nature and cost-effectiveness, making it a preferred "green" technology [4].

Furthermore, the inherent ability of these photo-activated processes to efficiently break down persistent and recalcitrant organic pollutants, emerges not merely as a

Metal Oxide Nanocatalysts for Sustainable Energy Production, First Edition. Edited by Abayomi Bamisaye et al.
Published 2026 by Wiley-VCH GmbH.

scientific advancement but as an indispensable tool for safeguarding global water security and public health in a sustainable manner [5].

Studies have shown that metal oxides (MOs) are at the forefront of photocatalytic research, distinguished by a unique combination of properties which includes exhibit high chemical stability, tunable band gaps, strong light absorption, low toxicity, coupled with remarkable magnetic and excellent electron transport properties [6–8]. This confers significant advantages over other photocatalyst materials, such as metal sulfides, oxynitrides, and oxysulfides [9]. Their superior stability, cost-effectiveness, favorable electronic structures, and inherent resistance to photocorrosion are also crucial attributes [10]. Furthermore, their high diffusion capacity, widespread availability, and abundance of active sites contribute to their overall high reactivity.

This extensive array of desirable properties is not merely a collection of isolated facts; it represents a compelling set of intrinsic material advantages that directly address and mitigate many of the practical limitations observed in alternative photocatalyst classes. For instance, MOFs are often hindered by their less stable nature [11], while metal sulfides are susceptible to severe photo-corrosion [12, 13]. This comparative superiority establishes MOs as a robust and enduring foundational platform for photocatalytic research and development. Their reliability provides a solid basis upon which further engineering and innovation can be built to enhance performance and broaden applications. Prominent examples of MOs extensively utilized in photocatalysis for dyes, antibiotics, phenol degradation, include titanium dioxide (TiO_2), SnO_2, MnO_2, Fe_2O_3, CuO, CeO_2, AgO, WO_3, ZnO, and bismuth oxide (Bi_2O_3) [4, 14–19].

7.2 Fundamental Principles and Mechanisms of Photocatalysis

Understanding the core scientific principles that govern photocatalytic reactions is essential to appreciating how light energy is transformed into chemical energy to drive degradation processes. This section elucidates the fundamental steps involved, from initial light absorption to the complex interplay of reactive species.

7.2.1 Photon Absorption and Electron-Hole Pair Generation

The fundamental process of photocatalysis initiates with the absorption of light by a semiconductor photocatalyst. When this material is irradiated with photons possessing energy equal to or greater than its characteristic bandgap, electrons (e^-) residing in the filled valence band (VB) absorb this energy [7, 20]. This absorption excites them, causing them to transition to the higher-energy, empty conduction band (CB). Simultaneously, this photoexcitation leaves behind positively charged holes (h^+) in the VB, thereby creating an electron-hole pair, this process is presented in Equations (1)–(7) as reported by [6, 7]. Figure 7.1 shows the mechanism of photodegradation process [21, 22].

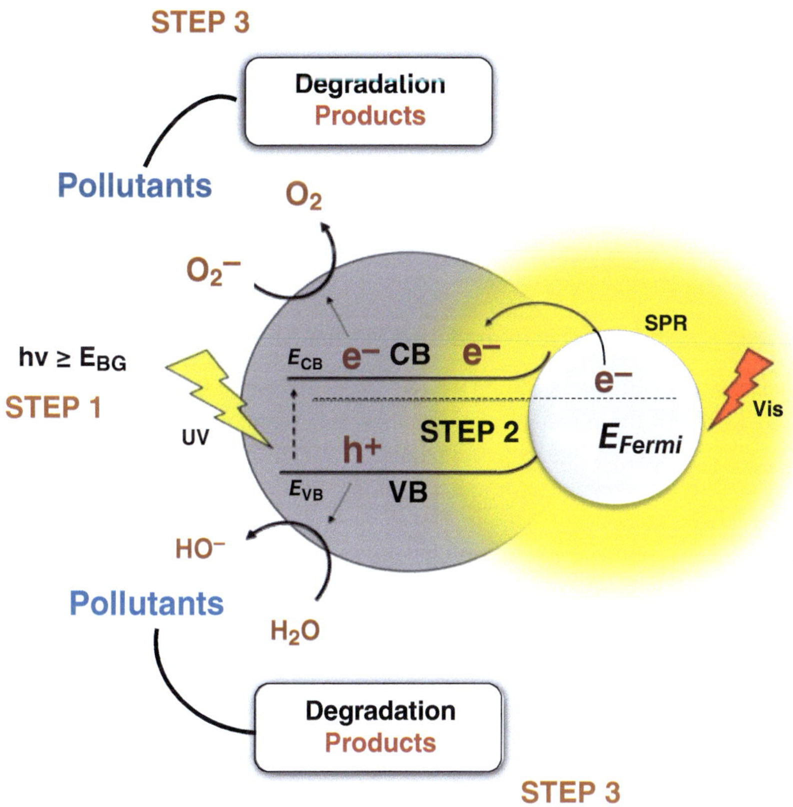

Figure 7.1 The mechanism of photodegradation process.

$$\text{Metal Oxide Photocatalysts} + h\nu \rightarrow h^+_{VB} + e^-_{CB} \tag{1}$$

Electron reaction

$$e^-_{CB} + O_2 \rightarrow O_2^- \tag{2}$$
$$O_2^- + H^+ \rightarrow HO_2 \tag{3}$$
$$2HO_2 \rightarrow H_2O_2 + O_2 \tag{4}$$
$$H_2O_2 + e^-_{CB} \rightarrow OH + OH^- \tag{5}$$

Hole reaction

$$h^+_{VB} + H_2O \rightarrow OH + H^+ \tag{6}$$
$$\frac{OH}{O_2^-} + \text{Organic pollutant} \rightarrow \text{Intermediate} \rightarrow CO_2 + H_2O + \text{Inorganic ion} \tag{7}$$

While the generation of these electron-hole pairs is the cornerstone of photocatalysis and represents the efficient conversion of light energy into chemical potential, a critical challenge arises from their inherent tendency to recombine rapidly [15].

This recombination is an undesirable process that significantly diminishes the overall photocatalytic efficiency by reducing the number of available charge carriers for chemical reactions. Consequently, this recombination phenomenon constitutes the primary kinetic bottleneck in photocatalytic systems. This implies that successful photocatalyst development must prioritize strategies for effective charge separation and efficient transfer of these charge carriers to the reactive surface sites, rather than solely focusing on maximizing light absorption. Overcoming this recombination is paramount for achieving high catalytic performance [23].

7.2.2 Direct and Indirect Dye Degradation Mechanisms

Photocatalytic dye degradation occurs via two main pathways: the direct and indirect mechanisms. In the direct mechanism, the MO photocatalyst absorbs light, generating reactive species like hydroxyl radicals (•OH) from water and superoxide radicals ($O_2^{\bullet-}$) from oxygen. These radicals, along with direct hole/electron reactions, break down dye molecules into harmless byproducts [14]. Meanwhile, in the indirect mechanism, the dye itself absorbs light, enters an excited state, and transfers electrons to the MO's CB, forming dye radical cations. The electrons react with oxygen to produce $O_2^{\bullet-}$, leading to further radical formation [24].

Research shows the direct mechanism dominates, highlighting the photocatalyst's key role in generating its own reactive species (• OH, $O_2^{\bullet-}$) from water and oxygen, rather than just accepting dye electrons. This emphasizes the importance of optimizing the catalyst's redox properties and surface chemistry to enhance radical production for efficient degradation. Thus, catalyst design should focus on maximizing these intrinsic photocatalytic capabilities.

7.2.3 Role of ROS

Several studies have reported that the photocatalytic degradation of organic pollutants relies on ROS, primarily photogenerated holes (h^+), which exhibit strong oxidative capability, even in a vacuum [25–27]. Hydroxyl radicals (•OH) contribute selectively, depending on the pollutant's structure, while superoxide ions ($O_2^{\bullet-}$) mainly target biological macromolecules [8]. The varying roles of ROS suggest that optimizing photocatalysis requires tailored ROS generation based on pollutant chemistry, moving beyond nonselective approaches for improved efficiency [20].

7.2.4 Evolution and Classification of MO Photocatalysts

The study and utilization of the chemistry of MOs photocatalysts in the field of material science has undergone a continuous evolutionary process, driven by the persistent demand for enhanced efficiency and broader applicability. This section gives an insight into the developmental trajectory of these materials, categorizing them into distinct generations based on their compositional complexity and mode of application, thereby highlighting the continuous innovation spurred by performance requirements [28, 29].

7.2.4.1 First-Generation: Single Component MOs

The first-generation of photocatalysts, primarily single transition MOs (TMOs) like TiO_2 and ZnO. These laid the foundation for photocatalytic applications such as pollutant degradation and water splitting [30, 31]. TiO_2 gained prominence due to its stability, nontoxicity, and strong oxidizing properties, while ZnO offered high exciton binding energy and ease of synthesis. However, these materials suffer from large bandgaps, restricting activity to UV light (only 5% of solar spectrum), and high electron-hole recombination rates which tends to limit their efficiency.

To address these issues, modifications are apt, which involve metal loading. This strategy has been explored in order to ameliorate the limitation of this novel material, as such enhancing charge separation and improving degradation rates. Studies also show that catalyst dosage and dye concentration influence efficiency, as seen with ZnO nanoparticles degrading azo dyes [9]. Additionally, green-synthesized Cu_2O nanoparticles demonstrated reduced recombination and effective dye degradation. Despite advancements, first-generation photocatalysts remain constrained by poor solar utilization, prompting the need for more complex, engineered designs.

7.2.4.2 Second-Generation: Doped and Binary MO Composites

The second-generation photocatalysts were developed to overcome the limitations of first-generation materials, such as wide bandgaps and high electron-hole recombination rates. Key strategies include doping single Mos with metals/nonmetals or creating binary composites to enhance photocatalytic activity. These modifications narrow bandgaps, improve light absorption, and reduce charge recombination.

Doped MOs show notable improvements: Fe-doped TiO_2 degraded 88.7% of acid orange 7, while Sn-doped ZnO fully degraded DR-31 dye in 60 minutes [32, 33]. Mn-doped ZnO enhanced electron-hole separation, and Au/TiO_2 demonstrated high reusability. Non-metal doping, like B-doped MnO_2, achieved 97% methylene orange degradation [34].

Binary composites further boosted performance: CuO/Bi_2O_3 degraded 87.87% MB, and ZnO/TiO_2 reached 98.13% methyl orange degradation, with pH playing a critical role [35, 36]. Sn-doped CeO_2–Fe_2O_3 nearly fully degraded MO solutions and remained stable over five cycles, aided by efficient charge transfer [37]. Despite these advances, second-generation catalysts are typically used in suspended form, making recovery and separation from treated water difficult and energy-intensive. This limitation spurred the development of third-generation photocatalysts to address practical application challenges. Table 7.1 depicts the literature review of second-generation MO photocatalysts.

7.2.4.3 Third-Generation: Supported/Immobilized Catalysts

The third-generation of photocatalysts addresses the recovery and recyclability issues of earlier suspended systems by immobilizing catalysts onto support materials. These supported systems offer high stability, easy regeneration, and simpler separation. They can be active or inactive supports. Examples of inactive support include TiO_2 on glass beads (80% dye decolorization), TiO_2/WO_3 on stainless steel (enhanced methylene blue degradation due to visible-light absorption

Table 7.1 Literature review of second-generation MO photocatalysts.

Material	Pollutant	Efficiency (%)	Notable features	References
Fe-doped TiO_2	Acid orange 7	88.7	Narrowed bandgap, reduced recombination	Singh et al. [32]
Sn-doped ZnO	DR-31 dye	100	Enhanced light absorption and reactivity	Bhatia et al. [33]
Mn-doped ZnO	Azo dye	96	Improved electron-hole separation	Aadnan et al. [18]
Au/TiO_2	Paracetamol	99.17	Noble metal enhanced charge separation	Edirisooriya et al. [38]
B-doped MnO_2	Methylene orange	97	Non-metal doping for improved oxidation efficiency	Gayathri et al. [34]
CuO/Bi_2O_3	Methylene blue	87.87	Binary composite performance	Kaliyaperumal et al. [36]
ZnO/TiO_2	Methyl orange	98.13	pH-dependent efficiency; strong synergy	Kahate et al. [35]
Sn-doped CeO_2–Fe_2O_3	Methyl orange	~100	Efficient charge transfer; good recyclability	Krishnan et al. [37]

and charge separation), and ZnO on glass/alumina (complete dye removal in 3 hours) [19, 39, 40]. Meanwhile, active supports include chitosan/TiO_2 composites for degrading malachite green under visible light [41]. While immobilization improves recovery, it reduces accessible surface area, limiting efficiency compared to suspended catalysts. This trade-off spurred advancements toward more complex, multicomponent systems.

7.2.4.4 Fourth-Generation: Ternary and Quaternary Composites

The fourth-generation of photocatalysts represents advanced MO systems designed to surpass earlier limitations, such as low surface area and inefficiency. These ternary or quaternary compositions feature multiple heterojunctions, active supports, and often magnetic properties for improved recoverability, offering high activity, stability, and reusability. This can be observed in a ZnO/Ag_2O/Ag_2S composite, enhancing methyl orange degradation via heterojunctions and silver's plasmonic effect [42] and in a Fe_3O_4/ZnO/Ag_3VO_4/AgI magnetic nanocomposite, degrading rhodamine B and methylene blue 120x and 50x faster than Fe_3O_4/ZnO, due to improved light absorption and charge separation [39, 43]. It can also be noticed in a $ZrCdPbO_4$ system excelling in degrading xanthene dyes, leveraging high surface area, visible light absorption, and reusability [44].

Despite their advantages, challenges include complex synthesis (e.g., creating multiple heterojunctions), identifying optimal material combinations, and potential

Table 7.2 Evolution and classification of MO photocatalysts.

Generation	Description	Examples/key features
First-generation	Single-component MOs	TiO_2, ZnO; limited to UV light; high recombination; improved by noble metal doping (e.g., Ag-TiO_2)
Second-generation	Doped and binary MO composites	Fe-TiO_2 (88.7% acid orange 7); Sn-ZnO; Mn-ZnO; ZnO/TiO_2 (98.13% methyl orange); improved visible light activity
Third-generation	Supported/immobilized catalysts for easier recovery	TiO_2 on glass, WO_3 on stainless steel, chitosan-TiO_2; better separation but lower surface area
Fourth-generation	Ternary/quaternary composites with complex heterojunctions and often magnetic properties	ZnO/Ag_2O/Ag_2S, Fe_3O_4/ZnO/Ag_3VO_4/AgI, $ZrCdPbO_4$; superior degradation and reusability

defects affecting performance. However, these systems hold great promise for efficient, recoverable photocatalysis in environmental applications. Table 7.2 shows the evolution and classification of MO photocatalysts.

7.3 Strategies for Enhancing Photocatalytic Activity

Several reports have shown that the pursuit of efficient photocatalysts focuses on maximizing light absorption and minimizing charge carrier recombination, particularly for visible and solar light applications [16]. Key strategies include:

7.3.1 Doping

This involves introducing impurities to modify electronic properties of the photoactive material, for example, Fe, Mn, Au, etc., which narrows bandgaps, introduces trap states, and enhances visible light absorption [45] or non-metal doping (N, C, S) redshifts absorption by elevating VBs. The likelihood of photocatalytic dye degradations is also enhanced by metal ion doping because of the chemical, physical, and electronic peculiarities of the same. Lanthanide ions f-orbitals associated have improved the absorption capability of the pollutants by the Lewis base bonding and use as a good electron trapper. Du et al. suggested that TiO_2 photocatalytic performance increased when stimulated to dope a rare earth metal ion [46].

In addition, there are also reports of concurrent depositions and doping, Ag/Ag(I)-TiO_2 where Ag(I) dopings are observed with the appearance of a distinct energy level of Ag 4d in the bandgap of TiO_2 [47]. This voltage was a result of the illumination, that is, the visible light meant that the isolated energy level of 4d got absorbed in the visible light. Quite to the contrary, the photogenerated carriers (electrons and holes) are postsynaptically due to the presence of the surface metallic Ag and offers more visible light activity as compared to the TiO_2 that has been modified with Ag.

7.3.2 Metal/MO Composites and Heterojunctions

Plasmonic nanostructures enhance light absorption through surface plasmon resonance (SPR), while core-shell structures (e.g., Au-Cu_2O) improve charge transfer and stability [48–50]. In heterojunctions, engineered interfaces separate charges, for example, Z-scheme (e.g., g-C_3N_4/TiO_2) preserves redox potential while reducing recombination [51, 52], p-n junctions create built-in electric fields for charge separation while Schottky junctions (e.g., Au/Cu_2O) use metal-semiconductor barriers to enhance charge transfer and S-scheme heterojunctions offer advanced charge separation mechanisms [53, 54].

A synergistic effect may arise when two disparate MO materials are combined to form heterojunctions to improve photocatalytic performance. ZnO-semiconductor heterojunction is beneficial in increasing charge separation and photocatalytic activity, such as with TiO_2 or Bi_2O_3 [55]. Such heterojunctions drain photo-induced electrons and holes. An integrated technique is a strategy that utilizes the combination of two semiconductors, and which usually entails the use of MOs mostly in the form M_xO_y/M_xO_t (whereby M and Me represent metal types and x, y, z, and t their distinguishable oxidation states in these MOs) [56]. This mixture of materials, a so-called nanocomposite, has some of the benefits, especially when used in photocatalytic technologies, because of its increased absorption of light, a facile recombination of photo-induced electron-hole pairs, and increased separation of photo-charges. As Lin and Chiang's study has revealed, the long lifespan of charge carriers, caused by the exchange of electrons that uses the conduction pathways at the particles that make up the nanocomposite structures, results in more electrons involved in the photodegrading cycle [57]. Such beneficial property is due to energy level structure of the composite materials [58].

7.3.3 Common Synthesis Methods

The synthesis method plays a crucial role in determining the morphology, size, crystallinity, and overall performance of MO photocatalysts. Common synthesis methods for MOs photocatalysts are presented in Table 7.3. Several common techniques are employed:

- **Sol-Gel Method:** A sol-gel process is recorded in the synthesis, properties, and applications of oxide nanomaterials because it is a method that is normally applied to manufacture MOs through the hydrolysis of metal reactive precursors (typically alkoxides in an alcohol solution). This forms corresponding hydroxide and the condensation of this hydroxide to form of a metal hydroxide polymer which resembles a tightly packed porous gel. The gel is evaporated and dried to result in a similar powder crystal of the target particle size MO. Depending on the heat treatment processes carried out, the produced MO can be presented as a form of a nanoparticle, bulk material or even finally oxygen-deficient [59]. This method was employed to synthesize copper oxide nanoparticles with a precursor solution of copper (II) acetate in ethanol [24, 60].

Table 7.3 Common synthesis methods for MO photocatalysts.

Method	Description	Key advantages	References
Sol-gel	Hydrolysis of metal alkoxides forms a gel, followed by drying and calcination	Fine control over chemical composition; uniform dopant dispersion	Joshi et al. [59], Aigbe and Osibote [71], Kasbaji et al. [72]
Hydrothermal	Reaction in sealed vessel at high temperature/pressure using water as a solvent/catalyst	Produces highly crystalline and pure materials; size and morphology control	Mukherji et al. [61], Shen et al. [62], Singh et al. [68]
Coprecipitation	Simultaneous precipitation of multiple ions in solution	Cost-effective; scalable; but difficult to control homogeneity	Singh et al. [68], Aigbe and Osibote [71], Goodarzi et al. [73]
In situ synthesis	Materials formed directly in the reaction context (e.g., composites or real-time property studies)	Enables complex structures and real-time analysis under practical/functional conditions	Shen et al. [62], Wang et al. [69], Park et al. [70]

- **Hydrothermal Method:** In essence, the term "hydrothermal" is used to describe a solvent that undergoes a transformation when it is composed of water. The agglomeration can occur and so the most appropriate capping agent or stabilizer is introduced in the reaction at the right time to hinder the growth of the particles. Dissolving of the particles in the various solvents is also aided by the stabilizers. Water acts as a catalyst and can influence the solid phases. It is favored for its simplicity, low cost, and ability to produce highly crystalline and pure materials with controlled size, structure, and morphology by adjusting parameters like pH, temperature, pressure, and time.

  In hydrothermal growth, in a medium of a dispersant (the oxide of a semiconductor is loaded onto graphene oxide [GO] after which it undergoes a subsequent reduction under higher temperature). The same procedure was applied to make $ZnFe_2O_4$/graphene composite using $Zn(NO_3)_2$ $6H_2O$ and $Fe(NO_3)_2$ $9H_2O$ as a precursor. Metal salts undergo oxidation, and GO is reduced in the process of growth. Mukherji et al. opted to grow graphene-$BiWO_6$ in-situ hydrothermal in the presence of GO [61]. Graphene TiO_2 composite synthesis was performed by Shen et al. using scalable approach and performed growth through the hydrothermal method with the use of water and glucose (reducing agent) [62]. The dispersion of graphene TiO_2 composite was also harnessed through a one-step solvothermal route in order to acquire dispersed MO uniformly on the graphene.
- **Coprecipitation Method:** In this technique, precipitation of an oxo-hydroxide form is done by a precipitating medium and a solution of a salt precursor (usually

a metal salt such as a nitrate or chloride) in a solvent (usually water). Their respective metal oxo-hydroxides are precipitated in water, adding a base to the solution. The most common precipitation media utilized in any of these methods are NaOH, KOH, and $(C_2H_5)_4NOH$, through which it was confirmed that morphologies and characteristics of the MOs are strongly determined by the pH of the and the type of the alkaline solution [63]. It is also common knowledge that the mode and rate of addition of such alkali solutions into the reaction mixture should be effectively put into considerations in this method so as to arrive at the required property of the MOs. They can also make use of surfactants, which is highly instrumental in the optimization of surface characteristics of the produced MO. The usage of surfactants reduces the size or width of acquired nanomaterials. The greatest advantage of this synthesis method is that it is low cost, and during the synthesis process, it has mild reaction conditions such as low temperatures, etc. A similar process was used to synthesize ZnO from Zn $(NO_3)_3$ [64].

- **In-situ Synthesis:** In this, metal salts and GO are employed. It is subsequently added to GO (dispersed solution), before getting transformed into MO by the presence of the oxidizing agent, after which GO was reduced to graphene [65]. The process of formation of graphene ZnO composite was based on the adsorption of Zn^{+2} ions to the surface of the GO sheet in ethanol solution using the presence of NaOH and $NaBF_4$ and drying at 150 °C [66, 67]. Comparison of these, SnO_2 and TiO_2 were also deposited on graphene as they used $SnCl_2$ and $TiCl_3$ as substrates [68]. The same method was used to obtain different morphological structures by reducing Sn and Ti salts in the presence of GO. This is as a result of varied rate of hydrolysis of the salt and their reduction capabilities. In the synthesis of various graphene-based composites using in-situ method, researchers have used various techniques of the dispersing GO. The 2D graphene has also been stabilized by anionic surfactants in an aqueous phase to grow in situ TiO_2 over graphene [69]. As compared to Park et al. described graphite TiO_2 composite after exfoliation reorganization of graphite oxide without any chemical agents in the aqueous slurry [70].

7.4 Applications of MO Photocatalysis

MO photocatalysis has emerged as a versatile technology with far-reaching applications across environmental remediation, energy production, and public health [59].

7.4.1 Pollutant Degradation

Photocatalysis is highly effective for the degradation of a wide array of organic pollutants. Organic dyes are frequently used as model pollutants due to their environmental prevalence from industries like textiles and the ease of monitoring their degradation. Common examples include anionic methyl orange (MO) and cationic methylene blue (MB), as well as rhodamine B (RhB), Congo red, and acid orange 7 [18, 19, 33, 37, 42, 43]. TiO_2 and ZnO are widely studied for this purpose, with innovations like fluorinated TiO_2 and Mn-doped ZnO showing enhanced visible light activity. Iron oxides have also been used for photodegrading azo dyes, converting them into simpler, mineralized products [18]. Similarly, photocatalysis addresses the growing concern of pharmaceutical residues in water systems. Examples include the degradation of clofibric acid, carbamazepine, tetracycline, sulfamethoxazole (SMX), and amoxicillin. Fe-based MOFs and MIL-100(Fe)/TiO_2 nanoarrays have shown high removal efficiencies for these persistent compounds [14, 28].

MO photocatalysts are effective against various pesticides, including isopropanol, diazinon, and perfluorooctanoic acid (PFOA). ZnO/$Na_2S_2O_8$ composites and FeO/carbon sphere systems have been utilized for accelerated degradation of these contaminants [74, 75]. This advancement extends to a broad spectrum of other organic compounds, such as phenols, formaldehyde, styrene, naphthalene, acetaldehyde, and nitrobenzene [76].

7.4.2 Energy Production

Photocatalytic and photoelectrochemical water splitting for hydrogen production is a key area of interest, aiming to generate H_2 and O_2 from water under illumination. Black titania and g-C_3N_4/Ti^{3+}-doped TiO_2 have demonstrated significant hydrogen production rates under solar and visible light [38, 77]. It has also been used in CO_2 reduction, which involves the conversion of carbon dioxide and water into valuable fuels like methane and carbon monoxide. Composites such as graphene: TiO_2O_2 and porous hyper cross-linked polymer-TiO_2-graphene have shown enhanced CO_2 conversion rates.

7.4.3 Antimicrobial Activity and Water Disinfections

Photocatalysis is a significant practical application for inactivating microorganisms and disinfecting water. This activity is often attributed to oxidative stress and cell membrane damage caused by ROS, as well as the release of metal ions (for certain MOs like ZnO) and direct interaction with cell walls [78]. TiO_2 and ZnO are common photocatalysts used for disinfection. Fe/ZnO nanoparticles have shown high efficiency in removing multidrug-resistant. The mechanisms of antibacterial activity are complex and can vary with the material and medium, often showing no direct correlation with ROS production or ion release alone.

7.4.4 Other Applications

Beyond these primary areas, MO photocatalysis finds utility in diverse fields. Figure 7.2 shows the diverse applications of MO photocatalysis. They have been used in solar cells and fuel cells to enhance efficiency, in gas sensors due to their photo-responsive electrical properties, as well as in biomedical imaging, drug delivery, and dental implants.

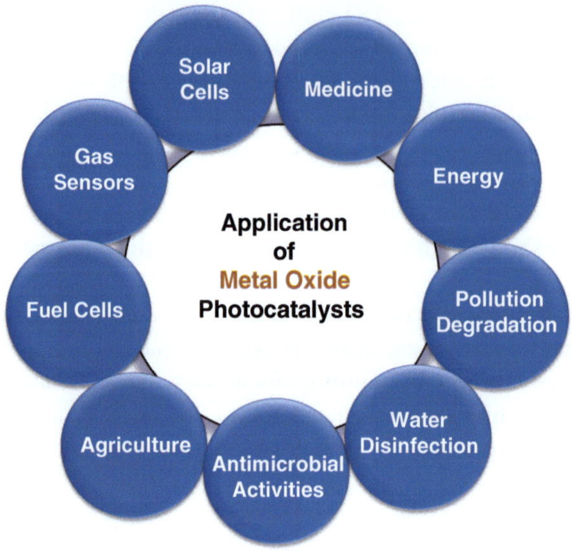

Figure 7.2 Diverse applications of MO photocatalysts.

7.5 Current Challenges in MO Photocatalysis

MO photocatalysis faces key challenges hindering its widespread use. Many photocatalysts like TiO_2 and ZnO have wide bandgaps, limiting light absorption to UV, leaving visible light unused [10, 79]. Rapid electron-hole recombination further reduces efficiency by wasting energy as heat. Photocorrosion (e.g., in ZnO, WO_3, CuO) degrades materials over time, impacting durability [63]. Poor understanding of defect roles, charge transport, and ROS interactions complicates optimization. Practical issues include separating nanoparticles from water, potential toxicity, and secondary pollutants. While solutions like magnetic separation exist, scalability and cost remain barriers. Overcoming these challenges is crucial for industrial viability [20].

7.6 Future Perspectives and Research Directions

The future of MO photocatalysis focuses on developing efficient, stable, and scalable solutions for environmental and energy challenges. Key research directions include enhancing light absorption, charge separation, and electron/hole pair lifetimes through advanced doping, heterostructures (e.g., Z-scheme, S-scheme), and defect engineering. Scalable, eco-friendly synthesis methods are needed for industrial adoption. Ecotoxicity and environmental impact must be addressed to prevent secondary pollution. Real-world testing under complex conditions such as varying pH and pollutant concentrations is essential for practical application. Standardized testing methods and stronger academia-industry collaboration will drive progress in water treatment and energy applications.

7.7 Conclusion

The upsurge in global water crisis, driven by unchecked industrialization and human activities, has led to severe ecological damage, resulting in a significant public health concern. MOs, with their robust photocatalytic properties, have proven invaluable in abating water and environmental pollution menace. Their intrinsic photoactive propensity, among other properties, enables the generation of ROS that is potent to activity degrade organic pollutants effectively. Recent innovations and modification techniques have significantly enhanced their photocatalytic efficiencies. However, there are certain limitations to the full-scale adoption of this technology in real-life applications. This piece provides an overview of the fundamental mechanisms, optimization strategies, and emerging applications of MO-based photocatalysts. Furthermore, it emphasizes their pivotal role in developing greener, safer, and more effective wastewater treatment technologies. As research continues to evolve, MOs stand poised as a sustainable cornerstone in the transition from conventional chemical treatments to more eco-benign alternatives.

References

1 Bamisaye A, Alli YA, Ige AR, Bamidele MO, Etafo NO, Manshor MR. Chitosan composites for wastewater treatment. In: Ngadi N, Ilyas RA, Ab Hamid NH, Sapuan SM, editors. *Advanced Composite Materials for Wastewater Treatment* [Internet]. Elsevier; 2025. p. 91–118. https://linkinghub.elsevier.com/retrieve/pii/B9780443336355000133

2 Ajala OA, Akinnawo SO, Bamisaye A, Adedipe DT, Adesina MO, Okon-Akan OA, et al. Adsorptive removal of antibiotic pollutants from wastewater using biomass/biochar-based adsorbents. *RSC Adv.* 2023; 13:4678–712.

3 Do HT, Phan Thi LA, Dao Nguyen NH, Huang CW, Le Q Van, Nguyen VH. Tailoring photocatalysts and elucidating mechanisms of photocatalytic degradation of perfluorocarboxylic acids (PFCAs) in water: a comparative overview. *J Chem Technol Biotechnol.* 2020;95(10):2569–78.

4 Mondal K, Sharma A. Recent advances in the synthesis and application of photocatalytic metal-metal oxide core-shell nanoparticles for environmental remediation and their recycling process. *RSC Adv.* 2016;6(87):83589–612.

5 Koe WS, Lee JW, Chong WC. An overview of photocatalytic degradation: photocatalysts, mechanisms, and development of photocatalytic membrane. *J Colloid Interface Sci.* 2019;44:2522–65.

6 Bamisaye A, Oluboyede OA, Etafo NO, Bamidele MO, Abiola-Kuforiji OT, Idowu MA. Synthesis of zinc oxide nanoparticles using *Syzygium malaccense* leaf extract: photocatalytic decomposition of ciprofloxacin and antimicrobial studies. *Nanotechnol Environ Eng.* 2025 Sep 30;10(3):54. https://link.springer.com/10.1007/s41204-025-00459-z

7 Ogunbiyi O, Bamisaye A, Abiodun AJ, Owoeye TF, Alli YA, Idowu MA. Biogenic synthesis of Zinc oxide nanoparticles for solar cell application and photodegradation of neomycin. *Mater Sci Eng B.* 2025 Sep 1;319:118324.

8 Etafo NO, Bamidele MO, Bamisaye A, Alli YA. Revolutionizing photocatalysis: unveiling efficient alternatives to titanium (IV) oxide and zinc oxide for comprehensive environmental remediation. *J Water Process Eng*. 2024;62:105369.

9 Raub AAM, Bahru R, Mohamed MA, Latif R, Haniff MASM, Simarani K, et al. Photocatalytic activity enhancement of nanostructured metal-oxides photocatalyst: a review. *Nanotechnology*. 2024;35(24):14047–59.

10 Arora AK, Jaswal VS, Singh K, Singh R. Applications of metal/mixed metal oxides as photocatalyst: a review. *Orient J Chem*. 2016;32(4):2035–42

11 Ding M, Cai X, Jiang HL. Improving MOF stability: approaches and applications. *Chem Sci*. 2019 Nov 13 [cited 2025 Jul 5];10(44):10209–30. Available from: https://pubs.rsc.org/en/content/articlehtml/2019/sc/c9sc03916c

12 Hlekelele L, Mdlalose L, Nomadolo N, Mtibe A, Chauke V. Carbon-metal sulfide nanomaterial photocatalysts for environmental remediation. In: *Metal Sulfide Nanomaterials for Environmental Applications* [Internet]. 2025 Jan 1 [cited 2025 Jul 5]. p. 235–63. Available from: https://www.sciencedirect.com/science/article/abs/pii/B978044313464700013X

13 Li WY, Liu J, Li ZR, Shen TT, Chen J, Hu ZY, et al. Probing the deactivation and regeneration of ZnIn2S4 in photocatalytic degradation of tetracycline. *J Colloid Interface Sci*. 2025 Sep 15 [cited 2025 Jul 5];694:137700. Available from: https://www.sciencedirect.com/science/article/pii/S0021979725010914

14 Baig MT, Zahid Z, Tauqeer M, Park CH, Park JP, Choi CH, et al. State-of-the-art developments in metal/metal oxide and graphene-based nano-hybrid systems for photocatalytic degradation of tetracycline antibiotics. *J Ind Eng Chem*. 2025;148:69–91.

15 Imran M, Abdullah AZ, Khan ME, Mohammad A. A focused review on photocatalytic potential of graphitic carbon nitride (g-C3N4) based metal oxide-nanostructures for effective remediation of most overused antibiotics. *J Environ Manage*. 2025;373:123759.

16 Lanjwani MF, Tuzen M, Khuhawar MY, Saleh TA. Trends in photocatalytic degradation of organic dye pollutants using nanoparticles: a review. *Inorg Chem Commun*. 2024;159:111613.

17 Srivastava SK. Recent advances in removal of pharmaceutical pollutants in wastewater using metal oxides and carbonaceous materials as photocatalysts: a review. *RSC Appl Interfaces*. 2024;1(3):340–429.

18 Aadnan I, Zegaoui O, El Mragui A, Daou I, Moussout H, Esteves da Silva JCG. Structural, optical and photocatalytic properties of Mn Doped ZnO nanoparticles used as photocatalysts for Azo-dye degradation under visible light. *Catalysts*. 2022;12(11):1382.

19 Daneshvar N, Salari D, Niaei A, Rasoulifard MH, Khataee AR. Immobilization of TiO2 nanopowder on glass beads for the photocatalytic decolorization of an azo dye CI direct red 23. *J Environ Sci Health*. 2005;40(8):1605–17.

20 Qumar U, Hassan JZ, Bhatti RA, Raza A, Nazir G, Nabgan W, et al. Photocatalysis vs adsorption by metal oxide nanoparticles. *J Mater Sci Technol*. 2022;131:122–66.

21 Danish MSS, Estrella LL, Alemaida IMA, Lisin A, Moiseev N, Ahmadi M, et al. Photocatalytic applications of metal oxides for sustainable environmental remediation. *Metals*. 2021;11(1):80.

22 Bamisaye A, Etafo NO, Bamidele MO, Awobifa OS. Targeted drug delivery with photoluminescent emitters: applications in diagnostic. In: Etafo NO, editor. *Tailored Light Emitters for Biomedical Applications. Engineering Materials*. Cham: Springer; 2025. p. 75–119. Available from: https://link.springer.com/10.1007/978-3-031-88731-4_4

23 Rakshitha R, Chethan R, Pallavi N. Photocatalytic removal of emerging contaminants from water using metal oxide-based nanoparticles. *Curr Nanosci*. 2024;20(3):339–55.

24 Geldasa FT, Kebede MA, Shura MW, Hone FG. Experimental and computational study of metal oxide nanoparticles for the photocatalytic degradation of organic pollutants: a review. *RSC Adv*. 2023;13(27):18404–42.

25 Ounis Dkhil Y, Peppel T, Sebek M, Strunk J, Houas A. Green synthesis of photocatalytically active ZnO nanoparticles using chia seed extract and mechanistic elucidation of the photodegradation of diclofenac and p-nitrophenol. *Catalysts* [Internet]. 2024 Dec 24 [cited 2025 Mar 21];15(1):4. https://www.mdpi.com/2073-4344/15/1/4/htm

26 Ghamarpoor R, Fallah A, Jamshidi M. A review of synthesis methods, modifications, and mechanisms of ZnO/TiO_2-based photocatalysts for photodegradation of contaminants. *ACS Omega*. 2024 Jun 18;9(24):25457–92. https://doi.org/10.1021/acsomega.3c08717

27 Tasleem F, Manzoor S, Tasleem S, Khan M, Khan SA, Nishan U, et al. Ultraviolet photodegradation of ciprofloxacin using zinc oxide and iron-doped zinc oxide (Fe-ZnO) nanoparticles (NPs): kinetic and isotherm measurements. *Anal Lett*. 2024;58(1):34–48.

28 Velempini T, Prabakaran E, Pillay K. Recent developments in the use of metal oxides for photocatalytic degradation of pharmaceutical pollutants in water—a review. *Mater Today Chem*. 2021;19:100380. https://doi.org/10.1016/j.mtchem.2020.100380

29 Gautam S, Agrawal H, Thakur M, Akbari A, Sharda H, Kaur R, et al. Metal oxides and metal organic frameworks for the photocatalytic degradation : a review. *J Environ Chem Eng*. 2020;8(3):103726. https://doi.org/10.1016/j.jece.2020.103726

30 Huong NTM, Thuy NT, Hoai PTT, Thi Huong P. Effective inhibition of antibiotic residues and bacterial pathogens in wastewater using TiO_2 metal oxide photocatalyst. *J Environ Sci Health B*. 2025;60(2):70–78.

31 Patwa R, Rohilla S, Goel N. Structural and spectroscopy analysis of nanocomposites of metal oxide ZnO/CuO/Ag by coprecipitation: potential application in photocatalysis. *Ceram Int*. 2025;51(11):14047–59.

32 Singh S, Parveen S, Clarizia L, Kumar P. An insight into photo-catalytic degradation mechanism of persistent pollutants with transition metal oxides and their composites: Photocatalysis mechanism, rate affecting parameters, and removal pathways. *Catal Rev*. 2025; 1–49. https://doi.org/10.1080/01614940.2024.2440664.

33 Bhatia S, Verma N, Bedi RK. Sn-doped ZnO nanopetal networks for efficient photocatalytic degradation of dye and gas sensing applications. *Appl Surf Sci*. 2017;407:495–502.

34 Gayathri M, Shanthi M, Satheeshkumar E, Jayaprakash N, Sundaravadivel E. Preparation and characterization of boron doped CN'S/MnO_2 and its photocatalytic application of dye degradation. *Mater Today Proc.* 2021;42:1506–12.

35 Kahate AS, Tripathy S, Madankar RS, Bhilkar P, Nagmote M, Singh NB, et al. Green nanomaterials for environmental mitigation. In: Desimone MF, editor. *Nanotechnology*. Springer; 2025. p. 49–73.

36 Kaliyaperumal R, Nagaraj K, Badgujar NP, Kasilingam T, Poovan V, Kumaravel T. Enhanced antimicrobial activity of bismuth oxide doped copper oxide nanocomposites synthesized via co-precipitation: influence of composition and morphology. *Bionanoscience* 2025;15(3):1–10.

37 Krishnan A, Vishwanathan PV, Mohan AC, Panchami R, Viswanath S, Krishnan AV. Tuning of photocatalytic performance of CeO_2-Fe_2O_3 composite by Sn-doping for the effective degradation of methlene blue (MB) and methyl orange (MO) dyes. *Surfaces Interfaces*. 2021;22:100808.

38 Edirisooriya EMNT, Senanayake PS, Xu P, Wang H. Recyclability and regeneration of Au/TiO_2 nanocomposite and Pt/TiO_2 atom-nano composite catalysts in photo-reforming plastics for hydrogen production. *J Environ Chem Eng.* 2025;13(3):116467.

39 Baruah S, Mahmood MA, Myint MTZ, Bora T, Dutta J. Enhanced visible light photocatalysis through fast crystallization of zinc oxide nanorods. *Beilstein J Nanotechnol.* 2010;1(1):14–20.

40 Anucha CB, Altin I, Bacaksiz E, Stathopoulos VN. Titanium dioxide (TiO_2)-based photocatalyst materials activity enhancement for contaminants of emerging concern (CECs) degradation: in the light of modification strategies. *Chem Eng J Adv.* 2022;10:100262.

41 Pérez-Obando J, Marin-Silva DA, Pinotti AN, Pizzio LR, Osorio-Vargas P, Rengifo-Herrera JA. Degradation study of malachite green on chitosan films containing heterojunctions of melon/TiO_2 absorbing visible-light in solid-gas interfaces. *Appl Catal B.* 2019;244:773–85.

42 Bhosale A, Kadam J, Gade T, Sonawane K, Garadkar K. Efficient photodegradation of methyl orange and bactericidal activity of Ag doped ZnO nanoparticles. *J Indian Chem Soc.* 2023;100(2):100920.

43 Hassanpour M, Safardoust-Hojaghan H, Salavati-Niasari M. Degradation of methylene blue and Rhodamine B as water pollutants via green synthesized Co_3O_4/ZnO nanocomposite. *J Mol Liq.* 2017;229:293–9.

44 do Nascimento CPG, Costa MSMA, Freire JMA, da Silva LTV, Coutinho LP, Monteiro NKV, et al. Degradation of xanthene-based dyes by photoactivated persulfate: experimental and computational studies. *Photochem Photobiol Sci.* 2023;22(12):2699–714.

45 Krishnan A, Swarnalal A, Das D, Krishnan M, Saji VS, Shibli SMA. A review on transition metal oxides based photocatalysts for degradation of synthetic organic pollutants. *J Environ Sci.* 2024;139:389–417.

46 Du P, Bueno-Lopez A, Verbaas M, Almeida AR, Makkee M, Moulijn JA, et al. The effect of surface OH-population on the photocatalytic activity of rare earth-doped P25-TiO_2 in methylene blue degradation. *J Catal.* 2008;260(1):75–80.

47 Liu R, Wang P, Wang X, Yu H, Yu J. UV-and visible-light photocatalytic activity of simultaneously deposited and doped Ag/Ag (I)-TiO_2 photocatalyst. *J Phys Chem.* 2012;116(33):17721–8.

48 Harish V, Ansari MM, Tewari D, Yadav AB, Sharma N, Bawarig S, et al. Cutting-edge advances in tailoring size, shape, and functionality of nanoparticles and nanostructures: a review. *J Taiwan Inst Chem Eng.* 2023 Aug;149:105010.

49 Ji X, Yu Y, Guan Y, Lai J, Wang L. Porous core-shell Cuau@Cu_2O catalyst for acidic C-N coupling toward urea electrosynthesis from Co_2 and nitrate. 2025 [cited 2025 Jul 13]. Available from: https://papers.ssrn.com/abstract=5309363

50 Nasiri H, Salahandish M, Sharifi H, Mohammadzadehasl N. Dependence of extinction cross-section, absorption and surface plasmon resonance of nanoparticle surface size and concentration of Silica@Gold nanoparticles. *Biomed Mat Dev* [Internet]. 2025 Sep 1 [cited 2025 Jul 13];3(2):1501–8. Available from: https://link.springer.com/article/10.1007/s44174-024-00272-3

51 Yang Y, Ma G, An Z, Wang W, Hu X, Wang Y, et al. Preparation of recyclable g-C_3N_4/TiO_2 heterojunction/alginate hydrogel microbeads and investigation of their adsorption-photocatalytic properties. *J Hazard Mater Advances.* 2025 May 1 [cited 2025 Jul 13];18:100650. https://www.sciencedirect.com/science/article/pii/S2772416625000622

52 Li Q, Zhou H, Li Z, Liu A, Wang E, Wu Y, et al. Efficient photocatalytic degradation of antibiotics using Z-scheme MIL-88(Fe)/Ti_3C_2/MoO_3: mechanistic insights and toxicity assessment. *J Hazard Mater* [Internet]. 2025 Mar 15 [cited 2025 Jul 13];486:137051. https://www.sciencedirect.com/science/article/pii/S030438942403632X

53 Ren H, Miao Z, Zhao Y, Ghasemi S, Feng X, Liu E, et al. Advances and challenges in multiple S-scheme heterojunction photocatalysts. *J Alloys Compd.* 2025 May 15 [cited 2025 Jul 13]; 1028:180646. https://www.sciencedirect.com/science/article/pii/S0925838825022078

54 Chu W, Sun G, Chen Z, Tian X, Wang S, Xie M, et al. Synergistic effects of shell-hollow confinement catalysis in Au@Cu_2O nanoreactors toward efficient electrocatalytic CO_2 to multi-carbon products. *Chem Eng J.* 2025 Sep 1 [cited 2025 Jul 13];519:165084. https://www.sciencedirect.com/science/article/pii/S1385894725059200

55 Huang Y, Wei Y, Wang J, Luo D, Fan L, Wu J. Controllable fabrication of Bi_2O_3/TiO_2 heterojunction with excellent visible-light responsive photocatalytic performance. *Appl Surf Sci.* 2017;423:119–30.

56 Vaezi MR. Coupled semiconductor metal oxide nanocomposites: types, synthesis conditions and properties. In: *Advances in Composite Materials for Medicine and Nanotechnology.* 2011. p. 365–400.

57 Lin CC, Chiang YJ. Preparation of coupled ZnO/SnO_2 photocatalysts using a rotating packed bed. *Chem Eng J.* 2012;181:196–205.

58 Johra FT, Jung WG. RGO–TiO_2–ZnO composites: synthesis, characterization, and application to photocatalysis. *Appl Catal A Gen.* 2015;491:52–7.

59 Joshi NC, Gururani P, Gairola SP. Metal oxide nanoparticles and their nanocomposite-based materials as photocatalysts in the degradation of dyes. *Biointerface Res Appl Chem.* 2022;12(5):6557–79.

60 Pouthika K, Madhumitha G. A mini review on recent advancements in metal oxide integrated nano-tubular inorganic clay mineral photocatalyst for organic pollutant degradation. *Inorg Chem*. 2025;45(1):30–62.

61 Mukherji A, Seger B, Lu GQ, Wang L. Nitrogen doped $Sr_2Ta_2O_7$ coupled with graphene sheets as photocatalysts for increased photocatalytic hydrogen production. *ACS Nano*. 2011;5(5):3483–92.

62 Shen J, Yan B, Shi M, Ma H, Li N, Ye M. One step hydrothermal synthesis of TiO_2-reduced graphene oxide sheets. *J Mater Chem*. 2011;21(10):3415–21.

63 Ahuja P, Ujjain SK, Kanojia R, Attri P. Transition metal oxides and their composites for photocatalytic dye degradation. *J Compos Sci*. 2021;5:82.

64 Albonetti C, Alkarsifi R, El Qacemi V, Dhuiege B, Ruani G, Seri M. Recycled or bio-based solvents for the synthesis of ZnO nanoparticles: characterization and validation in organic solar cells. *Materials*. 2024;17(6):1332.

65 Thakur S, Bi A, Mahmood S, Ruzimuradov O, Gupta R, Cho J, et al. Graphene oxide as an emerging sole adsorbent and photocatalyst: chemistry of synthesis and tailoring properties for removal of emerging contaminants. *Chemosphere*. 2024;352:141483.

66 Du J, Lai X, Yang N, Zhai J, Kisailus D, Su F, et al. Hierarchically ordered macro–mesoporous TiO_2– graphene composite films: improved mass transfer, reduced charge recombination, and their enhanced photocatalytic activities. *ACS Nano*. 2011;5(1):590–6.

67 Chen JS, Wang Z, Dong XC, Chen P, Lou XWD. Graphene-wrapped TiO_2 hollow structures with enhanced lithium storage capabilities. *Nanoscale*. 2011;3(5):2158–61.

68 Singh A, Dhau J, Kumar R, Badru R, Singh P, Mishra YK, et al. Tailored carbon materials (TCM) for enhancing photocatalytic degradation of polyaromatic hydrocarbons. *Prog Mater Sci*. 2024;144:101289.

69 Wang D, Choi D, Li J, Yang Z, Nie Z, Kou R, et al. Self-assembled TiO_2–graphene hybrid nanostructures for enhanced Li-ion insertion. *ACS Nano*. 2009;3(4):907–14.

70 Park Y, Kang SH, Choi W. Exfoliated and reorganized graphite oxide on titania nanoparticles as an auxiliary co-catalyst for photocatalytic solar conversion. *Phys Chem Chem Phys*. 2011;13(20):9425–31.

71 Aigbe UO, Osibote OA. Green synthesis of metal oxide nanoparticles, and their various applications. *J Hazard Mater Advances*. 2024;13:100401.

72 Kasbaji M, Ibrahim I, Mennani M, Mohamed MM, Salama TM, Moneam IA, et al. Future trends in dye removal by metal oxides and their nano/composites: a comprehensive review. *Inorg Chem Commun*. 2023;158:111546.

73 Goodarzi N, Ashrafi-Peyman Z, Khani E, Moshfegh AZ. Recent progress on semiconductor heterogeneous photocatalysts in clean energy production and environmental remediation. *Catalysts*. 2023;13(7):1102.

74 Ahmad F, Mehmood M. A critical review of photocatalytic degradation of organophosphorus pesticide "parathion" by different mixed metal oxides. *Adv J Chem A*. 2022;5(4):287–310.

75 Ahmadi E, Shokri B, Mesdaghinia A, Nabizadeh R, Khani MR, Yousefzadeh S, et al. Synergistic effects of α-Fe_2O_3-TiO_2 and $Na_2S_2O_8$ on the performance of a non-thermal plasma reactor as a novel catalytic oxidation process for dimethyl phthalate degradation. *Sep Purif Technol*. 2020;250:117185.

76 Yoo TH, Ryou H, Lee IG, Cho J, Cho BJ, Hwang WS. Comparison of Ga_2O_3 and TiO_2 nanostructures for photocatalytic degradation of volatile organic compounds. *Catalysts*. 2020;10(5):545.

77 Fan Y, Liu X, Li K, Zhang X, Xu L, Li B. Defect engineering of transition metal oxides and synergistic interaction of two-dimensional conducting metal–organic frameworks for efficient photocatalytic hydrogen evolution. *Sep Purif Technol*. 2025;354:128832.

78 Etafo NO, Bamisaye A, Bamidele MO, Renteria EV, Alli YA, Bakare OC, et al. Beyond the swipe: a review of photocatalytic antimicrobial biocompatible touchscreen technology. *Appl Mater Today*. 2025 Jun;44:102697.

79 Riaz U, Ashraf SM, Kashyap J. *Enhancement of Photocatalytic Properties of Transitional* [Internet]. Elsevier; 2015. http://doi.org/10.1016/j.materresbull.2015.06.035

Part 3

Applications of Metal Oxide Nanocatalysts in Energy Production for a Sustainable Environment

8

Metal Oxides for Carbon Dioxide Capture and Conversion

*Aash Mohammad[1], Rohit Kumar Singh[2], Niyamat Ullah Khan[1], Mohammad Asif[3], Wasim Khan[4], and Sujeet Kumar Pandey[1]**

[1] *Department of Chemical and Biochemical Engineering, Rajiv Gandhi Institute of Petroleum Technology, Jais, Uttar Pradesh, India*
[2] *Department of Mechanical Engineering, GL Bajaj Institute of Technology and Management, Greater Noida, Uttar Pradesh, India*
[3] *Centre for Risk, Integrity and Safety Engineering (C-RISE), Faculty of Engineering and Applied Science, Memorial University of Newfoundland, St. John's, NL, Canada*
[4] *Department of Chemical Engineering, College of Engineering, Taibah University, Yanbu, Kingdom of Saudi Arabia*

*Corresponding author: 20ce0012@rgipt.ac.in

8.1 Introduction

The rapid escalation of anthropogenic carbon dioxide (CO_2) emissions—primarily from the combustion of fossil fuels—has been recognized as the primary contributor to global climate change. Elevated atmospheric CO_2 levels have led to rising global temperatures, ocean acidification, and extreme weather events, posing significant risks to ecosystems, human health, and economic stability [1]. According to the Intergovernmental Panel on Climate Change (IPCC), limiting global warming to below 1.5 °C above preindustrial levels will require drastic reductions in CO_2 emissions and the deployment of negative emission technologies [2]. In this context, carbon management strategies that involve both the capture and conversion of CO_2 have garnered substantial scientific and industrial attention [3].

CO_2 capture and conversion represent a dual-purpose approach to mitigate emissions while enabling the valorization of waste carbon into valuable fuels and chemicals [4]. While traditional CO_2 capture technologies—such as amine scrubbing—are well-developed, they are energy-intensive and face limitations in scalability and sustainability [3]. On the other hand, CO_2 conversion technologies aim to utilize captured CO_2 as a carbon feedstock for producing syngas, hydrocarbons, formic acid, methanol, and other industrially relevant products [5]. The coupling of capture and conversion in an

Metal Oxide Nanocatalysts for Sustainable Energy Production, First Edition. Edited by Abayomi Bamisaye et al.
Published 2026 by Wiley-VCH GmbH.

integrated system can help close the carbon loop and move toward a circular carbon economy [3].

Among the various materials explored for CO_2 capture and conversion, metal oxides have emerged as promising candidates due to their thermal and chemical stability, tunable surface properties, and catalytic versatility [6]. Metal oxides offer diverse functionalities—from basic sites that chemisorb CO_2 to redox-active surfaces that facilitate catalytic transformations [7]. Their structural flexibility allows for morphological and compositional engineering at the nanoscale, which is critical for enhancing activity, selectivity, and recyclability [6].

This chapter presents a comprehensive review of the role of metal oxides in CO_2 capture and conversion. It begins by examining the mechanisms of physisorption, chemisorption, and carbonate formation during CO_2 capture. Subsequently, the chapter explores major conversion technologies, including thermochemical, electrochemical, and photocatalytic processes. Special emphasis is placed on nanostructured metal oxides, their synthesis and performance. The chapter also highlights current challenges, recent advancements, and future directions, aiming to inform ongoing research and the development of sustainable CO_2 utilization pathways.

8.2 Classification and Properties of Metal Oxides

Metal oxides are a chemically diverse group of inorganic compounds characterized by varying oxidation states, surface chemistries, and crystalline structures. They are broadly classified into three categories based on their acid-base behavior: basic oxides, amphoteric oxides, and acidic oxides. Among these, basic metal oxides (such as CaO, MgO, and BaO) readily react with acidic gases like CO_2 to form carbonates, making them effective for chemisorption and long-term storage. In contrast, amphoteric oxides like Al_2O_3 and ZnO exhibit dual behavior depending on the reacting species, while acidic oxides such as SiO_2 and V_2O_5 are typically inert toward CO_2 capture under mild conditions. Mixed metal oxides, often engineered by doping or compositional tuning, combine multiple functionalities and enhance performance under various capture and conversion process [6].

The surface area and porosity of metal oxides are critical to their performance in CO_2 capture and conversion. Porous metal oxides are categorized as microporous (<2 nm), mesoporous (2–50 nm), and macroporous (>50 nm). Mesoporous structures, in particular, offer an optimal balance between surface accessibility and structural stability. High surface areas (e.g., >150 m^2/g) improve adsorption kinetics by exposing more active sites for CO_2 interaction. For instance, mesoporous Fe_2O_3 prepared using a biomass-derived template achieved a surface area of ~134 m^2/g, enabling high gas diffusion and reactivity [6].

A fundamental feature that distinguishes reactive metal oxides is their ability to accommodate oxygen vacancies—missing lattice oxygen atoms that act as electron-rich sites for CO_2 activation. These defect sites enhance CO_2 adsorption through increased local basicity and can serve as reaction centers for redox processes. The presence of oxygen vacancies in materials like CeO_2, TiO_2, and ZrO_2 has been confirmed via XPS studies, which reveal shifts in binding energies associated with reduced oxidation states [7, 8]. These vacancies not only promote CO_2 binding but also facilitate electron transfer during catalytic reduction.

In terms of redox behavior, transition metal oxides (TMOs) such as MnO_x, Fe_2O_3, Co_3O_4, and CeO_2 are notable for their reversible oxidation states [9, 10]. This property allows them to function in cyclic redox reactions such as thermochemical CO_2 splitting, where the oxide alternates between oxidized and reduced states (e.g., Ce^{4+}/Ce^{3+}, Fe^{3+}/Fe^{2+}). The redox-active nature of TMOs is intimately linked to their d-orbital configurations, enabling controlled electron transfer essential for catalytic applications [11].

Thermal and chemical stability are additional properties that make metal oxides robust candidates for high-temperature CO_2 capture or conversion applications. For instance, Yttria-stabilized zirconia (YSZ) and CeO_2 retain their structural integrity at temperatures above 800 °C, allowing prolonged operation without sintering or phase degradation. These characteristics are vital in processes such as solar thermochemical CO_2 splitting and catalytic hydrogenation [6].

Finally, tunable surface properties, including Lewis acidity/basicity, polarity, and hydrophilicity, can be engineered by compositional doping, morphology control, or hybridization with other functional materials. For example, doping TiO_2 with Zr or W enhances its surface basicity and creates additional sites for CO_2 chemisorption [12–14]. These modifications not only improve adsorption capacity but also influence product selectivity during CO_2 reduction reactions (CO_2RRs).

Altogether, the classification and physicochemical properties of metal oxides provide a tunable platform for CO_2 capture and conversion. Their structural adaptability, redox functionality, and stability across diverse operating conditions enable broad applicability in integrated carbon management systems.

8.3 Mechanisms of CO_2 Capture by Metal Oxides

Metal oxides exhibit diverse interactions with CO_2, primarily governed by surface chemistry, crystallinity, and defect structures. The capture of CO_2 by metal oxides typically occurs through three key mechanisms: physisorption, chemisorption, and carbonate formation. Understanding these interactions is critical for designing materials with enhanced CO_2 uptake and stability under realistic process conditions.

Physisorption involves weak van der Waals interactions between CO_2 molecules and the surface of the oxide. It is generally favored at low temperatures and is characterized by reversible binding without significant electronic rearrangement [4]. Though physisorption offers high recyclability and rapid kinetics, its low binding energy limits capacity at elevated temperatures, making it less ideal for industrial flue gas capture.

In contrast, chemisorption entails the formation of stronger chemical bonds between CO_2 molecules and surface-active sites such as oxygen vacancies, metal cations, or hydroxyl groups. This process leads to the formation of surface-bound species like bicarbonates or monodentate/bidentate carbonates, especially on basic metal oxides such as MgO, CaO, and CeO_2 [15, 16]. These interactions are highly temperature-dependent and often exhibit irreversibility, necessitating regeneration under mild thermal cycling.

The third mechanism, carbonate formation, involves the reaction of CO_2 with surface or lattice oxygen atoms to form bulk or surface carbonates. This pathway is particularly relevant for alkaline earth metal oxides like CaO and SrO, where CO_2 reacts exothermically to form stable carbonate salts [17, 18]. These materials show high CO_2 uptake capacities but often suffer from sintering and structural degradation during repeated thermal cycling, especially above 700 °C [19].

The reactivity of metal oxides towards CO_2 is influenced by factors such as surface area, porosity, and the presence of oxygen vacancies or defect sites. Nanostructuring metal oxides significantly improves their CO_2 uptake by increasing active surface sites and enabling faster gas diffusion [20, 21]. For instance, MgO nanocubes and ceria nanoparticles with controlled crystallographic facets have demonstrated enhanced chemisorption due to their high surface energy and defect density [22, 23].

Additionally, the basicity of the oxide surface plays a decisive role in determining CO_2 affinity. Strongly basic oxides such as CaO and BaO preferentially form carbonates, while TMOs like TiO_2 or Fe_2O_3 interact via weaker binding modes [24]. Tuning the basicity through doping or surface modification has emerged as a strategy to optimize capture efficiency and selectivity.

The CO_2 capability of metal oxides depends on the interplay between physical and chemical interactions, surface structure, and thermal behavior. While physisorption allows reversible capture, chemisorption and carbonate formation are better suited for long-term storage or integrated conversion applications. Advancements in nanoscale engineering, defect tailoring, and hybrid composite formation continue to expand the versatility of metal oxides in next-generation CO_2 capture technologies [2].

8.4 CO_2 Conversion Technologies Using Metal Oxides

Metal oxides play vital roles in the catalytic conversion of CO_2 into value-added chemicals and fuels through thermochemical, electrochemical, and photocatalytic pathways. These processes not only help mitigate CO_2 emissions but also enable its integration into sustainable chemical cycles. This section discusses the mechanisms, materials, and performance metrics associated with each method, with emphasis on the functional role of metal oxides [25].

8.4.1 Thermochemical CO_2 Reduction

Thermochemical CO_2 reduction involves the redox cycling of metal oxides such as CeO_2 and Fe_2O_3 to split CO_2 at high temperatures. This process typically follows a two-step looping scheme. In the first step, the metal oxide is reduced under an inert or reducing environment. In the second step, the reduced oxide reacts with CO_2 to form CO and reoxidize the oxide. This mechanism is well-established for ceria-based systems, which exhibit excellent redox stability and oxygen exchange capacity [6].

Nanostructuring metal oxides enhances thermal stability and redox performance by facilitating oxygen vacancy generation and minimizing sintering during cycling. For example, nanocrystalline Fe_2O_3-based oxides supported on YSZ have shown high CO evolution rates due to better dispersion and increased active surface area [6, 26]. These oxides are promising for solar thermochemical CO_2 splitting using concentrated thermal energy.

8.4.2 Electrochemical CO_2 Reduction Reaction

Electrochemical CO_2 reduction is carried out in electrolyzers, where CO_2 is converted into chemical products using electricity, often powered by renewables. Metal oxides such as Cu_2O, SnO_2, and ZnO have been extensively studied for their catalytic activity and product selectivity. For example, Cu_2O-based catalysts exhibit excellent selectivity for hydrocarbons and alcohols due to their favorable binding with CO_2 reduction intermediates [27–29].

SnO_2 and ZnO catalysts favor the formation of CO and formate due to their strong affinity for the *COOH intermediate* while suppressing the hydrogen evolution reaction (HER). When electrodeposited onto carbon cloth or integrated into gas diffusion layers, these oxides achieve Faradaic efficiencies exceeding the CO production at moderate overpotentials [26]. Surface defect engineering, oxygen vacancies, and facet control have all been employed to optimize selectivity and long-term stability.

Recent studies highlight that mixed oxide systems (e.g., CuO-ZnO) supported on conductive substrates further enhance performance by synergistically combining reaction pathways and improving electron transport [30].

8.4.3 Photocatalytic CO_2 Reduction

Photocatalytic conversion uses solar photons to excite electrons in semiconducting oxides, driving CO_2 reduction on the surface. TiO_2 remains the most widely used photocatalyst due to its stability and abundance, though its large bandgap limits visible light absorption. ZnO, WO_3, and $BiVO_4$ have been investigated as alternatives or cocatalysts to enhance light harvesting and charge separation [31, 32].

Defect engineering, especially the introduction of oxygen vacancies, has been shown to enhance photocatalytic performance by extending charge carrier lifetimes and providing CO_2 activation sites. Black TiO_2 with engineered surface disorder and sub-bandgap states shows significantly improved CO_2-to-methanol conversion under visible light [33, 34]. High-entropy oxides (HEOs) with tunable band structures have also demonstrated promising performance under simulated solar irradiation [35, 36].

Composite systems that integrate TiO_2 with plasmonic nanoparticles (e.g., Ag or Au) or carbon-based materials (e.g., graphene) further improve light absorption and charge mobility, making them strong candidates for practical applications.

8.4.4 Hybrid and Tandem CO_2 Conversion Systems

Emerging technologies integrate thermochemical, electrochemical, and photocatalytic mechanisms to overcome limitations of standalone processes. For example, solar thermochemical looping can be coupled with low-temperature electrochemical conversion in tandem systems that use solar heat for oxide reduction and electricity for CO_2 activation [4].

Such hybrid systems improve overall energy efficiency, offer flexible operation modes, and align well with intermittent renewable energy sources. Challenges remain in synchronizing the kinetics and durability of each subsystem, but recent demonstrations show promise for large-scale deployment.

8.5 Nanostructuring and Morphological Effects

The design and engineering of nanostructured metal oxides have a significant impact on their ability to capture and convert CO_2. At the nanoscale, materials exhibit unique surface phenomena, increased defect densities, and enhanced electronic properties—all of which directly influence gas–solid interactions. By tailoring morphology, porosity, and particle size, researchers have successfully enhanced both adsorption kinetics and catalytic efficiency.

Particle size reduction increases surface-to-volume ratios and exposes more active sites. Nanoparticles, nanosheets, and hollow nanospheres offer improved CO_2 accessibility and shortened diffusion paths. For example, nanosized mesoporous Fe_2O_3 synthesized using waste wood as a template exhibited an enhanced surface area enabling greater interaction with gaseous species during catalysis [37, 38]. These structures also reduce sintering at elevated temperatures, a critical limitation in conventional bulk oxides.

Morphological engineering through core-shell structures, hierarchical pores, and hollow architecture facilitates efficient gas transport and adsorption. Core-shell oxides like TiO_2-ZrO_2 and ZnO-Al_2O_3 integrate the surface reactivity of the shell with the redox capacity of the core [39]. Moreover, hierarchically porous materials that combine micropores, mesopores, and macropores show improved catalytic activity due to superior molecular transport and active site accessibility. Literature reported that dual-surfactant templating creates such hierarchical pores in TMOs, dramatically improving CO_2 uptake rates and reducing diffusion resistance [40].

The presence of oxygen vacancies is strongly influenced by nanostructure. For instance, doping TiO_2 with Zr^{4+} was shown to enhance surface oxygen vacancies while maintaining structural integrity during calcination at above 400 °C. This modification increased the anatase phase, resulting in better adsorption of CO_2 and electron transfer efficiency [41, 42]. These oxygen-deficient sites act as nucleophiles, promoting the binding and activation of CO_2 molecules.

Synthesis techniques play a pivotal role in determining morphology and defect profiles. Soft-templating and sol-gel routes are widely used to produce mesoporous oxides with controlled pore sizes and phase purity. A one-pot method involving hydrogen peroxide-based peroxo complexes to produce mixed-valent oxides of Ti, Mn, Cu, Fe, and Zn with uniform pore distribution and tailored porosity [43, 44]. Mechanical nanocasting, an alternative to sol-gel synthesis, offers better scalability and reaction time control while minimizing precursor waste and contamination [45, 46].

The catalytic properties of nanostructured oxides are also influenced by crystalline facets and polymorphic phase composition. High-energy facets expose more coordinatively unsaturated sites, which are favorable for CO_2 adsorption and reduction. In the case of hexagonal WO_3, selective exposure of (002) facets enhanced charge transfer and oxygen vacancy generation, ultimately improving gas-phase photocatalytic reactions [47].

The nanostructuring of metal oxides offers a multifaceted strategy to enhance CO_2 capture and conversion. Morphology, defect engineering, and synthetic control converge to deliver materials with high surface areas, tailored porosity, and superior electronic and catalytic behavior. Continued advancements in nanoscale design will enable more efficient and scalable carbon management technologies.

8.6 Challenges in Metal Oxide-Based CO_2 Capture and Conversion Systems

Despite their advantages, metal oxide-based systems for CO_2 capture and conversion face several technical, thermodynamic, and scalability challenges that hinder their widespread implementation in industrial carbon management strategies.

One of the primary challenges is energy efficiency. Many thermochemical CO_2 splitting cycles operate at temperatures exceeding 800 °C, demanding high thermal inputs and complex heat management infrastructure. Although redox-active oxides like CeO_2 and Fe_2O_3 are stable at such conditions, the energy cost per mole of CO produced remains relatively high [48]. In electrochemical systems, overpotentials required to initiate CO_2 reduction often lead to significant parasitic energy losses and low system-wide energy efficiencies. Faradaic efficiencies for many oxide-based electrocatalysts are often below 60% under industrially relevant conditions [49, 50].

Material degradation is another limiting factor, particularly under cyclic thermal or redox conditions. Sintering and phase transitions during repeated heating and cooling cycles can reduce surface area, collapse porosity, and deactivate catalytic sites. For example, α-MoO_3 used in battery and catalytic systems undergoes irreversible phase change under cycling, severely limiting long-term performance [51]. In photocatalytic systems, prolonged UV exposure and thermal cycling can degrade the nanostructure and reduce activity. Compositional modifications such as doping or forming core-shell structures have been proposed to enhance structural robustness, though often at the cost of added complexity and reduced scalability.

A persistent issue in CO_2 reduction systems is product selectivity. Most metal oxide catalysts produce mixtures of products, such as CO, CH_4, methanol, or formate, depending on potential, pH, and surface composition. Achieving precise control over selectivity is challenging, especially when competing reactions such as the HER occur simultaneously [52]. Moreover, oxide catalysts often exhibit sluggish reaction kinetics, requiring higher overpotentials that further reduce energy efficiency and selectivity.

Scalability and synthesis reproducibility are also major concerns. While sol-gel, hydrothermal, and nanocasting techniques yield highly porous and active materials at the laboratory scale, these methods often lack uniformity, suffer from poor batch-to-batch consistency, and are difficult to translate to industrial-scale production. Precise control of particle size, dopant dispersion, and pore structure is difficult to achieve outside of controlled lab environments [53, 54]. High-temperature continuous-flow reactors and templating techniques are being investigated for scalable manufacturing, but challenges in thermal management and contamination persist.

The cost and lifetime of oxide materials also impact deployment. TMOs such as MnO_x and Fe_2O_3 are low-cost and earth-abundant but often suffer from fast deactivation. More efficient systems based on ZnO, Cu_2O, or CeO_2 involve rare elements or expensive synthetic routes. Strategies such as hybridization with conductive supports (e.g., graphene or CNTs) or cocatalyst deposition (e.g., Pt or Ru) improve activity but introduce economic and material recovery concerns [55, 56].

Ultimately, overcoming these challenges requires the convergence of material design, reactor engineering, and process integration. Solutions include multifunctional composite oxides, tandem processes that couple capture with conversion, and advanced characterization to understand degradation pathways. Innovations that reduce thermal budgets, improve selectivity, and enable long-term cycling without material fatigue will be essential for real-world deployment.

8.7 Recent Advancements and Future Directions

Recent advancements in metal oxide-based systems for CO_2 capture and conversion have focused on enhancing efficiency, tuning surface chemistry, and broadening product selectivity. Novel synthesis strategies, advanced in-situ characterization, and machine learning (ML) are now central to improving catalyst performance and developing next-generation materials.

One major development is the rise of HEOs—materials composed of five or more metal cations that provide a multicomponent reaction environment. Their configurational entropy enables structural stability and defect formation, critical for CO_2 adsorption and activation. These materials demonstrate enhanced oxygen vacancy formation and robust thermal performance in redox cycles. For instance, high-entropy Ta-based oxides have shown promise for thermal barrier applications and tunable surface chemistry relevant to CO_2 catalysis [57].

Tandem and multifunctional catalysts are another emerging class. These systems integrate CO_2 capture and conversion in a single material platform. Examples include CeO_2-CoO-Co heterointerfaces, which show dual functionality in water-gas shift and CO_2 hydrogenation reactions by synergistically modulating oxidation states and minimizing catalyst poisoning [58, 59]. Similarly, Ni-NiO-Y_2O_3 composites have shown high stability and activity due to interfacial promotion of hydrogen evolution and intermediate stabilization.

Advances in photocatalytic heterojunction engineering have enabled efficient visible-light utilization. Z-scheme architectures like TiO_2/CdS thin films facilitate directional charge transfer and reduce recombination losses, thereby improving CO_2-to-CH_4 selectivity. The direct evidence of charge migration using in-situ XPS and FTIR confirms the mechanistic advantage of such architectures [6, 26].

In-situ and operando spectroscopy now play a pivotal role in capturing real-time dynamics at the catalyst surface. Soft X-ray absorption, XPS under illumination, and transient photoluminescence have been used to monitor oxidation states, intermediate formation, and defect behavior. For example, in-situ XPS tracking of Ti^{2+}-Ti^{4+} transitions in doped TiO_2 confirmed the active role of oxygen vacancies in CO_2 binding and reduction under visible light [60, 61].

The use of ML and density functional theory (DFT) is accelerating catalyst discovery. ML models trained on DFT datasets predict CO_2 adsorption energies and selectivity trends across large metal oxide libraries. These tools help identify optimal compositions and surface terminations before experimental synthesis, thereby reducing trial-and-error efforts [62, 63].

Future directions include the development of scalable green synthesis methods like soft templating and mechanochemical routes, integration of solar-driven tandem systems, and the deployment of AI-assisted screening to design multifunctional oxide catalysts with atomic precision. Equally important is the need for lifetime assessments, technoeconomic analysis, and modular reactor designs that support real-world implementation under variable energy inputs.

Continued integration of experimental advances with predictive modeling and scalable synthesis will be crucial for realizing practical and energy-efficient CO_2 capture and conversion technologies.

8.8 Conclusions

Metal oxides have emerged as versatile and effective materials for the dual roles of CO_2 capture and conversion. Their inherent properties, such as thermal and chemical stability, tunable surface functionalities, redox behavior, and structural diversity, enable them to operate under a wide range of conditions relevant to industrial carbon management. From basic oxides like CaO and MgO with high

CO_2 chemisorption capacity to TMOs such as TiO_2, Fe_2O_3, and CeO_2 that support photocatalytic, thermochemical, and electrochemical transformations, these materials are central to achieving a circular carbon economy.

Mechanistic pathways of CO_2 capture via physisorption, chemisorption, and carbonate formation are well established. Defect engineering, particularly the creation of oxygen vacancies, significantly enhances CO_2 activation by modifying electronic environments and increasing surface reactivity. Simultaneously, advancements in photocatalytic heterojunctions, tandem catalysts, and HEOs have improved product selectivity and system robustness across multiple CO_2 conversion strategies.

Despite these advances, the scalability of metal oxide-based systems is constrained by factors such as high thermal energy demands, catalyst deactivation, limited Faradaic efficiency, and synthesis reproducibility. Addressing these concerns requires continued optimization of morphology, synthesis routes, and compositional engineering. Hybrid and multifunctional catalysts, including dual-function oxide-metal systems, have shown promise for simultaneous CO_2 capture and conversion with minimized energy losses.

Looking forward, future research will benefit from deeper integration of operando spectroscopy and in-situ characterization techniques to monitor real-time surface transformations and reaction intermediates. These tools are essential for understanding deactivation pathways and guiding rational catalyst design. Furthermore, the use of ML and computational modeling such as DFT will accelerate the discovery of novel oxide systems with optimized band structures, vacancy concentrations, and reaction energetics.

To transition from lab-scale innovation to industrial implementation, efforts must also focus on process integration, modular reactor design, and techno-economic evaluations. By coupling metal oxide-based capture with catalytic conversion in tandem systems powered by renewable energy, the CO_2 lifecycle can be effectively closed. Such integrated approaches represent a feasible path toward large-scale decarbonization of chemical and energy sectors.

References

1 Jackson RB, Friedlingstein P, Quéré CL, Abernethy S, Andrew RM, Canadell JG, et al. Global fossil carbon emissions rebound near pre-COVID-19 levels. *Environ Res Lett*. 2022;17:031001.

2 Kazlou T, Cherp A, Jewell J. Feasible deployment of carbon capture and storage and the requirements of climate targets. *Nat Clim Chang*. 2024;14(10):1047–55.

3 Omodolor IS, Otor HO, Andonegui JA, Allen BJ, Alba-Rubio AC. Dual-function materials for CO_2 capture and conversion: a review. *Ind Eng Chem Res*. 2020;59(40):17612–31.

4 Zanatta M. Materials for direct air capture and integrated CO_2 conversion: advancement, challenges, and prospects. *ACS Mater Au*. 2023;3(6):576–83.

5 Rony ZI, Mofijur M, Ahmed SF, Kabir Z, Chowdhury AA, Almomani F. Recent advances in the solar thermochemical splitting of carbon dioxide into synthetic fuels. *Front Energy Res.* 2022;10:982269.

6 Sahoo S, Wickramathilaka KY, Njeri E, Silva D, Suib SL. A review on transition metal oxides in catalysis. *Front Chem.* 2024;12:1374878.

7 Reddy BM, Khan A. Nanosized CeO_2–SiO_2, CeO_2–TiO_2, and CeO_2–ZrO_2 mixed oxides: influence of supporting oxide on thermal stability and oxygen storage properties of ceria. *Catal Surv Asia.* 2005;9(3):155–71.

8 Kim M, Park G, Jo K, Lee H. Enhanced redox properties of Gd-doped CeO_2–TiO_2 induced by oxygen vacancies and disordered structure. *Mater Today Chem.* 2023;29:101440.

9 Yuan X, Wu X, Wu Y, van Ree T. *Metal Oxides for Emission Control.* Elsevier eBooks; 2018 Jan 1. p. 391–414.

10 Vidal H, Kašpar J, Pijolat M, Colon G, Bernal S, Cordón A, et al. Redox behavior of CeO_2–ZrO_2 mixed oxides. *Appl Catal B-Environ.* 2000 Jun 1;27(1):49–63.

11 Kaeffer N, Leitner W. Electrocatalysis with molecular transition-metal complexes for reductive organic synthesis. *JACS Au.* 2022 May 31;2(6):1266–89.

12 Li Y, Walsh AG, Li D, Do D, Ma H, Wang C, et al. W-Doped TiO_2 for photothermocatalytic CO_2 reduction. *Nanoscale.* 2020 Jan 1;12(33):17245–52.

13 Ruibin LV, Zhang M, Huang H, Yuan S, Liu L, Li K, et al. Self-doped TiO_2 nanotubes with surface modification by ionic liquids for enhanced photoreduction of CO_2 to acetic acid. *Appl Surf Sci.* 2023 Mar 2;621:156897.

14 Bolbol AM, Abd-Elkader OH, Elshimy H, Zaki ZI, Shata SA, Kamel M, et al. The effect of Zr (IV) doping on TiO_2 thin film structure and optical characteristics. *Results Phys.* 2022 Sep 5;42:105955.

15 Chen H, Nanayakkara CE, Grassian VH. Titanium dioxide photocatalysis in atmospheric chemistry. *Chem Rev.* 2012;112(11):5919–48.

16 Burghaus U. Surface chemistry of CO_2—adsorption of carbon dioxide on clean surfaces at ultrahigh vacuum. *Prog Surf Sci.* 2014 May;89(2):161–217.

17 Dunstan MT, Donat F, Bork AH, Grey CP, Müller CR. CO_2 capture at medium to high temperature using solid oxide-based sorbents: fundamental aspects, mechanistic insights, and recent advances. *Chem Rev.* 2021;121(20):12681–745.

18 Suksumrit K, Kleiber S, Lux S. The role of carbonate formation during CO_2 hydrogenation over MgO-supported catalysts: a review on methane and methanol synthesis. *Energies.* 2023 Jan 1;16(7):2973.

19 Krödel M, Leroy C, Kim SM, Naeem MA, Kierzkowska A, Wu YH, et al. Of glasses and crystals: mitigating the deactivation of CaO-Based CO_2 sorbents through calcium aluminosilicates. *JACS Au.* 2023 Nov 10;3(11):3111–26.

20 Chavhan MP, Kryeziu A, Ganguly S, Parmentier J. Monolithic metal-based/porous carbon nanocomposites made from dissolved cellulose for use in electrochemical capacitor. *Green Carbon.* 2024 Jan 8;2(1):109–17.

21 Du Y, Meng Q, Wang J, Yan J, Fan H, Liu Y, et al. Three-dimensional mesoporous manganese oxides and cobalt oxides: high-efficiency catalysts for the removal of toluene and carbon monoxide. *Micropor Mesopor Mat.* 2012 Nov;162:199–206.

22 Sharma N, Kakkar R. Adsorption of sarin on MgO nanotubes: role of doped and defect sites. *J Comput Sci.* 2015 Sep;10:225–36.
23 Chakraborty S, Pal A. Microporous assembly of MnO_2 nanosheets for malachite green degradation. *Sep Purif Technol.* 2014 Sep 25;134:26–36.
24 Yu X, Shi M, Fan Y, Yang L, Zhang J, Liu W, et al. Activation or passivation: influence of halogen dopant (F, Cl, Br) on photothermal activity of Mn_2O_3 in degrading toluene. *Appl Catal B Environ Energy.* 2022;309:121236.
25 Diao K, Jiang Q, Yang M, Lu J, Zhu Y. Enhancing the dynamic thermal stability of metal oxide oil-based nanofluids through synchronous chemical modification and physical coating. *Surf Interfaces.* 2024;46:104119.
26 Li C, Guo R, Zhang Z, Wu T, Pan W. Converting CO_2 into value-added products by Cu_2O-based catalysts: from photocatalysis, electrocatalysis to photoelectro catalysis. *Small.* 2023;19(19):e2207875.
27 Wang J, Lou Y, Xu C, Song S, Liu W. Magnetic lanthanide oxide catalysts: an application and comparison in the heterogeneous catalytic ozonation of diethyl phthalate in aqueous solution. *Sep Purif Technol.* 2016 Feb 1;159:57–67.
28 Aizaz U, Ul Hassan I, Onaizi SA. MXenes-based materials for CO_2 capture and conversion: a comprehensive review. *Renew Sustain Energy Rev.* 2025 May;214:115524.
29 Krishnan U, Babu C, Jivanlal SA. Effective tri-metallic TiO_2 supported catalyst for hydrocarbon production from carbon dioxide. *J Indian Chem Soc.* 2022 Dec 19;100(1):100859.
30 Nguyen TN, Dinh CT. Gas diffusion electrode design for electrochemical carbon dioxide reduction. *Chem Soc Rev.* 2020;49(21):7488–504.
31 Packiaraj R, Mahendraprabhu K, Devendran P, Nallamuthu N, Palanivel B, Venkatesh KS, et al. Electrochemical performances of ZNO–NIO–CUO mixed metal oxides as smart electrode material for Solid-State asymmetric device fabrication. *Energy Fuels.* 2021;36(1):603–17.
32 Alam S, Yamashita H, Verma P. Unveiling the critical role of High/Low-Index facets in nanostructured energy materials for enhancing the photoelectrochemical water splitting. *ChemCatChem.* 2024 Dec 25; 17(5):e202401672.
33 Madima N, Chauke NM, Ngqoloda S, Mmelesi OK, Raphulu M. Recent advances in the development of defective black TiO_2 nanomaterials for application in energy and environmental sustainability: a review. *Results Eng.* 2025 Apr 16;26:104868.
34 Tripathy SR, Baral SS. Defect engineering in semiconductor photocatalysts: enhancing photocatalytic activity for green energy production. *Adv Energy Sustainability Res.* 2025 May 19; 2500110.
35 Guo HX, Wang WM, He CY, Liu BH, Yu DM, Liu G, et al. Entropy-assisted high-entropy oxide with a spinel structure toward high-temperature infrared radiation materials. *ACS Appl Mater Interfaces.* 2021 Dec 27;14(1):1950–60.
36 Edalati P, Itagoe Y, Ishihara H, Ishihara T, Emami H, Arita M, et al. Visible-light photocatalytic oxygen production on a high-entropy oxide by multiple-heterojunction introduction. *J Photochem Photobiol A Chem.* 2022 Dec 1;433:114167.

37 Liang H, Liu K, Ni Y. Synthesis of mesoporous α-Fe_2O_3 via sol–gel methods using cellulose nano-crystals (CNC) as template and its photo-catalytic properties. *Mater Lett.* 2015 Jul 3;159:218–20.

38 Bao C, Serrano-Lotina A, Niu M, Portela R, Li Y, Lim KH, et al. Microwave-associated chemistry in environmental catalysis for air pollution remediation: a review. *Chem Eng J* [Internet]. 2023 Jun 15;466:142902.

39 Swadener JG. Strain engineering of ZrO_2@TiO_2 Core@shell nanoparticle photocatalysts. *Solar*. 2023 Jan 4;3(1):15–24.

40 Yang XY, Chen LH, Li Y, Rooke JC, Sanchez C, Su BL. Hierarchically porous materials: synthesis strategies and structure design. *Chem Soc Rev.* 2017;46(2):481–558.

41 Li Y, Fang Z, Feng L, Liu F, Shi Y, Li J, et al. Study of mesoporous Zr-TiO_2 catalyst with rich oxygen vacancies for N-methylmorpholine oxidation to N-methylmorpholine-N-oxide. *Molecules.* 2024 Aug 11;29(16):3812.

42 Khlyustova A, Sirotkin N, Kusova T, Kraev A, Titov V, Agafonov A. Doped TiO_2: the effect of doping elements on photocatalytic activity. *Mater Adv.* 2020 Aug 17;1(5):1193–201.

43 Oaki Y, Imai H. One-pot synthesis of manganese oxide nanosheets in aqueous solution: chelation-mediated parallel control of reaction and morphology. *Angew Chem.* 2007 Jun 15;46(26):4951–5.

44 Lu H, Wright DS, Pike SD. The use of mixed-metal single source precursors for the synthesis of complex metal oxides. *Chem Commun.* 2020;56(6):854–71.

45 Khan NR, Sharmin T, Bin Rashid A. Exploring the versatility of aerogels: broad applications in biomedical engineering, astronautics, energy storage, biosensing, and current progress. *Heliyon* [Internet]. 2024 Jan 15; 10(1):e23102.

46 Xiao W, Yang S, Zhang P, Li P, Wu P, Li M, et al. Facile synthesis of highly porous metal oxides by mechanochemical nanocasting. *Chem Mater.* 2018 Apr 13;30(9):2924–9.

47 Xie Y, Liu G, Yin L, Cheng HM. Crystal facet-dependent photocatalytic oxidation and reduction reactivity of monoclinic WO_3 for solar energy conversion. *J Mater Chem.* 2012 Jan 1;22(14):6746–51.

48 Marxer D, Furler P, Takacs M, Steinfeld A. Solar thermochemical splitting of CO_2 into separate streams of CO and O_2 with high selectivity, stability, conversion, and efficiency. *Energy Environ Sci.* 2017;10(5):1142–9.

49 Li F, Thevenon A, Rosas-Hernández A, Wang Z, Li Y, Gabardo CM, et al. Molecular tuning of CO_2-to-ethylene conversion. *Nature.* 2019 Nov 20;577(7791):509–13.

50 Zhao K, Liu Y, Quan X, Chen S, Yu H. CO_2 electroreduction at low overpotential on oxide-derived cu/carbons fabricated from metal organic framework. *ACS Appl Mater Interfaces.* 2017 Feb 2;9(6):5302–11.

51 Li Y, Sun H, Cheng X, Zhang Y, Zhao K. In-situ TEM experiments and first-principles studies on the electrochemical and mechanical behaviors of α-MoO_3 in Li-ion batteries. *Nano Energy.* 2016 Sep;27:95–102.

52 Ait Ahsaine H, BaQais A. Metal and metal oxide electrocatalysts for the electrochemical reduction of CO_2 -to-C1 chemicals: are we there yet? *Green Chem Lett Rev.* 2023 Jan 2;16(1):2160215.

53 Yarbrough R, Davis K, Dawood S, Rathnayake H. A sol–gel synthesis to prepare size and shape-controlled mesoporous nanostructures of binary (II–VI) metal oxides. *RSC Advances*. 2020;10(24):14134–46.

54 Mushtaq M, Sathyakam PU. Optimizing hematite photoanodes: a critical review of hydrothermal, sol–gel, ALD, and electrodeposition techniques. *Clean Technol Environ Policy*. 2024 Dec 6;27(2):887–918.

55 Nguyen D, Lee CW, Na J, Kim MC, Tu NDK, Lee SY, et al. Mass transport control by surface graphene oxide for selective CO production from electrochemical CO_2 reduction. *ACS Catal*. 2020 Feb 6;10(5):3222–31.

56 Bramhaiah K, Bhattacharyya S. Challenges and future prospects of graphene-based hybrids for solar fuel generation: moving towards next generation photocatalysts. *Mater Adv*. 2022;3:142–72.

57 Sang W, Zhang H, Zhang H, Liu X, Chen X, Xie W, et al. Novel $(Sm_{0.2}Lu_{0.2}Yb_{0.2}Y_{0.2}Dy_{0.2})_3TaO_7$ high-entropy ceramic for thermal barrier coatings. *Ceram Int*. 2022;49(6):9052–9.

58 Yu Y, Xia W, Yu A, Simakov DSA, Ricardez-Sandoval L. Transition-metal-doped CeO_2 for the reverse water-gas shift reaction: an experimental and theoretical study on CO_2 adsorption and surface vacancy effects. *ChemSusChem*. 2025 Jan 14; 18(2):e202400681.

59 Zhang Q, Bown M, Pastor-Pérez L, Duyar MS, Reina TR. CO_2 conversion via reverse water gas shift reaction using fully selective Mo–P multicomponent catalysts. *Ind Eng Chem Res*. 2022 Aug 19;61(34):12857–65.

60 Jing L, Xin B, Yuan F, Xue L, Wang B, Fu H. Effects of surface oxygen vacancies on photophysical and photochemical processes of Zn-doped TiO_2 nanoparticles and their relationships. *J Phys Chem B*. 2006 Aug 23;110(36):17860–5.

61 Zhou Y, Zhang Z, Fang Z, Qiu M, Ling L, Long J, et al. Defect engineering of metal–oxide interface for proximity of photooxidation and photoreduction. *Proc Natl Acad Sci USA*. 2019 May 7;116(21):10232–7.

62 Iwama R, Takizawa K, Shinmei K, Baba E, Yagihashi N, Kaneko H. Design and analysis of metal oxides for CO_2 reduction using machine learning, transfer learning, and Bayesian optimization. *ACS Omega*. 2022 Mar 17;7(12):10709–17.

63 Yohannes AG, Lee C, Talebi P, Mok DH, Karamad M, Back S, et al. Combined high-throughput DFT and ML screening of transition metal nitrides for electrochemical CO_2 reduction. *ACS Catal*. 2023 Jun 22;13(13):9007–17.

9

Green Hydrogen from Water Splitting Using Metal Oxide Nanocatalysts

Sujeet Kumar Pandey[1], Rohit Kumar Singh[2], Aash Mohammad[1], Mohammad Asif[3], and Wasim Khan[4]*

[1]*Department of Chemical and Biochemical Engineering, Rajiv Gandhi Institute of Petroleum Technology, Jais, Uttar Pradesh, India*
[2]*Department of Mechanical Engineering, GL Bajaj Institute of Technology and Management, Greater Noida, Uttar Pradesh, India*
[3]*Centre for Risk, Integrity and Safety Engineering (C-RISE), Faculty of Engineering and Applied Science, Memorial University of Newfoundland, St. John's, NL, Canada*
[4]*Department of Chemical Engineering, College of Engineering, Taibah University, Yanbu, Kingdom of Saudi Arabia*

*Corresponding author: 20ce0012@rgipt.ac.in

9.1 Introduction

The escalating concerns over climate change, energy security, and environmental sustainability, the demand for clean energy alternatives has surged globally [1]. Among available alternative energy sources, hydrogen fuel, particularly green H_2 has emerged as a promising solution for sustainable energy systems. Green H_2, produced via the electrolysis of water using renewable energy sources, represents an environmental benefits, energy security, economic opportunities and technological advancement (Figure 9.1) pathway to decarbonize sectors such as transportation, heavy industry, and power generation [2].

Hydrogen's appeal lies in its high energy content per unit mass (about 120 MJ/kg), its clean combustion product (water), and its potential role in balancing intermittent renewable energy supplies. However, the primary challenge remains the development of cost-effective and efficient H_2 production methods [3]. Traditional methods like steam methane reforming (SMR) generate significant carbon emissions, while water electrolysis powered by renewables offers a clean alternative [4].

Green H_2 refers specifically to hydrogen gas generated through the electrolysis of water using electricity sourced from renewable energy (solar, wind, and hydro) [5]. This process contrasts with grey hydrogen (from fossil fuels) and blue hydrogen (from fossil fuels with carbon capture) [6]. The clean nature of green hydrogen

Metal Oxide Nanocatalysts for Sustainable Energy Production, First Edition. Edited by Abayomi Bamisaye et al.
Published 2026 by Wiley-VCH GmbH.

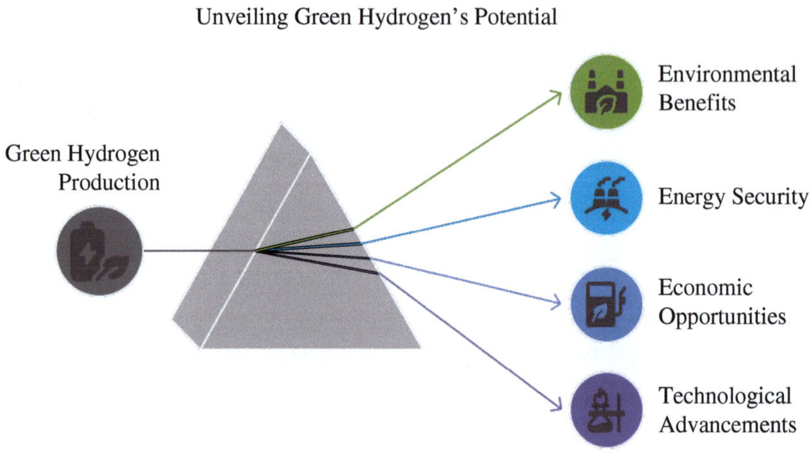

Figure 9.1 Potential of green hydrogen.

makes it an attractive candidate for achieving net-zero carbon goals [7]. Water splitting involves two half-reactions [8]:

$$\text{HER: } 2H^+ + 2e^- \rightarrow H_2$$

$$\text{OER: } 2H_2O \rightarrow O_2 + 4H^+ + 4e^-$$

Although theoretically simple, these reactions are kinetically sluggish, especially the oxygen evolution reaction (OER), necessitating the use of catalysts to reduce overpotentials and enhance reaction rates [9]. Conventional catalysts like platinum and iridium offer high efficiency but are expensive and scarce [10]. In contrast, metal oxide nanocatalysts present a viable alternative due to their earth abundance, low cost, and tunable catalytic properties [11].

There are three main approaches to water splitting:

Photocatalytic Water Splitting: Utilizes semiconducting materials that absorb solar photons to generate electron-hole pairs, which drive the hydrogen evolution reaction (HER) and OER [12, 13].

Electrocatalytic Water Splitting: Employs external electrical energy to split water with the aid of electrocatalysts [14].

Photoelectrochemical (PEC) Water Splitting: Combines features of photocatalysis and electrocatalysis using light-absorbing electrodes [15].

Among these, electrocatalytic and photocatalytic methods are gaining traction due to their scalability and compatibility with renewable energy systems [16]. Metal oxide nanostructures such as TiO_2, Fe_2O_3, NiO, and Co_3O_4 have demonstrated remarkable performance in both HER and OER due to their large surface areas, variable oxidation states, and ability to stabilize reaction intermediates [17].

This chapter delves into the current landscape of green H_2 production through water splitting, with a particular emphasis on the role of metal oxide nanocatalysts. It explores synthesis strategies, reaction mechanisms, recent advancements, and future directions, aiming to provide a comprehensive understanding of the field.

This foundational introduction serves to contextualize the role of advanced nanomaterials in overcoming the practical barriers to scalable and sustainable hydrogen energy systems.

9.2 Overview of Water Splitting Processes and Role of Metal Oxide Nanostructures in Enhancing Efficiency

Water splitting is a pivotal process in green hydrogen production that involves breaking down H_2O into H_2 and O_2 using energy inputs [18, 19]. This reaction can be driven through various methods, primarily photocatalysis and electrocatalysis processes (Figure 9.2) [20]. Each technique presents unique advantages and challenges but shares the common need for efficient catalysts to accelerate the sluggish HER and OER [21].

9.2.1 Photocatalytic Water Splitting

Photocatalysis is a solar-driven water-splitting method where semiconductor materials absorb photons to generate electron-hole pairs [22, 23]. These charge carriers then migrate to the surface and participate in redox reactions electrons reduce protons to hydrogen (HER), and holes oxidize water to oxygen (OER) [24]. The efficiency of this process hinges on several factors: the light absorption capacity of the photocatalyst, the separation efficiency of charge carriers, and the surface catalytic activity [25].

Metal oxide semiconductors such as TiO_2, ZnO, and Fe_2O_3 are widely used photocatalysts due to their chemical stability, cost-effectiveness, and ability to absorb UV light [26]. However, their relatively wide band gaps limit their absorption to UV spectra, which constitutes only a small portion of solar energy [27]. To address this, researchers have engineered metal oxide nanostructures doped with nonmetals or transition metals, creating heterostructures, and designing core-shell morphologies to extend light absorption into the visible range and improve charge separation [28].

For instance, nitrogen-doped TiO_2 nanostructures have shown enhanced photocatalytic activity by narrowing the band gap [29]. Similarly, heterostructures like

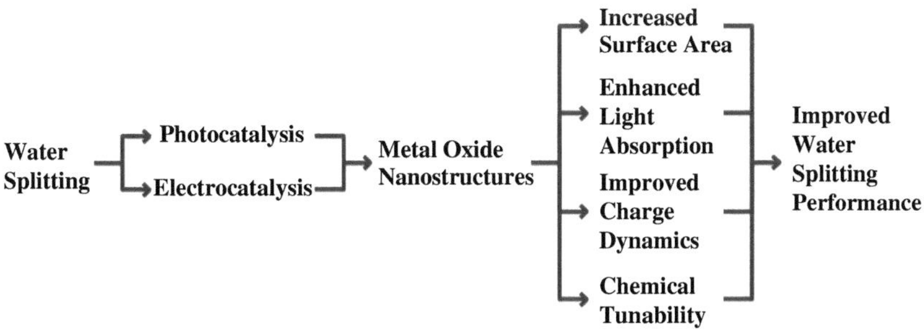

Figure 9.2 Different methods of water splitting and factors affecting water splitting.

TiO_2/ZnO and $BiVO_4$/Fe_2O_3 utilize the complementary electronic properties of each material to boost charge transfer and suppress recombination [30].

9.2.2 Electrocatalytic Water Splitting

Electrocatalytic water splitting involves applying an external electric potential to drive HER and OER in an electrolyzer [31, 32]. This method is modular, scalable, and can be powered directly by renewable electricity [33]. The challenge lies in reducing the overpotentials of both half-reactions to minimize energy input and maximize efficiency [34].

In this context, metal oxide nanostructures are particularly attractive due to their robustness and tunable electronic properties [35]. Transition metal oxides (TMOs) like NiO, Co_3O_4, and MnO_2 have been extensively explored as OER electrocatalysts [36]. These materials can exhibit multiple oxidation states, enabling efficient adsorption and transformation of oxygen-containing intermediates [37].

Nickel-based nanostructures doped with iron or cobalt have demonstrated significantly reduced overpotentials and increased stability in alkaline media [38]. Furthermore, nanostructuring increases the electrochemically active surface area, offering more catalytic sites and enhancing mass transport.

On the HER side, although metal oxides are generally less active than noble metals like Pt, they can be combined with conductive substrates (e.g., carbon nanotubes or graphene) to improve electrical conductivity and HER performance [39]. Some hybrid metal oxide systems, such as MoO_3/MoS_2 and $NiFe_2O_4$/rGO, offer synergistic effects that enhance HER kinetics [40].

9.2.3 Enhancing Efficiency Through Nanostructuring

Metal oxide nanostructures contribute to improved water-splitting performance through:

Increased Surface Area: Nanostructures provide a higher density of active sites for reaction [41–43].

Enhanced Light Absorption: Nanoscale engineering allows for bandgap tuning to absorb more solar spectrum [44].

Improved Charge Dynamics: Reduced dimensions shorten charge diffusion paths and minimize recombination [45].

Chemical Tunability: Doping and surface modifications adjust electronic structure and catalytic activity [46].

Both photocatalytic and electrocatalytic water-splitting processes benefit greatly from the strategic design of metal oxide nanostructures. Through band engineering, heterostructure formation, and nano-architecture control, researchers have significantly enhanced the efficiency and viability of green H_2 production. The continued development of advanced nanomaterials will be key to overcoming the economic and technical barriers to large-scale deployment [47].

9.3 Role of Metal Oxide Nanostructures in OER and HER

Water splitting comprises two critical half-reactions: the HER at the cathode and the OER at the anode [53]. While HER involves the reduction of protons to hydrogen gas, OER pertains to the oxidation of water molecules to oxygen gas. Both reactions are intrinsically slow due to high activation barriers and require efficient catalysts to proceed at appreciable rates under practical conditions. Comparison of HER and OER important points in Figure 9.3. Metal oxide nanostructures have emerged as leading candidates in this field due to their unique electronic structures, abundant active sites, and tunable surface properties (Table 9.1) [54–56].

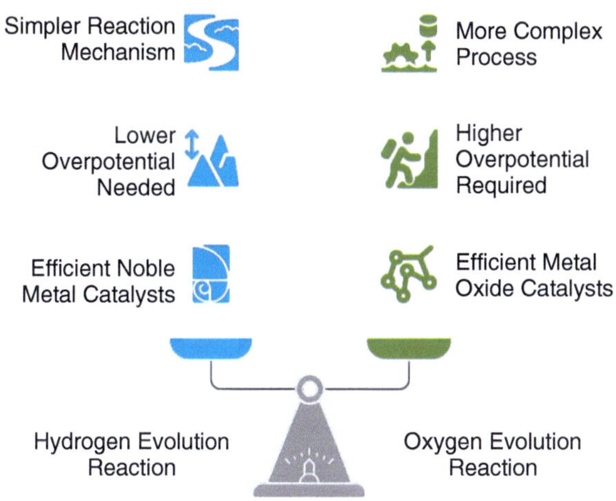

Figure 9.3 Comparing HER and OER in water splitting.

Table 9.1 Comparison of nanocatalysts in terms of band gap, HER, and OER.

Nanocatalyst	Bandgap (eV)	HER activity	OER activity	Notes	References
TiO_2	3.0–3.2	Moderate	Low	UV active, chemically stable	Khan et al. [48]
ZnO	3.2	Moderate	Low	Structurally similar to TiO_2	Zahra et al. [49]
Fe_2O_3	2.1	Low	Moderate	Absorb visible light	Ahmad et al. [50]
Co_3O_4	1.6–2.1	Good	Good	High conductivity and activity	Gorylewski et al. [51]
NiO	3.6	High	Good	Abundant and cost-effective	Gorylewski et al. [52]

9.3.1 Fundamentals of HER and OER

The HER typically follows two possible mechanisms [57–61]:

Volmer–Heyrovsky mechanism: $H^+ + e^- \rightarrow H_{ads}$,
followed by $H_{ads} + H^+ + e^- \rightarrow H_2$.
Volmer–Tafel mechanism: two H_{ads} species recombine to form H_2 (Tafel step).

The OER process is more complex, typically requiring the transfer of four electrons [62–64]:

$2H_2O \rightarrow O_2 + 4H^+ + 4e^-$ (in acidic conditions), or
$4OH^- \rightarrow O_2 + 2H_2O + 4e^-$ (in alkaline conditions).

Due to the multi-electron nature and sluggish kinetics of OER, the overpotential required for oxygen generation is typically higher than that for HER, thus making OER the rate-determining step in water electrolysis [65].

9.3.2 Metal Oxide Nanostructures for OER

TMOs, especially those containing nickel, cobalt, iron, and manganese, are widely recognized as efficient OER catalysts in alkaline media. These metal oxides can exhibit multiple oxidation states, which facilitate the redox reactions required for water oxidation. For example:

$NiFe_2O_4$ and $CoFe_2O_4$: The incorporation of iron into nickel or cobalt oxide matrices improves electrical conductivity and modifies the electronic structure to promote adsorption of OER intermediates (e.g., OH^-, OOH^-) [66, 67].
MnO_2 and Co_3O_4: These materials are known for their abundance and stability in alkaline electrolytes, with nanostructured forms (e.g., nanorods, nanosheets) providing high surface areas and enhanced catalytic activity [68, 69].

Metal oxide nanostructures such as nanowires, nanotubes, and mesoporous frameworks enhance mass transport and electron transfer, further boosting OER performance. Additionally, oxygen vacancies and defect engineering can be used to increase the number of active sites and alter the binding energy of oxygen intermediates.

Recent research by S. Marimuthu et al. demonstrated that nanoclusters of transition metals (Fe, Co, Ni, Cu) supported on nickel oxide nanosheets exhibited a significantly reduced OER overpotential (nearly 380 mV), highlighting the synergistic effects between nanostructuring and compositional tuning [70].

9.3.3 Metal Oxide Nanostructures for HER

While noble metals such as platinum are the most effective HER catalysts, their high cost and scarcity necessitate alternatives. Metal oxide nanostructures offer a promising solution, particularly when combined with conductive additives or modified to enhance electronic conductivity.

MoO_3, WO_3, and CuO: These oxides exhibit moderate HER activity, which can be improved through defect engineering, doping with nonmetals (e.g., nitrogen), or forming heterojunctions [71–73].

NiO and Co_3O_4: Though more commonly associated with OER, these oxides can be tuned for HER by modifying their morphology and electronic properties [74, 75].

Hybrid materials such as MoO_3/MoS_2 and $NiFe_2O_4$/graphene combine the benefits of both metal oxides and conductive phases. The oxide component provides stability and catalytic sites, while the carbonaceous matrix ensures efficient charge transport [76, 77].

Furthermore, metal oxide-based nanocomposites like ZnO/rGO and TiO_2/CuO demonstrate enhanced HER activity due to improved electron mobility and better dispersion of active centers [78, 79].

9.3.4 Synergistic Effects in Bifunctional Catalysts

A major advantage of some metal oxide nanostructures is their bifunctionality, being effective for both HER and OER. This property is highly desirable for overall water splitting applications using a single catalyst system.

Perovskite oxides (e.g., $LaNiO_3$, $Ba_{0.5}Sr_{0.5}Co_{0.8}Fe_{0.2}O_3{-}\delta$) have demonstrated promising bifunctional activity [80, 81].

Spinel oxides (e.g., $NiCo_2O_4$, $ZnCo_2O_4$) exhibit high conductivity, stability, and surface reactivity [82, 83].

Such bifunctional catalysts simplify electrolyzer design and reduce system complexity. Nanostructuring these materials further improves performance by enabling fine-tuning of surface chemistry and defect concentrations.

Catalyst system	Overpotential (HER/OER)	Electrolyte	Stability
NiO@Fe-Co cluster	62/380 mV	1 M KOH	>24 h
ZnO-In_2O_3 hybrid	220/410 mV	0.5 M Na_2SO_4	Good
Co_3O_4 LDH	170/320 mV	1 M NaOH	Excellent

9.4 Mechanisms of Hydrogen Production via Water Splitting: Photocatalytic and Electrocatalytic Water Splitting

Green hydrogen, generated via the electrolysis of water using renewable energy sources, represents a pivotal solution in the shift towards carbon-neutral energy systems. The production of hydrogen through water splitting involves cleaving the H—O—H bonds to release hydrogen and oxygen gases. This can be achieved primarily through two methods: photocatalytic water splitting and electrocatalytic water splitting. Each method capitalizes on distinct physicochemical principles and materials, particularly emphasizing the role of nanostructured catalysts, especially metal oxide nanomaterials, to improve performance [84, 85].

9.4.1 Overview of Water Splitting Reactions

Water splitting is thermodynamically uphill, requiring an input of 1.23 V under standard conditions ($\Delta G = +237.2$ kJ/mol). The overall water-splitting reaction [84, 86]:

$$H_2O \rightarrow H_2 + \frac{1}{2}O_2 \ (\Delta G > 0)$$

Comprises two half-reactions:

Cathodic reaction (HER): $2H^+ + 2e^- \rightarrow H_2$
Anodic reaction (OER): $2H_2O \rightarrow O_2 + 4H^+ + 4e^-$

Both reactions must proceed efficiently to ensure high rates of hydrogen generation. Catalysts play a crucial role in reducing the energy barrier and increasing the reaction kinetics.

9.4.2 Photocatalytic Water Splitting

Photocatalysis mimics natural photosynthesis, utilizing semiconductors to absorb light and generate electron–hole pairs that drive redox reactions. The basic mechanism includes:

Light Absorption: Photons with energy equal to or greater than the band gap excites electrons from the valence band (VB) to the conduction band (CB), creating electron-hole pairs [87].

Charge Separation and Migration: Electrons migrate to the surface-active sites where reduction occurs, and holes drive oxidation [88].

Surface Reactions: Electrons reduce protons to hydrogen, while holes oxidize water to oxygen [36].

TiO_2 and ZnO are classic examples of photocatalytic metal oxides due to their favorable band positions and high stability. However, their large band gaps (>3.0 eV) limit activity under visible light. Strategies to improve photocatalytic efficiency include [89, 90]:

- Doping with nonmetals (e.g., N, S) or metals (e.g., Fe, Cu) to narrow the band gap.
- Creating heterojunctions (e.g., TiO_2/CdS) for efficient charge separation.
- Surface modification with cocatalysts like Pt, RuO_2, or NiO to promote redox reactions.

A recent study by M. S. Quezada et al. demonstrated that Zn-Al-In mixed metal oxide (MMO) nanocomposites triggered by visible light achieved high H_2 generation rates (nearly 372 μmol h^{-1} g^{-1}), showcasing the potential of multi-metal systems to broaden light absorption and increase photocatalytic activity [61].

9.4.3 Electrocatalytic Water Splitting

Electrocatalysis uses externally applied potential to drive HER and OER in water electrolysis cells. Electrocatalytic systems are often divided into two types:

Alkaline Electrolysis (KOH, NaOH): Widely used due to the stability of non-precious metal catalysts [91, 92].
Proton Exchange Membrane (PEM) Electrolysis: Uses solid polymer electrolyte for compact systems but often requires noble metal catalysts [93, 94].

Mechanistically, the HER and OER processes involve intermediate adsorbed species (H_{ads}, OH^-, OOH^-), with the energy of adsorption and desorption controlling catalytic activity. For optimal performance, catalysts must balance the adsorption energy to avoid trapping or repelling intermediates.

9.4.4 Role of Metal Oxide Nanostructures

Metal oxide nanostructures serve as both light absorbers and catalytic centers. Their high surface area, tunable morphology, and electronic structure facilitate effective HER and OER in both photocatalytic and electrocatalytic systems [95–97]. Notable examples include:

NiO and Co_3O_4: Spinel-type oxides that exhibit high bifunctional activity [98, 99].
Fe_2O_3 (Hematite): An abundant photoanode material with a suitable bandgap (nearly 2.1 eV), though often hindered by short hole diffusion lengths [100].
TiO_2: Extremely stable but limited to UV absorption; often modified via doping or compositing [101].

Nanostructuring improves charge carrier mobility, increases the number of active sites, and facilitates better light absorption. Additionally, nanomaterials can be engineered to contain oxygen vacancies, which enhance catalytic activity by modifying the electronic environment and lowering the energy barrier for intermediate formation.

9.4.5 Comparative Assessment

Feature	Photocatalysis	Electrocatalysis
Energy source	Solar energy	Electrical energy
Reaction mechanism	Light-induced electron-hole pairs	Potential-induced redox reactions
Catalysts used	Semiconducting oxides (TiO_2, ZnO)	Metal oxides, phosphides, and sulfides
Efficiency challenges	Low quantum yield and recombination	Overpotential and electrode stability
Key advantage	Direct use of sunlight	Precise control and high efficiency
Limitation	Poor visible light absorption	High cost of noble metals (in PEM)

9.4.6 Enhancing Water Splitting Performance

To maximize hydrogen output and catalyst lifespan, researchers are focusing on the following innovations:

Band Engineering: Adjusting the band gap and band alignment of photocatalysts for better solar absorption [102].
Core-Shell Structures: For example, $Fe_2O_3@TiO_2$ nanocomposites combine visible light absorption with surface stability [103, 104].
Conductive Supports: Integration of graphene or carbon nanotubes with metal oxides reduces charge recombination [105–108].
Dual-Function Nanocatalysts: Materials that facilitate both HER and OER reduce system complexity [109, 110].

Moreover, hybrid systems combining photocatalysis with electrocatalysis, known as PEC water splitting, are gaining attention. These devices integrate semiconductor light absorbers with electrode setups to enhance overall energy conversion efficiency.

9.5 Recent Progress in Developing High-Efficiency Catalysts for H_2 Production

The development of high-efficiency catalysts for hydrogen production via water splitting is a cornerstone of advancing green hydrogen technologies. Recent research has intensified in discovering and optimizing nanostructured materials, particularly metal oxides, that can catalyze the HER and OER efficiently. This section outlines some of the key breakthroughs, emerging material classes, and strategic innovations in catalyst engineering.

9.5.1 Design Strategies for Efficient Catalysts

To create high-performance catalysts, researchers focus on enhancing surface activity, improving charge transfer, reducing overpotentials, and maximizing the utilization of abundant, low-cost materials. Major strategies include:

Doping and Alloying: Introducing heteroatoms into the catalyst lattice to modulate electronic properties and enhance reaction kinetics [111].
Morphological Engineering: Creating nanostructures (e.g., nanosheets, nanorods, hollow spheres) to expose more active sites [112].
Interface Engineering: Forming heterostructures or composites to facilitate charge separation and increase stability [113].
Defect Engineering: Generating oxygen vacancies or dislocations to boost adsorption and activation of reactants [114].

9.5.2 TMOs as Catalysts

TMOs like NiO, Co_3O_4, MnO_2, Fe_2O_3, and CuO have been widely studied due to their multiple oxidation states and robust structural properties.

Nickel Oxide (NiO): Known for high catalytic activity in alkaline HER. Recent studies report the creation of NiO nanosheets doped with Fe and Co, achieving low overpotentials (nearly 60 mV) and enhanced stability [112].

Cobalt Oxide (Co_3O_4): A spinel oxide with bifunctional OER and HER capabilities. Cobalt-based materials, especially when combined with conductive supports like graphene, show significant promise in PEM electrolysis [115].

Manganese Oxide (MnO_2): Economical and eco-friendly, with applications in photocatalytic OER. MnO_2 in layered or tunnel structures facilitates better ion transport and catalytic efficiency [116].

9.5.3 Multi-Metal and Mixed Oxide Nanocomposites

MMOs harness the synergistic effects between multiple metal cations to improve conductivity and charge mobility. Recent advances include:

Zn-Al-In Nanocomposites: These mixed oxides have demonstrated enhanced visible light absorption and impressive H_2 evolution rates [61].

$NiFeO_x$ and $CoFeO_x$: These materials are widely regarded as top-performing OER electrocatalysts in alkaline solutions. Their excellent conductivity and synergistic charge transfer contribute to low Tafel slopes and high turnover frequencies [117, 118].

9.5.4 Perovskite and Spinel Structures

Perovskite oxides (ABO_3) and spinels (AB_2O_4) are gaining traction due to their tunable electronic structures and stability.

$LaNiO_3$ and $Ba_{0.5}Sr_{0.5}Co_{0.8}Fe_{0.2}O_3-\delta$ (BSCF): These perovskites offer high OER performance in alkaline media. Oxygen vacancies in perovskites play a vital role in facilitating charge transport [119, 120].

Co_3O_4 Spinels: These are among the best bifunctional catalysts, with unique redox properties and high surface activity, particularly when structured as hollow or porous nanospheres [121].

9.5.5 Photocatalyst Innovations

In the domain of photocatalysis, considerable effort has gone into improving visible light absorption and minimizing recombination of photogenerated charges. Noteworthy developments include:

TiO_2-Based Hybrids: TiO_2 combined with narrow bandgap semiconductors like CdS or $BiVO_4$ has shown significant improvement in solar-driven hydrogen production [122, 123].

Doped ZnO and $BiFeO_3$ Nanoparticles: Doping with transition metals (Fe, Cu, Mn) allows for better light harvesting and catalytic activity. Cu-doped ZnO has demonstrated increased carrier density and HER efficiency [124].

2D Materials and Quantum Dots: Materials like MoS_2 nanosheets and carbon quantum dots enhance the photocatalytic activity when integrated with metal oxides, thanks to superior charge transfer and optical properties [125–127].

9.5.6 Noble-Metal-Free Catalysts

A major thrust of recent research is the move away from precious metals like Pt, Ir, and Ru. Some effective noble-metal-free alternatives include:

Ni-Mo Alloys: Exhibiting Pt-like HER performance in alkaline media [128].
Cobalt Phosphides (CoP) and Nickel Sulfides (NiS): Show excellent bifunctional catalytic behavior and durability [129, 130].
Iron-based Catalysts: Fe-Ni layered double hydroxides (LDHs) and FeCo oxide composites have emerged as low-cost yet efficient OER catalysts [131, 132].

9.5.7 Role of Nanostructuring and Support Materials

Nano-architectures such as nanotubes, nanorods, and mesoporous frameworks allow precise control over catalytic surface area and reaction interfaces. Key trends include:

Graphene and CNT Supports: Provide superior electrical conductivity and prevent catalyst agglomeration [133].
Metal-Organic Frameworks (MOFs): Used as precursors or templates for porous metal oxide catalysts [134].
Aerogels and Hydrogels: Porous, lightweight networks that enhance catalyst dispersion and reactant accessibility [135].

9.6 Challenges and Potential in Scaling Up Hydrogen Production

The transition from laboratory-scale hydrogen production to industrial-scale implementation presents numerous challenges and opportunities. While the technological feasibility of hydrogen generation through water splitting is well-established, its commercial viability hinges on economic, technical, infrastructural, and policy-driven factors. This section explores the multifaceted barriers to scaling up hydrogen production and the potential pathways for overcoming them.

9.6.1 Technical and Material Challenges

A primary technical challenge lies in the performance and durability of the catalysts used for the OER and HER. Although metal oxide nanostructures show promise in enhancing catalytic activity and stability, many remain susceptible to degradation under prolonged operational conditions, particularly in harsh pH environments.

Additionally, issues like poor electrical conductivity, sluggish charge carrier mobility, and limited photostability of photocatalysts restrict efficiency. The synthesis of uniformly active nanostructures at scale with reproducible performance characteristics remains a significant bottleneck.

9.6.2 Economic Constraints

Cost is a major barrier to the large-scale deployment of green hydrogen. While efforts to replace noble metals with earth-abundant alternatives are ongoing, the cost of advanced materials, fabrication techniques, and supporting infrastructure remains high. Water electrolysis, especially using PEM technology, is currently more expensive compared to SMR, which, despite its carbon-intensive nature, dominates global hydrogen production.

Photocatalytic systems, though promising for decentralized solar-to-hydrogen (STH) production, often suffer from low STH conversion efficiencies and high capital expenditure.

9.6.3 Infrastructure and Distribution

The existing infrastructure for hydrogen production, storage, and transportation is insufficient to support a large-scale green hydrogen economy. Hydrogen has a low volumetric energy density and requires high-pressure or cryogenic storage, both of which demand specialized equipment and safety measures.

Moreover, the absence of a dedicated hydrogen pipeline network complicates distribution. Relying on compressed gas cylinders or liquefied hydrogen tanks introduces logistical complexities and additional costs.

9.6.4 Water and Energy Requirements

Water splitting requires pure water, ideally deionized, which poses a challenge in arid regions. Large-scale deployment may put pressure on freshwater resources unless seawater electrolysis becomes viable. Furthermore, the energy input needed for water splitting must come from renewable sources to ensure the process remains carbon-neutral. Scaling green hydrogen production will, therefore, demand parallel investments in solar, wind, or other renewable energy infrastructure.

9.6.5 Standardization and Safety

The absence of standardized protocols for evaluating catalyst performance, long-term stability, and safety presents difficulties in benchmarking new technologies. Hydrogen, being highly flammable, also requires strict safety regulations and robust engineering controls to prevent leaks, explosions, or occupational hazards.

9.6.6 Policy and Market Integration

Despite increasing global interest in hydrogen, supportive regulatory frameworks, subsidies, and market incentives are still in nascent stages in many regions. Policies that prioritize fossil fuel usage or impose high tariffs on renewable technologies can hinder green hydrogen's growth.

Effective policy frameworks should encourage public-private partnerships, R&D funding, and clear long-term commitments to decarbonization strategies. Additionally, integration with existing energy markets, such as coupling hydrogen with electric grids or transportation networks, is essential for large-scale viability.

9.6.7 Potential Pathways Forward

Despite these challenges, several potential strategies could accelerate the scaling of green hydrogen production:

Decentralized Systems: Small-scale, localized hydrogen production units, particularly those powered by off-grid renewables, can reduce dependence on central infrastructure.

Modular and Scalable Electrolyzers: Development of plug-and-play electrolyzers that can be scaled in parallel.

Seawater Electrolysis: Advances in corrosion-resistant catalysts and membranes could make direct seawater electrolysis practical.

Hybrid Systems: Integration of photovoltaic-electrochemical (PV-EC) or PEC devices for more efficient STH conversion.

Circular Economy Models: Utilizing industrial waste heat or integrating hydrogen production with CO_2 capture can create synergistic energy systems.

9.7 Conclusion

In conclusion, the utilization of metal oxide nanocatalysts for green H_2 production via water splitting represents a transformative opportunity in the pursuit of sustainable energy solutions. Through both photocatalytic and electrocatalytic processes, metal oxide nanostructures have demonstrated significant potential in enhancing the kinetics and efficiency of the HER and OER.

Research advancements have led to the development of high-performance catalysts that address many of the fundamental limitations of traditional water-splitting systems, including stability, cost, and low overpotential requirements. Despite these achievements, numerous challenges remain, particularly in scaling these technologies to meet global energy demands.

Future work must focus on improving catalyst longevity, reducing reliance on critical raw materials, optimizing system integration with renewable energy sources, and developing efficient, scalable manufacturing techniques. Policy frameworks, investment in infrastructure, and international collaboration will be crucial in driving green hydrogen from pilot-scale research to commercial reality.

Overall, metal oxide nanocatalysts play a central role in the evolving hydrogen economy and stand as key enablers of climate change mitigation, energy security, and a cleaner global energy landscape.

References

1 Sivula K, Le Formal F, Grätzel M. Solar water splitting: progress using hematite (α-Fe_2O_3) photoelectrodes. *ChemSusChem*. 2011 Apr 18;4(4):432–49.

2 Yin Y, Zhang X, Sun C. Transition-metal-doped Fe_2O_3 nanoparticles for oxygen evolution reaction. *Prog Nat Sci Mater*. 2018 Aug;28(4):430–6.

3 Pawar SM, Pawar BS, Hou B, Kim J, Aqueel Ahmed AT, Chavan Harish S, et al. Self-assembled two-dimensional copper oxide nanosheet bundles as an efficient oxygen evolution reaction (OER) electrocatalyst for water splitting applications. *J Mater Chem A Mater*. 2017;5(25):12747–51.

4 Kuang M, Han P, Wang Q, Li J, Zheng G. CuCo hybrid oxides as bifunctional electrocatalyst for efficient water splitting. *Adv Funct Mater*. 2016 Dec 26;26(46):8555–61.

5 Lu C, Wang J, Czioska S, Dong H, Chen Z. Hierarchically structured cu-based electrocatalysts with nanowires array for water splitting. *J Phys Chem C*. 2017 Nov 22;121(46):25875–81.

6 Zheng Z, Morgan M, Maji P, Xia X, Zu X, Zhou W. CuO nanorod arrays by gas-phase cation exchange for efficient photoelectrochemical water splitting. *RSC Adv*. 2023;13(6):3487–93.

7 Seabold JA, Choi KS. Efficient and stable photo-oxidation of water by a bismuth vanadate photoanode coupled with an iron oxyhydroxide oxygen evolution catalyst. *J Am Chem Soc*. 2012 Feb 1;134(4):2186–92.

8 Liu B, Wang X, Zhang Y, Xu L, Wang T, Xiao X, et al. A $BiVO_4$ photoanode with a VO_x layer bearing oxygen vacancies offers improved charge transfer and oxygen evolution kinetics in photoelectrochemical water splitting. *Angew Chem Int Ed*. 2023 Mar 31; 62(10):e202217346.

9 Shabdan Y, Markhabayeva A, Bakranov N, Nuraje N. Photoactive tungsten-oxide nanomaterials for water-splitting. *Nanomaterials*. 2020 Sep 18;10(9):1871.

10 Thangamuthu M, Vankayala K, Xiong L, Conroy S, Zhang X, Tang J. Tungsten oxide-based Z-scheme for visible light-driven hydrogen production from water splitting. *ACS Catal*. 2023 Jul 7;13(13):9113–24.

11 Villafán Vidales HI, Jiménez-González A, Bautista-Orozco A, Arancibia-Bulnes CA, Estrada CA. Solar production of WO_3: a green approach. *Green Process Synth*. 2015 Jan 1;4(3):167–177.

12 Dymerska A, Kukułka W, Biegun M, Mijowska E. Spinel of Nickel-cobalt oxide with rod-like architecture as electrocatalyst for oxygen evolution reaction. *Materials*. 2020 Sep 4;13(18):3918.

13 Santhosh Kumar R, Karthikeyan SC, Ramakrishnan S, Vijayapradeep S, Rhan Kim A, Kim JS, et al. Anion dependency of spinel type cobalt catalysts for efficient overall water splitting in an acid medium. *Chem Eng J.* 2023 Jan;451:138471.

14 Han B, Risch M, Lee YL, Ling C, Jia H, Shao-Horn Y. Activity and stability trends of perovskite oxides for oxygen evolution catalysis at neutral pH. *Phys Chem Chem Phys.* 2015;17(35):22576–80.

15 Zhao JW, Li Y, Luan D, Lou XW (David). Structural evolution and catalytic mechanisms of perovskite oxides in electrocatalysis. *Sci Adv.* 2024 Sep 27; 10(39):eadq4696.

16 Beall CE, Fabbri E, Schmidt TJ. Perovskite oxide based electrodes for the oxygen reduction and evolution reactions: the underlying mechanism. *ACS Catal.* 2021 Mar 5;11(5):3094–114.

17 Wang Q, O'Hare D. Recent advances in the synthesis and application of layered double hydroxide (LDH) nanosheets. *Chem Rev.* 2012 Jul 11;112(7):4124–55.

18 Chandrasekaran S, Kim EJ, Chung JS, Bowen CR, Rajagopalan B, Adamaki V, et al. High performance bifunctional electrocatalytic activity of a reduced graphene oxide–molybdenum oxide hybrid catalyst. *J Mater Chem A Mater.* 2016;4(34):13271–9.

19 Wolf S, Roschger M, Genorio B, Garstenauer D, Hacker V. Mixed transition-metal oxides on reduced graphene oxide as a selective catalyst for alkaline oxygen reduction. *ACS Omega.* 2023 Mar 28;8(12):11536–43.

20 Liu W, Ni C, Gao M, Zhao X, Zhang W, Li R, et al. Metal–organic-framework-based nanoarrays for oxygen evolution electrocatalysis. *ACS Nano.* 2023 Dec 26;17(24):24564–92.

21 Li C, Zhang H, Liu M, Lang FF, Pang J, Bu XH. Recent progress in metal–organic frameworks (MOFs) for electrocatalysis. *Ind Chem Mater.* 2023;1(1):9–38.

22 Zhou L, Lu S, Guo S. Recent progress on precious metal single atom materials for water splitting catalysis. *SusMat.* 2021 Jun 23;1(2):194–210.

23 Alam N, Noor T, Iqbal N. Catalyzing sustainable water splitting with single atom catalysts: recent advances. *Chem Rec.* 2024 Mar 19; 24(3):e202300330.

24 Tosoni S, Di Liberto G, Matanovic I, Pacchioni G. Modelling single atom catalysts for water splitting and fuel cells: a tutorial review. *J Power Sources.* 2023 Feb;556:232492.

25 Gupta A, Ghosh S, Bhalothia D, Thangarasu S, Ghosh B, Urkude R, et al. Stabilizing a Ru single atom catalyst through electronic metal–support interaction with a $NiCo_2O_4$ support for overall water splitting and urea electrolysis. *J Mater Chem A Mater.* 2024;12(35):23819–36.

26 Yuan CZ, Huang S, Zhao H, Li J, Zhang L, Weng Y, et al. Vacancy defect tuning of electronic structures of transition metal (hydr)oxide-based electrocatalysts for enhanced oxygen evolution. *Energy Adv.* 2023;2(1):73–85.

27 Hu M, Zhu Q, Zhao Y, Zhang G, Zou C, Prezhdo O, et al. Facile removal of bulk oxygen vacancy defects in metal oxides driven by hydrogen-dopant evaporation. *J Phys Chem Lett.* 2021 Oct 7;12(39):9579–83.

28 Yu K, Lou L, Liu S, Zhou W. Asymmetric oxygen vacancies: the intrinsic redox active sites in metal oxide catalysts. *Adv Sci.* 2020 Jan 5;7(2):1901970.

29 Bi F, Meng Q, Zhang Y, Chen H, Jiang B, Lu H, et al. Engineering triple O-Ti-O vacancy associates for efficient water-activation catalysis. *Nat Commun*. 2025 Jan 20;16(1):851.

30 Yang Q, Lu Z, Li T, Sun X, Liu J. Hierarchical construction of core–shell metal oxide nanoarrays with ultrahigh areal capacitance. *Nano Energy*. 2014 Jul;7:170–8.

31 Ahsan MT, Usman M, Ali Z, Javed S, Ali R, Farooq MU, et al. 3D hierarchically mesoporous zinc-nickel-cobalt ternary oxide ($Zn_{0.6}Ni_{0.8}Co_{1.6}O_4$) nanowires for high-performance asymmetric supercapacitors. *Front Chem*. 2020 Jun 15;8:487.

32 Sheng X, He D, Yang J, Zhu K, Feng X. Oriented assembled TiO_2 hierarchical nanowire arrays with fast electron transport properties. *Nano Lett*. 2014 Apr 9;14(4):1848–52.

33 Zhang F, Qi L. Recent progress in self-supported metal oxide nanoarray electrodes for advanced lithium-ion batteries. *Adv Sci*. 2016 Sep 15;3(9):1600049.

34 Yang Y, Niu S, Han D, Liu T, Wang G, Li Y. Progress in developing metal oxide nanomaterials for photoelectrochemical water splitting. *Adv Energy Mater*. 2017 Oct 12;7(19):1700555.

35 Wang J, Cui W, Liu Q, Xing Z, Asiri AM, Sun X. Recent progress in cobalt-based heterogeneous catalysts for electrochemical water splitting. *Adv Mater*. 2016 Jan 9;28(2):215–30.

36 Jafari T, Moharreri E, Amin A, Miao R, Song W, Suib S. Photocatalytic water splitting—the untamed dream: a review of recent advances. *Molecules*. 2016 Jul 9;21(7):900.

37 Wang Q, Lei Y, Wang D, Li Y. Defect engineering in earth-abundant electrocatalysts for CO_2 and N_2 reduction. *Energy Environ Sci*. 2019;12(6):1730–50.

38 Loni E, Shokuhfar A, Siadati MH. Cobalt-based electrocatalysts for water splitting: an overview. *Catal Surv Asia*. 2021 Jun 11;25(2):114–47.

39 Wang Y, Zhang J. Structural engineering of transition metal-based nanostructured electrocatalysts for efficient water splitting. *Front Chem Sci Eng*. 2018 Dec 13;12(4):838–54.

40 Mugheri AQ, Tahira A, Aftab U, Abro MI, Chaudhry SR, Amaral L, et al. Facile efficient earth abundant NiO/C composite electrocatalyst for the oxygen evolution reaction. *RSC Adv*. 2019;9(10):5701–10.

41 Zhang H, Bai X. Photocatalytic production of hydrogen peroxide over Z-scheme Mn3O4/Co9S8 with p-n heterostructure. *Appl Catal B*. 2021 Dec;298:120516.

42 Bai L, Wen X, Guan J. Amorphous Fe Co Ni oxide for oxygen evolution reaction. *Mater Today Energy*. 2019 Jun;12:311–7.

43 Frites M, Simpson A, Gautam R, Khan SUM. Mixed transition metal oxides for efficient electrochemical splitting of water to hydrogen and oxygen. *WIT Trans Ecol Environ*. 2014;186:139–46.

44 Wang G, Chen T, Liu S, Wang F, Li M, Xie M, et al. Construction of TiO_2–$MnO_2$0D–2D nanostructured heterojunction for enhanced photocatalytic hydrogen production. *Dalton Trans*. 2021;50(25):8711–7.

45 Etemadi H, Soltani T, Yoshida H, Zhang Y, Telfer SG, Buchanan JK, et al. Synergistic Effect of redox dual PdO_x/MnO_x cocatalysts on the enhanced H_2 production potential of a SnS/α-Fe_2O_3 heterojunction via ethanol photoreforming. *ACS Omega*. 2022 Nov 22;7(46):42347–58.

46 Zheng D, Xue Y, Wang J, Varbanov PS, Klemeš JJ, Yin C. Nanocatalysts in photocatalytic water splitting for green hydrogen generation: challenges and opportunities. *J Clean Prod*. 2023 Aug;414:137700.

47 Qadeer MA, Zhang X, Farid MA, Tanveer M, Yan Y, Du S, et al. A review on fundamentals for designing hydrogen evolution electrocatalyst. *J Power Sources*. 2024 Sep;613:234856.

48 Khan MM, Ansari SA, Pradhan D, Ansari MO, Han DH, Lee J, et al. Band gap engineered TiO_2 nanoparticles for visible light induced photoelectrochemical and photocatalytic studies. *J Mater Chem A*. 2014;2(3):637–44.

49 Zahra T, Ahmad KS, Thomas AG, Zequine C, Malik MA, Gupta RK. Organic template-based ZnO embedded Mn_3O_4 nanoparticles: synthesis and evaluation of their electrochemical properties towards clean energy generation. *RSC Adv*. 2020;10(17):9854–67.

50 Ahmad I, Ahmed J, Batool S, Zafar MN, Hanif A, Zahidullah, et al. Design and fabrication of Fe_2O_3/FeP heterostructure for oxygen evolution reaction electrocatalysis. *J Alloys Compd*. 2022 Feb;894:162409.

51 Gorylewski D, Zasada F, Słowik G, Lofek M, Grzybek G, Tyszczuk-Rotko K, et al. Modulation of the electronic properties of Co_3O_4 through Bi octahedral doping for enhanced activity in the oxygen evolution reaction. *ACS Catal*. 2025 Mar 21;15(6):4746–58.

52 Mannaa MA, Qasim KF, Alshorifi FT, El-Bahy SM, Salama RS. Role of NiO nanoparticles in enhancing structure properties of TiO_2 and Its applications in photodegradation and hydrogen evolution. *ACS Omega*. 2021 Nov 16;6(45):30386–400.

53 Robinson DM, Go YB, Greenblatt M, Dismukes GC. Water oxidation by λ-MnO_2: catalysis by the cubical Mn_4O_4 subcluster obtained by delithiation of spinel $LiMn_2O_4$. *J Am Chem Soc*. 2010 Aug 25;132(33):11467–9.

54 Zheng KW, Li B, Li X, Gao W. Understanding the oxygen-evolution-reaction catalytic activity of metal oxides based on the intrinsic descriptors. *Phys Chem Chem Phys*. 2022;24(46):28632–40.

55 Shi G, Tano T, Tryk DA, Yamaguchi M, Iiyama A, Uchida M, et al. Temperature dependence of oxygen evolution reaction activity in alkaline solution at Ni–Co oxide catalysts with amorphous/crystalline surfaces. *ACS Catal*. 2022 Nov 18;12(22):14209–19.

56 Li L, Liu Y, Chen Y, Zhai W, Dai Z. Research progress on layered metal oxide electrocatalysts for an efficient oxygen evolution reaction. *Dalton Trans*. 2024;53(21):8872–86.

57 Rusinque B, Escobedo S, de Lasa H. Hydrogen production via Pd-TiO_2 photocatalytic water splitting under near-UV and visible light: analysis of the reaction mechanism. *Catalysts*. 2021 Mar 23;11(3):405.

58 Cao W, Wei B, Fu X, Ma N, Gao H, Xu L. Colored TiO_2 hollow spheres for efficient water-splitting photocatalysts. *RSC Adv*. 2016;6(110):108969–73.

59 Ong YP, Ho LN, Ong SA, Banjuraizah J, Ibrahim AH, Thor SH, et al. A highly sustainable hydrothermal synthesized MnO2 as cathodic catalyst in solar photocatalytic fuel cell. *Chemosphere*. 2021 Jan;263:128212.

60 Sekar S, Lee S, Vijayarengan P, Kalirajan KM, Santhakumar T, Sekar S, et al. Upcycling of wastewater via effective photocatalytic hydrogen production using MnO_2 nanoparticles—decorated activated carbon nanoflakes. *Nanomaterials.* 2020 Aug 17;10(8):1610.

61 Suárez-Quezada M, Suárez-Quezada VM, Tobola-Inchaurregui F, Oros-Ruiz S, Cipagauta-Díaz S. Hydrogen production through water splitting reactions using Zn-Al-In mixed metal oxide nanocomposite photocatalysts induced by visible light. *Catalysts.* 2024 Nov 20;14(11):835.

62 Santos HLS, Mascaro LH. Enhanced photoelectrochemical water splitting using nanostructured films: p-CuO sensitized with polyhedral n-Cu_2O particles and CuS as photocathode. *Mater Adv.* 2024;5(16):6596–604.

63 Yang P, Shi H, Wu H, Yu D, Huang L, Wu Y, et al. Manipulating the surface states of $BiVO_4$ through electrochemical reduction for enhanced PEC water oxidation. *Nanoscale.* 2023;15(9):4536–45.

64 Ho-Kimura S. Experimental evidence for photoactivated $BiVO_4$ anodes with enhanced photoelectrochemical water oxidation. *ACS Appl Energy Mater.* 2024 Mar 11;7(5):1902–13.

65 Jafari S, Shaghaghi Z. Engineering active sites in ternary CeO_2-CuO-Mn_3O_4 heterointerface embedded in reduced graphene oxide for boosting water splitting activity. *Sci Rep.* 2025 Feb 3;15(1):4145.

66 Avcı ÖN, Sementa L, Fortunelli A. Mechanisms of the oxygen evolution reaction on $NiFe_2O_4$ and $CoFe_2O_4$ inverse-spinel oxides. *ACS Catal.* 2022 Aug 5;12(15):9058–73.

67 Chen Q, Wang R, Lu F, Kuang X, Tong Y, Lu X. Boosting the oxygen evolution reaction activity of $NiFe_2O_4$ nanosheets by phosphate ion functionalization. *ACS Omega.* 2019 Feb 28;4(2):3493–9.

68 Paudel N, Ale Magar B, Acharya K, Lambert TN, Vasiliev I. Influence of defects and surfaces on the electrochemical performance of MnO_2 cathodes in rechargeable alkaline Zn/MnO_2 batteries: a first-principles study. *ACS Appl Energy Mater.* 2024 Apr 8;7(7):2767–78.

69 Fuchigami T, Kimata R, Haneda M, Kakimoto K. Complex three-dimensional Co_3O_4 nano-raspberry: highly stable and active low-temperature CO oxidation catalyst. *Nanomaterials.* 2018 Aug 26;8(9):662.

70 Marimuthu S, Maduraiveeran G. Tailoring the heterointerfaces of earth-abundant transition-metal nanoclusters on nickel oxide nanosheets for enhanced overall water splitting through electronic structure optimization. *Langmuir.* 2024 Oct 29;40(43):22549–60.

71 Zhang L, Jin Z, Tsubaki N. Activating and optimizing the MoS_2@MoO_3 S-scheme heterojunction catalyst through interface engineering to form a sulfur-rich surface for photocatalyst hydrogen evolution. *Chem Eng J.* 2022 Jun;438:135238.

72 Zhang C, Liu N, Wang H, Chen B, Huang C. Hydrogen-assisted defect engineering of Pt/MoO3-x catalysts with promoted Pt dispersion to achieve ultra-low Pt loading for hydrogen evolution reaction. *Int J Hydrogen Energy.* 2025 Feb;104:114–21.

73 Zhao Y, Balasubramanyam S, Sinha R, Lavrijsen R, Verheijen MA, Bol AA, et al. Physical and chemical defects in WO_3 thin films and their impact on photoelectrochemical water splitting. *ACS Appl Energy Mater*. 2018 Nov 26;1(11):5887–95.
74 Peng L, Liang Y, Wu S, Li Z, Sun H, Jiang H, et al. Nanoporous Ni/NiO catalyst for efficient hydrogen evolution reaction prepared by partial electro-oxidation after dealloying. *J Alloys Compd*. 2022 Aug;911:165061.
75 Ma J, Wei H, Liu Y, Ren X, Li Y, Wang F, et al. Application of Co_3O_4-based materials in electrocatalytic hydrogen evolution reaction: a review. *Int J Hydrogen Energy*. 2020 Aug;45(41):21205–20.
76 Aftab U, Ansari HM, Abro MI, Baloch MM, Sirajuddin, Ibupoto ZH. Mixed MoS_2/MoO_3 nanostructures for hydrogen evolution reaction. *J Nanosci Nanotechnol*. 2021 Apr 1;21(4):2500–10.
77 Zhang X, Zhang X, Wang XG, Xie Z, Zhou Z. $NiFe_2O_4$–CNT composite: an efficient electrocatalyst for oxygen evolution reactions in Li–O_2 batteries guided by computations. *J Mater Chem A Mater*. 2016;4(24):9390–3.
78 Pitiphattharabun S, Auewattanapun K, Htet TL, Thu MM, Panomsuwan G, Techapiesancharoenkij R, et al. Reduced graphene oxide/zinc oxide composite as an electrochemical sensor for acetylcholine detection. *Sci Rep*. 2024 Jun 20;14(1):14224.
79 Yang K, Cheng G, Chen R, Zhao K, Liang Y, Li W, et al. Simply coupling TiO_2nanospheres with Cu_2O particles to boost the photocatalytic hydrogen evolution through p–n heterojunction-induced charge transfer. *Energy Technol*. 2022 Jan 9;10(1):2100259.
80 Shen TH, Spillane L, Vavra J, Pham THM, Peng J, Shao-Horn Y, et al. Oxygen evolution reaction in $Ba_{0.5}$ $Sr_{0.5}$ $Co_{0.8}$ $Fe_{0.2}$ $O_{3-\delta}$ aided by intrinsic Co/Fe spinel-like surface. *J Am Chem Soc*. 2020 Sep 16;142(37):15876–83.
81 Zhou W, Sunarso J. Enhancing Bi-functional electrocatalytic activity of perovskite by temperature shock: a case study of $LaNiO_{3-\delta}$. *J Phys Chem Lett*. 2013 Sep 5;4(17):2982–8.
82 Boyjoo Y, Jin Y, Mao X, Zhao G, Gengenbach T, Du A, et al. Crystal facet engineering of spinel $NiCo_2O_4$ with enhanced activity and water resistance for tuneable catalytic methane oxidation. *EES Catalysis*. 2024;2(2):638–46.
83 Xin H, Li D, Shi L, Ji M, Lin Y, Yu J, et al. A simple approach to fabricate of Ni-$NiCo_2O_4$@$ZnCo_2O_4$ yolk-shell nano-tetrahedron composite as high-performance anode material for lithium-ion batteries. *Chem Eng J*. 2018 Jun;341:601–9.
84 Turner JA. Sustainable hydrogen production. *Science*. 2004 Aug 13;305(5686):972–4.
85 Nikolaidis P, Poullikkas A. A comparative overview of hydrogen production processes. *Renew Sustain Energy Rev*. 2017 Jan;67:597–611.
86 Grätzel M. Photoelectrochemical cells. *Nature*. 2001 Nov;414(6861):338–44.
87 Fujishima A, Honda K. Electrochemical photolysis of water at a semiconductor electrode. *Nature*. 1972 Jul;238(5358):37–8.

88 Chen X, Mao SS. Titanium dioxide nanomaterials: synthesis, properties, modifications, and applications. *Chem Rev*. 2007 Jul 1;107(7):2891–959.

89 Wang ZL. Zinc oxide nanostructures: growth, properties and applications. *J Phys Condens Matter*. 2004 Jun30; 16(25):R829–58.

90 Rafique M, Hajra S, Irshad M, Usman M, Imran M, Assiri MA, et al. Hydrogen production using TiO_2-based photocatalysts: a comprehensive review. *ACS Omega*. 2023 Jul 25;8(29):25640–8.

91 Wang J, Gao Y, Kong H, Kim J, Choi S, Ciucci F, et al. Non-precious-metal catalysts for alkaline water electrolysis: *operando* characterizations, theoretical calculations, and recent advances. *Chem Soc Rev*. 2020;49(24):9154–96.

92 Kawaguchi K, Goto K, Konno A, Nohira T. Novel high-temperature alkaline water electrolysis using molten KOH–H_2O system. *J Electrochem Soc*. 2023 Aug 1;170(8):084507.

93 Belami D, Lindley M, Jonnalagadda US, Goncalves Bullock AM, Fan Q, Liu W, et al. Active and highly durable supported catalysts for proton exchange membrane electrolysers. *EES Catalysis*. 2024;2(5):1139–51.

94 Yang Y, Chen D, Hu S, Pei P, Xu X. Advanced Ir-based alloy electrocatalysts for proton exchange membrane water electrolyzers. *Small*. 2025 Feb 3;21(8):2410372.

95 Ganguly A, Anjaneyulu O, Ojha K, Ganguli AK. Oxide-based nanostructures for photocatalytic and electrocatalytic applications. *CrystEngComm*. 2015;17(47):8978–9001.

96 Tahir MB, Sohaib M, Sagir M, Rafique M. Role of nanotechnology in photocatalysis. In: Olabi A-G, editor. *Encyclopedia of Smart Materials*. Elsevier; 2022. p. 578–89.

97 Pascariu P, Gherasim C, Airinei A. Metal oxide nanostructures (MONs) as photocatalysts for ciprofloxacin degradation. *Int J Mol Sci*. 2023 May 31;24(11):9564.

98 Li J, Li J, Ren J, Hong H, Liu D, Liu L, et al. Electric-field-treated Ni/Co_3O_4 film as high-performance bifunctional electrocatalysts for efficient overall water splitting. *Nanomicro Lett*. 2022 Dec 22;14(1):148.

99 Xu H, Zhang D, Liu M, Ye D, Huo S, Chen W, et al. Self-supporting hierarchical Co_3O_4-nanowires@NiO-nanosheets core-shell nanostructure on carbon foam to form efficient bifunctional electrocatalyst for overall water splitting. *J Colloid Interface Sci*. 2024 Jan;654:1293–302.

100 Kim JY, Magesh G, Youn DH, Jang JW, Kubota J, Domen K, et al. Single-crystalline, wormlike hematite photoanodes for efficient solar water splitting. *Sci Rep*. 2013 Sep 17;3(1):2681.

101 El Mragui A, Zegaoui O, Daou I, Esteves da Silva JCG. Preparation, characterization, and photocatalytic activity under UV and visible light of Co, Mn, and Ni mono-doped and (P,Mo) and (P,W) co-doped TiO_2 nanoparticles: a comparative study. *Environ Sci Pollut Res*. 2021 May 26;28(20):25130–45.

102 Ontiveros D, Viñes F, Sousa C. Bandgap engineering of MXene compounds for water splitting. *J Mater Chem A Mater*. 2023;11(25):13754–64.

103 Wang M, Pyeon M, Gönüllü Y, Kaouk A, Shen S, Guo L, et al. Constructing Fe_2O_3/TiO_2 core–shell photoelectrodes for efficient photoelectrochemical water splitting. *Nanoscale*. 2015;7(22):10094–100.

104 Han H, Riboni F, Karlicky F, Kment S, Goswami A, Sudhagar P, et al. α-Fe_2O_3/$TiO_2$3D hierarchical nanostructures for enhanced photoelectrochemical water splitting. *Nanoscale*. 2017;9(1):134–42.

105 Perera DC, Rasaiah JC. Computational study of H_2O adsorption, hydrolysis, and water splitting on $(ZnO)_3$ nanoclusters deposited on graphene and graphene oxides. *ACS Omega*. 2023 Sep 5;8(35):32185–203.

106 Wang J, Peng Z, Zhu K, Xu M, Tao T, Jia D, et al. The fate of oxygen on graphene-catalyst in the photocatalytic water splitting reaction. *Catal Sci Technol*. 2021;11(21):7083–90.

107 Gao X, Zhang H, Li Q, Yu X, Hong Z, Zhang X, et al. Hierarchical $NiCo_2O_4$ hollow microcuboids as bifunctional electrocatalysts for overall water-splitting. *Angew Chem Int Ed*. 2016 May 17;55(21):6290–4.

108 Bilal M, Altaf A, Bint-E-Khalid E, Zafar HK, Tahir N, Nafady A, et al. $NiCo_2O_4$ nano-needles as an efficient electro-catalyst for simultaneous water splitting and dye degradation. *RSC Adv*. 2023;13(34):23547–57.

109 Kumar M, Meena B, Subramanyam P, Suryakala D, Subrahmanyam C. Recent trends in photoelectrochemical water splitting: the role of cocatalysts. *NPG Asia Mater*. 2022 Dec 11;14(1):88.

110 Chen X, Zhang Z, Chi L, Nair AK, Shangguan W, Jiang Z. Recent advances in visible-light-driven photoelectrochemical water splitting: catalyst nanostructures and reaction systems. *Nanomicro Lett*. 2016 Jan 28;8(1):1–12.

111 Jin H, Liu X, Chen S, Vasileff A, Li L, Jiao Y, et al. Heteroatom-doped transition metal electrocatalysts for hydrogen evolution reaction. *ACS Energy Lett*. 2019 Apr 12;4(4):805–10.

112 Zhou Z, Han FX, Yao HH, Li YZ, Yang YG, Guo XY, et al. Novel fe-based amorphous composite coating with a unique interfacial layer improving thermal barrier application. *ACS Appl Mater Interfaces*. 2021 May 19;13(19):23057–66.

113 Zhang X, Liu K, Zhang S, Miao F, Xiao W, Shen Y, et al. Enabling remarkable cycling performance of high-loading MoS_2@Graphene anode for sodium ion batteries with tunable cut-off voltage. *J Power Sources*. 2020 May;458:228040.

114 Nath I, Chakraborty J, Heynderickx PM, Verpoort F. Engineered synthesis of hierarchical porous organic polymers for visible light and natural sunlight induced rapid degradation of azo, thiazine and fluorescein based dyes in a unique mechanistic pathway. *Appl Catal B*. 2018 Jul;227:102–13.

115 Chen F, Zhang R, Zhang Y, Hao X, Liu Y, Xu M, et al. Special atmosphere annealed Co_3O_4 bifunctional catalysts with oxygen defects for high-efficient water splitting. *Appl Surf Sci*. 2024 Jun;657:159839.

116 Wang P, Zhang S, Wang Z, Mo Y, Luo X, Yang F, et al. Manganese-based oxide electrocatalysts for the oxygen evolution reaction: a review. *J Mater Chem A Mater*. 2023;11(11):5476–94.

117 Ehsan MA, Ullah Z, Nazar MF, Younas M, Hakeem AS. Rapid synthesis of nickel–iron-oxide (NiFeOx) solid-solution nano-rods thin films on nickel foam as advanced electrocatalyst for sustained water oxidation. *Mater Today Sustain*. 2023 Sep;23:100451.

118 Zhu M, Xu H, Dai J, Guan D, Hu Z, She S, et al. A dynamically stable self-assembled CoFe (oxy)hydroxide-based nanocatalyst with boosted electrocatalytic performance for the oxygen-evolution reaction. *J Mater Chem A Mater.* 2024;12(36):24308–17.

119 Petrie JR, Cooper VR, Freeland JW, Meyer TL, Zhang Z, Lutterman DA, et al. Enhanced bifunctional oxygen catalysis in strained $LaNiO_3$ perovskites. *J Am Chem Soc.* 2016 Mar 2;138(8):2488–91.

120 Sun J, Zhang Z, Gong Y, Wang H, Wang R, Zhao L, et al. Plasma engraved $Bi_{0.1}(Ba_{0.5}Sr_{0.5})_{0.9}Co_{0.8}Fe_{0.2}O_{3-\delta}$ perovskite for highly active and durable oxygen evolution. *Sci Rep.* 2019 Mar 12;9(1):4210.

121 Tan Ys, Shi X, Han W, Kuang D, Wang C. Spiny Co_3O_4@Hollow carbon spheres—polyacrylonitrile/carbon black fiber-based bifunctional air electrodes. *Langmuir.* 2025 Feb 4;41(4):2629–38.

122 El Hattab A, Mirzaei A, Nada AA, Shayegan Z, Roualdes S, Chaker M. TiO_2/$BiVO_4$ dual photoanodes: extending light harvesting and addressing band edge misalignment for photoelectrochemical water splitting. *J Power Sources.* 2025 Jul;643:237070.

123 Rafiq K, Sabir M, Abid MZ, Jalil M, Nadeem MA, Iqbal S, et al. Tuning of TiO_2/CdS Hybrid Semiconductor with Au cocatalysts: state-of-the-art design for sunlight-driven H_2generation from water splitting. *Energy Fuels.* 2024 Mar 7;38(5):4625–36.

124 Chen G, Yang M, Tian B, Yao J, Chen S, Li D, et al. Cu-doped ZnO nanoparticles and its application for the photocatalytic degradation of Rhodamine B. *Sci Rep.* 2025 May 25;15(1):18246.

125 Wang B, Deng Z, Fu X, Li Z. MoS_2/CQDs obtained by photoreduction for assembly of a ternary MoS_2/CQDs/$ZnIn_2S_4$ nanocomposite for efficient photocatalytic hydrogen evolution under visible light. *J Mater Chem A Mater.* 2018;6(40):19735–42.

126 Chen L, Hsieh SL, Kuo CH, Hsieh S, Chen WH, Chen CW, et al. Novel MoS_2 quantum dots as a highly efficient visible-light driven photocatalyst in water remediation. *RSC Adv.* 2020;10(53):31794–9.

127 Liang Z, Guo Y, Xue Y, Cui H, Tian J. 1T-phase MoS_2 quantum dots as a superior co-catalyst to Pt decorated on carbon nitride nanorods for photocatalytic hydrogen evolution from water. *Mater Chem Front.* 2019;3(10):2032–40.

128 Park SH, To DT, Myung NV. A review of nickel-molybdenum based hydrogen evolution electrocatalysts from theory to experiment. *Appl Catal A Gen.* 2023 Feb;651:119013.

129 Ji L, Wang J, Teng X, Meyer TJ, Chen Z. CoP Nanoframes as bifunctional electrocatalysts for efficient overall water splitting. *ACS Catal.* 2020 Jan 3;10(1):412–9.

130 Jiang N, Tang Q, Sheng M, You B, Jiang D, Sun Y. Nickel sulfides for electrocatalytic hydrogen evolution under alkaline conditions: a case study of crystalline NiS, NiS_2, and Ni_3S_2 nanoparticles. *Catal Sci Technol.* 2016;6(4):1077–84.

131 Lin Q, Nan G, Fu D, Xie L. Oxygen evolution reaction on NiFe-LDH/(Ni,Fe)OOH: theoretical insights into the effects of electronic structure and spin-state evolution. *Phys Chem Chem Phys*. 2025;27(9):4926–33.

132 Zhang J, Chen J, Jing Y, Xu X, Liu C, Jia J. Fe, Co, N-doped composite as an efficient catalyst for oxygen reduction and oxygen evolution reaction. *Mater Today Chem*. 2025 Mar;44:102531.

133 Qin X, Ola O, Zhao J, Yang Z, Tiwari SK, Wang N, et al. Recent progress in graphene-based electrocatalysts for hydrogen evolution reaction. *Nanomaterials*. 2022 May 25;12(11):1806.

134 Wang X, Ying J, Mai Y, Zhang J, Chen J, Wen M, et al. MOF-derived metal oxide composite $Mn_2Co_1O_x$/CN for efficient formaldehyde oxidation at low temperature. *Catal Sci Technol*. 2019;9(20):5845–54.

135 Chen J, Xie Z, Tang Y, Tang Z, Wang X. Co@Ir core–shell nanochain aerogels for the hydrogen evolution reaction and the oxygen evolution reaction in alkaline media. *New J Chem*. 2024;48(31):13950–6.

10

Innovation of Metal Oxide Nanocatalysts for Air Pollution Control

Abdulazeez Muhammed[1], Mustapha Abdulsalam[2], Innocent Ojeba Musa[3], Miracle Uwa Livinus[4], Ismail Rabiu[5], Abdulhakeem Idris Abdulhakeem[6], Emmanuel Olabi Moses[7], and Issa Sheriffdeen Bale[8]*

[1,2,3] *Department of Microbiology, Skyline University Nigeria, Kano, Nigeria*
[4] *Department of Biochemistry, Skyline University Nigeria, Kano, Nigeria*
[5] *Department of Biomedical Science and Environmental Biology, Kaohsiung Medical University, Kaohsiung, Taiwan*
[6] *Department of Biodiversity and Bioresources, Al-Farabi Kazakh National University, Almaty, Kazakhstan*
[7] *Department of Public Health, Birmingham City University, Birmingham, UK*
[8] *Department of Microbiology, Federal University of Health Sciences, Ila-Orangun, Osun, Nigeria*

*Corresponding author: mdabdulazeez814@gmail.com

10.1 Introduction

The worldwide pollution problem is a key environmental issue [1]. Research on human development and pollution impact has developed its volume while enhancing its scholarly standards after the mid-20th century [2]. The detrimental health effects of air contaminants within the environment have encouraged numerous researchers to focus on multiple research domains [3].

The utilization of fossil fuels across industries throughout the 21st century world has led to environmental decline through atmospheric contamination [4]. Environmental pollution creates obstacles to human survival and gives rise to difficulties in sustaining human reproduction on Earth [1]. Climate change originates from environmental degradation through two manifestations that involve industrial production of greenhouse gases causing global warming and subsequent changes in weather patterns [1].

The rapidly escalating environmental issue affects worldwide survival and ecological sustainability for all living organisms [2]. Various human-initiated activities under the guise of development and civilization, and natural disasters trigger the quality deterioration of environmental resources and features [3].

The removal of airborne pollutants along with their tracking represents an excellent use of nanotechnology that has revolutionized environmental cleanup

Metal Oxide Nanocatalysts for Sustainable Energy Production, First Edition. Edited by Abayomi Bamisaye et al.
Published 2026 by Wiley-VCH GmbH.

operations by offering transformational options to address complicated pollution problems [2].

Problems of environmental contamination [4]. Nano adsorbents form a part of the developed nanotechnology techniques in combination with nanofiltration, along with nano photocatalysts and magnetic nanoparticles to create nano sensors for air pollution detection [5].

According to [6], nanocatalysts exhibit one-of-a-kind characteristics along with high surface area, together with reactivity and selectivity. The combination of properties that characterize nanocatalysts makes them a superior choice for destroying environmental pollutants through catalysis [5].

Nanocatalysts attract increasing use for environmental remediation because they achieve higher process efficiency while reducing energy needs and offering environmentally sustainable solutions for pollution reduction.

10.2 Mechanisms of Nanocatalyst in Environmental Remediation

The nanometer range characterizes nanocatalysts whose materials possess catalyst functions [4]. Environmental remediation benefits greatly from their exceptional physical and chemical properties, which include large surface area-to-volume ratios together with reactivity and easy functionalization potential [7].

The main objective of environmental remediation is to eliminate or decontaminate pollutants in contaminated air [5]. The breakdown of toxic substances occurs faster with nanocatalysts because they provide an efficient, sustainable method to combat environmental pollution [5].

[8] presented a detailed explanation of the nanocatalysts' pollution cleanup operation principles in her work.

1. Photocatalysis
2. Fenton-like Reactions
3. Electrocatalysis
4. Adsorption and Desorption
5. Oxidation-Reduction Reactions

10.2.1 Photocatalysis

The acceleration of chemical reactions through light exposure represents photocatalysis. UV light usually drives photocatalysis, but visible light can be effective based on the catalyst material. Nanocatalysts like titanium dioxide (TiO_2) and zinc oxide (ZnO) are among the most widely used photocatalysts due to their strong UV absorption and excellent stability [9].

10.2.1.1 Mechanism of Photocatalysis

UV light exposure induces TiO_2, along with related photocatalysts to both absorb energy and undergo states of excitation [10]. The process generates electron-hole pairs because of the energy absorption [10]. The unique energy of UV light exceeds the threshold needed to lift electrons from valence to conduction bands and creates positively charged holes (h^+) [7].

After trapping electrons and holes, the material can initiate chemical reactions with nearby water and oxygen molecules that produce reactive oxygen species (ROS), which include hydroxyl radicals (•OH) and superoxide anions ($O_2^{\bullet -}$). The holes (h^+) generated by the photogeneration oxidize surface-adsorbed water or hydroxide ions to produce •OH, and the photogenerated electrons (e^-) reduce dissolved oxygen to make $O_2^{\bullet -}$. The ROS are then involved in further oxidation reactions that weaken or modify the molecules in the environment [11].

10.2.2 Fenton-like Reactions

The basis of Fenton-like reactions includes using hydrogen peroxide (H_2O_2), which reacts with iron-based nanomaterials to create hydroxyl radicals (•OH) [7]. The conventional Fenton reaction uses iron ions (Fe^{2+}) to trigger the production of hydroxyl radicals through a reaction with hydrogen peroxide, which degrades organic pollutants [12].

Fenton-like reactions utilize iron-based nanoparticles such as nano-zero-valent iron (nZVI) or iron oxide (Fe_2O_3) to boost the destruction of pollutants through a related method [12].

10.2.2.1 Mechanism of Fenton-like Reactions

The combination of iron nanoparticles (nZVI) or iron oxide catalysts together with hydrogen peroxide occurs in Fenton-like systems. The reaction sequence between H_2O_2 and Fe^{2+} produces Fe^{3+} and $\bullet OH + OH^-$ radicals: $Fe^{2+} + + H_2O_2 \rightarrow Fe^{3+} + + \bullet OH + OH^-$ [13].

The reactive hydroxyl radicals (•OH) effectively decompose organic substances and toxic components (Cr, As, Pb) as well. Pollutants undergo oxidation through initiating reactions where radicals turn the contaminants into beneficial substances such as CO_2 and water or simpler organic compounds [12].

10.2.3 Electrocatalysis

Energy-based reactions combining nanocatalysts for enhanced electrochemical reactions occur when the catalyst transfers electrons to pollutants [12]. The environmental degradation process receiving improved efficiency through electrochemical treatment benefits from nanocatalysts consisting of platinum (Pt), palladium (Pd), or gold (Au) [14].

10.2.3.1 Mechanism of Electrocatalysis

Redox reaction efficiency increases in electrocatalysis through nanocatalysts, which support electron movements between pollutants and electrodes [13]. Nanocatalysts help electrochemical pollutant treatments by enabling the conversion of nitrogen oxides (NO_x) and SO_2 pollutants as well as volatile organic compounds (VOCs) into safe end products [13].

Nanocatalysts show exceptional value in metal contaminant removal by enabling chemical reactions that transform dissolved metals into solid forms that evacuation becomes feasible.

10.2.4 Adsorption and Desorption

Nanocatalysts help environmental pollution remediation through their ability to adsorb and desorb pollutants. The surface of nanocatalysts recognizes pollutants through adsorption mechanisms until the catalyst releases these pollutants through desorption processes under specific circumstances [13].

10.2.4.1 Mechanism of Adsorption and Desorption

Nanocatalysts, which include activated carbon and graphene along with carbon nanotubes, exhibit extensive surface areas that act as adsorbents for heavy metals, organic chemicals, and dye contaminants [13]. Pollutants attach to nanomaterial surfaces by physical and chemical forces, which encompass electrostatic forces and hydrogen bonds as well as van der Waals attractions [15].

The adsorbed pollutants enable filtration separation or receive chemical reactions on the catalyst surface before removal [5]. The nanocatalyst can be reused when desorption occurs under pressure or temperature and pH variations [9].

10.2.5 Oxidation-Reduction Reactions

Reduction-oxidation (redox) reactions serve as the core mechanism for various catalytic operations that use nanomaterials [12]. Nanocatalysts help perform oxidation reactions (despite electron loss) while concurrently performing reduction reactions (electron gain) for breaking down polluting substances [16].

10.2.5.1 Mechanism of Oxidation-Reduction Reactions

The nanocatalysts Pd, Pt, and copper (Cu) provide catalyst power to initiate redox reactions, which convert pollutants through reduction or oxidation processes [17]. Educts and products within chromium (VI) conversion to chromium (III) react according to $Cr^{6+} + 3e^- \rightarrow Cr^{3+}$.

The mechanism of nanocatalysts for oxidation reactions includes the production of hydroxyl radicals (•OH) and superoxide anions ($O_2^{\bullet -}$) to achieve compound breakdown through oxidation processes [13] (Figure 10.1).

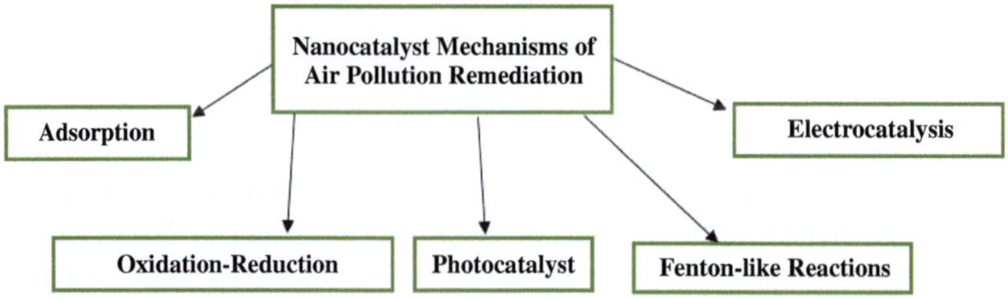

Figure 10.1 Diagrammatic representation of nanocatalyst mechanisms of air pollution remediation.

10.3 Nanotechnology and Air Pollution

Nanotechnology functions by applying nanoscience to materials, which exist at dimensions falling in the range of 10^{-9} m [17]. Both physical characteristics and chemical properties change in materials that measure 10^{-9} m and exhibit unique electrical networking and enhanced strength and colored properties [10].

The properties of strength and color, along with others, differ between bulk materials and those scaled at nanometer dimensions [1]. Metallic materials show electrical properties of semiconductors or insulators when sized at the nanometer level [18].

Scientists attribute the changes in nanoscale properties to four factors: size-dependent surface area ratios, improved mechanical properties, stronger thermal resistance, and modified catalytic response and electrical, magnetic, and optical characteristics [1].

Users can prepare lightweight yet strong nanosystems through these distinctive nanomaterial characteristics for device production that includes indoor and outdoor nano-purifiers according to [10].

The industrial scale development of human civilization leads to a growing production of dangerous air contaminants NO_x, SO_x, and carbon monoxide (CO) [14]. The current pollution control methods show various shortcomings, making several procedures inefficient and generating hazardous waste materials [16].

Most environmental cleanup techniques fail to eliminate small contaminants present in the environment [19]. A new affordable solution must be developed to address this problem, according to [20]. Nanofilters and nanobiomaterials represent two nanotechnology applications for environmental air pollutant reduction [10].

10.4 Nanocoatings

Nanostructured coatings provide superior functionalities compared to standard coatings and achieve better adhesive properties while enabling the development of distinctive surface characteristics [13]. Nanocoatings consist of multiple microns of thickness, which enables them to become resistant to chemical/mechanical damage and heat while also performing as self-cleaning surfaces [9].

The improved structural properties allow the replacement of fewer raw materials in structure construction [21]. Nanocoatings lead to decreased energy usage through their application while simultaneously decreasing the use of detergent chemicals [13]. Nanocoatings result in decreased pollutant emissions because of their application [18].

10.5 Nanosensor

The continuous control of air pollution stands as an essential air pollution requirement [13]. Nanosensors represent contemporary technological solutions that detect physical stimuli within nanometer measurement boundaries [13]. Nanosensors exist in two categories that include nanoparticles and nanostructured materials [12]. Nanoparticles in the form of nanoscale spherical materials function as both receptors and visual-spatial sensors [19].

Scientists have built nanosensors using porous silicon nanomaterials for identifying chemical and biological interactions [21]. The industrial sector faces the major risk of toxic gas leaks through its facilities [18]. The key industrial sensors presently operate effectively at measuring low concentrations of gases according to [17].

The need to develop faster and more accurate sensors requires employing modern technologies for their manufacturing process [22]. Single- or multilayered nanotubes form this type of sensor that functions by absorbing toxic gas molecules [19]. These sensors can detect small quantities of fatal gas molecules. Nanosensors identify multiple sorts of gaseous compounds, such as dioxin, while surpassing traditional sensor performance [23].

Through their development, smart dust nanosensors achieve a novel application method. A successful first production of smart dust led these sensors to near scientific applications [12]. The creation of smart dust serves to make advanced sensors possible through nanocomputer sensors produced in very light forms [12].

The particles have a silicon composition and employ a wireless mechanism to transmit data from their structure to a central database [22]. The nanosensors hover in air networks for prolonged periods thanks to their capability of extracting energy from solar power [22].

Smart dust transmits environmental information, including air temperature and pressure data, humidity ranges, and chemical readings, extending up to 20 km before sending data to control air pollution evolution in specific areas [6].

10.6 Degradation as a Nano Purifier

The research shows promise regarding air quality improvement through reduced air pollutant levels using nanocatalysts [9]. A nanocatalytic system performs rapid selective chemical transformations that yield high-quality products while permitting catalyst regeneration better than traditional catalyst systems [6].

Photocatalytic properties can be found in materials that include TiO_2, ZnO, and tungsten oxide (WO_3) [9]. Photocatalysts serve as effective environmental remediators because they break down organic pollutants into harmless end products [22]. The underlying semiconductor principle permits organic molecules to achieve photo-oxidation [21]. When sufficient light intensity prevails, the valence band to conduction band charge transfer begins, resulting in the oxidation of substances in the surrounding area [6].

The advancement of nanotechnology enables researchers to enhance the properties of semiconductor photocatalysts regarding both reactivity and selectivity [22].

10.7 Adsorption as a Nano Purifier

The industrial use of nano-adsorbents relies on carbon nanostructures because of their exceptional physical characteristics, which include average pore diameter, pore volume, and surface area with high selectivity, affinity, and capacity [24].

Highly reactive surface sites together with structural bonds function significantly in adsorption processes [4]. Nanomaterials serve as adsorbents that enable different treatments for air contaminants according to [24].

Carbon nanotubes on quartz filters function to eliminate air pollutants, including VOCs, for monitoring purposes through interactions between π electrons [15]. Nano-adsorptive materials exhibit super adsorbent characteristics that capture aerial pollutants, including hazardous gases and VOCs as described by [9] hydrogen bonding, electrostatic forces, π—π interactions, and van der Waals forces.

10.8 Filtration as an Air Purifier

The filtration-based air purifier functions through membranes which possess nanometer-sized pores between 1 and 10 nm [9]. The production of drinking water using nanofiltration technology relies on the use of polyethylene terephthalate along with aluminum and its variants [11].

The main force behind nanofiltration operations is maintained through pressure. Nanofiltration membranes outperform standard methods through their strong ability to reject multivalent ions and pesticides and heavy metals while functioning as efficient particulate air filters [1].

The application of nanofiltration with silver nanoparticles occurs because these nanoparticles demonstrate antibacterial properties while exhibiting minimal harm towards mammalian cells [6]. Nanosilver finds its place in filtering membranes through different impregnation methods according to [9]. Air filtration devices now commonly use silver nanoparticles filters due to their antibacterial capabilities together, with minimal harm to mammalian cells [21].

Different means of nanosilver incorporation exist for materials, including spraying methods, product painting, and plastic incorporation [1]. Nanotechnology provides treatment methods which enhance air pollution performance through adsorption, alongside degradation and filtration techniques [11].

10.9 Design a Mega Machine Device Consisting of Nanotechnology-Based Cascaded Filters to Clean Polluted Air

The published research findings in nanotechnology enable dangerous pollutants to be extracted from contaminated air streams [3]. The creation of an air cleaning solution exists for home use and industrial applications at various capacity levels [1].

The implementation of cascaded nanotechnology filters through a specific design structure allows additive functionalities that purify contaminated air for clean air production [11]. A cascaded mega air-cleaner machine employs various filters containing distinct nanomaterials that sequence serially according to airflow direction [1]. The design relies on an imaginative perfect environmental cleaning system with full operational efficiency that uses an ideal fan to process air with any degree of impurities through multiple nanotechnology-based filtration stages to produce completely clean output [11]. An ideal fan or ideal pump operates at the first stage of the system through perfectly renewable energy sources [11]. The energy source known as the ideal energy source [3] should operate similarly in ideal air purifiers as it functions in ideal water purifiers [1].

10.10 Challenges and Future Directions

Several issues need to be resolved before nanocatalysts can achieve mass deployment for environmental treatment, according to [12]. The main problem at present involves nanocatalyst reliability combined with their potential for reuse [15]. Time exposes nanomaterials to degradation mechanisms that cause both aggregation and loss of functioning capacity. The advancement of more stable and reusable nanocatalysts continues through modifications and functionalization of their surfaces [25].

Environmental challenges arise because nanoparticles generate risks to their surrounding environment [2]. Scientists need to conduct extensive research about the environment-friendly properties of nanomaterials, specifically nanocatalysts, because their long-term impacts on ecosystems along with human health remain unclear [12].

Future research will dedicate efforts to improving sustainable nanocatalysts and scalability and their integration within existing pollution control systems [9].

10.11 Conclusion

Nanocatalysts made from metal oxides demonstrate great potential as air pollution control agents because they provide high catalytic performance alongside durable structure and adjustable usage across environmental processes. The physical and chemical properties of these materials provide strong redox activity and adjustable electronic structures together with a high surface area that effectively degrades VOCs, NO_x, CO, and particulate matter. The progress in nanotechnology research enabled scientists to make innovative synthesis techniques which boost the performance quality and selectivity of these catalysts. The research needs additional attention to tackle catalyst deactivation issues and improve the cost-effectiveness and scale-up abilities of this method. Future work needs to develop sustainable metal oxide nanocatalysts that both reduce production costs and improve their capacity for reusability as well as extended operational lifetime. A reduction of air pollution needs research to establish catalyst-powered advanced air purification systems and hybrid nanomaterials development. Ongoing innovation and interdisciplinary work between researchers help metal oxide nanocatalysts develop into promising tools for creating a sustainable environment.

References

1 Muhammed A, Aransiola SA, Bamisaye A, Abioye OP, Maddela NR, editors. Blue economy in Nigeria and the African continent. In: *Marine Bioprospecting for Sustainable Blue-Bioeconomy*. Cham: Springer; 2024. https://doi.org/10.1007/978-3-031-68911-6_18

2 Aitha S. *Principles of Environmental Studies*. Srinivas Publication; 2019. p. 6–21.

3 Gupta VK, Saleh TA. Sorption of pollutants by porous carbon, carbon nanotubes and fullerene—an overview. *Environ Sci Pollut Res*. 2013;20:2828–43.

4 Aithal PS, Aithal S. The concept and characteristics of ideal energy system and its realization constraints. *Int J Appl Eng Manag Lett*. 2018;2(3):127–37.

5 Aithal S, Aithal PS. Cleaning the environment using nanotechnology – a review-based mega-machine design. In: Paul PK, Aithal PS, Saavedra R, Aremu B, Sinha RR, editors. *Environmental Information Sciences: With Aspects on Allied Areas and Other Emerging Interdisciplinary Environmental Concerns*. New Delhi: New Delhi Publishers; 2020. p. 13–40.

6 Amade R, Hussain S, Ocana IR, Bertran E. Growth and functionalization of carbon nanotubes on quartz filter for environmental applications. *J Environ Eng Ecol Sci*. 2014;3:1–7.

7 Baraton MI, Merhari L. Advances in air quality monitoring via nanotechnology. *J Nanopart Res*. 2004;6:107–17.

8 Chen B, Kan H. Air pollution and population health: a global challenge. *J Environ Health Prev Med*. 2008;13:94–101.

9 Eranna G. *Metal Oxide Nanostructures as Gas Sensing Devices*. CRC Press; 2011.

10 Fulekar MH, Pathak B, Kale RK. Nanotechnology: perspective for environmental sustainability. In: Fulekar MH, Pathak B, Kale RK, editors. *Environment and Sustainable Development*. New Delhi: Springer; 2014. p. 87–114.

11 Griffin RD. *Principles of Hazardous Materials Management*. 2nd ed. CRC Press; 2009.

12 Jimenez-Cadena G, Riu J, Rius FX. Gas sensors based on nanostructured materials. *Analyst* 2007;132:1083–99.

13 Mammadova S, Nasibova A, Khalilov R, Mehraliyeva S, Valiyeva M, Anar S, et al. Nanomaterials application in air pollution remediation. *Eurasian Chem Commun*. 2022;4:1–7.

14 Mohammed EF. Revolution of nanotechnology in air pollution. *Int Conf Chem Environ Eng*. 2018;9:3–9.

15 Nasrollahi F, Moradi E, Rafiei M. Nano coating and reducing the energy consumption in city buildings. *Majlesi J Energy Manag*. 2014;3:17–22.

16 Rani NU, Sharma P, Sharma RK. The role of biotechnology in environmental remediation: opportunities and challenges. *Afr J Biol Sci.* 2024;6:5–18.
17 Schell LM, Gallo MV, Denham M, Ravenscroft J. Effects of pollution on human growth and development: an introduction. *J Physiol Anthropol.* 2006;25(1):103–12.
18 Sharma R, Bandichhor R, Mishra V, Sharma S, Yadav S, Mehta S, et al. Advanced metal oxide-based nanocatalysts for the oxidative synthesis of fine chemicals. *Mater Adv.* 2023;4(10):1795–830.
19 Singh SB, Tandon PK. Catalysis: a brief review on nano-catalyst. *J Environ Chem Eng.* 2014;2:106–15.
20 Sofian I, Harwin Y, Kurniawan A, Adityawarman D, Indarto A. Nanotechnologies in water and air pollution treatment. *J Environ Technol Innov.* 2012;1:136–48.
21 Taran M, Safaei M, Karimi N, Almasi A. Benefits and application of nanotechnology in environmental science: an overview. *Biointerface Res Appl Chem.* 2021;11(1):7860–70.
22 Theodore L, Kunz RG. *Nanotechnology: Environmental Implications and Solutions.* John Wiley & Sons; 2005.
23 Wang S, Sun H, Ang HM, Tadé MO. Adsorptive remediation of environmental pollutants using novel graphene-based nanomaterials: reviews in chem. *Eng. J Chem Eng.* 2013;226:336–47.
24 Yuan W, Shi G. Graphene-based gas sensors. *J Mater Chem.* 2013;1:10078–91.
25 Yunus IS, Harwin, Kurniawan A, Adityawarman D, Indarto A. Nanotechnologies in water and air pollution treatment. *Environ Technol Rev.* 2012;1(1):136–48.

11

Metal Oxide Nanocatalysts for Water Purification Processes

Ikechukwu P. Ejidike[1, 2, 3], Olayinka S. Okoh[1, 2], Temitope O. Fakoya[4], Solomon A. Olaleru[5], Mercy O. Bamigboye[6], and Ameen O. Adiru[7]*

[1]*Biological and Analytical Application of Synthetic and Nanomaterials Research (BAASNMR) Group, Department of Chemical Sciences, Faculty of Science, Anchor University, Lagos, Nigeria*
[2]*Center for Global Health, Anchor University, Lagos, Nigeria*
[3]*Department of Chemistry, College of Science, Engineering and Technology, University of South Africa, Pretoria, South Africa*
[4]*Department of Industrial Chemistry, Faculty of Science, University of Ilesa, Ilesa, Nigeria*
[5]*Department of Physical Science, School of Science, Yaba College of Technology, Lagos, Nigeria*
[6]*Department of Industrial Chemistry, Faculty of Physical Sciences, University of Ilorin, Ilorin, Nigeria*
[7]*Department of Chemistry and Industrial Chemistry, Faculty of Pure and Applied Sciences, Kwara State University, Malete, Nigeria*

*Corresponding author: pejidike30@gmail.com; iejidike@aul.edu.ng

11.1 Introduction

Water contamination is one of the attendant effects of rising population, urbanization, and industrialization/industrial growth to match the rising population, and this is becoming a global concern as it affects global public health. Human activities, not limited to farming, poor waste management, and industrial by-products, litter the land with hazardous substances, including heavy metals, organic compounds, pharmaceuticals, and pesticides. These end up polluting surface water and groundwater. These pollutants, besides compromising the aquatic ecosystem, also pose a threat to public health via bioaccumulation and potential carcinogenic or endocrine-disrupting effects [1]. The traditional methods of water treatment, which include biological, mechanical, and physical techniques, have failed in sufficiently/adequately eliminating these contaminants from water. While advanced technologies like activated carbon adsorption and membrane filtration have been identified to offer some advantages over the traditional methods, they have been marred by high-cost, energy demands, and inadequate degradation of pollutants [2]. Search for alternative water purification methods to address these challenges birthed metal oxide nanocatalysts

Metal Oxide Nanocatalysts for Sustainable Energy Production, First Edition. Edited by Abayomi Bamisaye et al.
Published 2026 by Wiley-VCH GmbH.

(MONCs), and these are gaining recognition being more efficient and sustainable methods of water purification.

High surface area, photoactivity, and redox capabilities are some characteristics of MONCs like cerium oxide, manganese dioxide, iron oxide, titanium oxide, and zinc oxide. MONCs further possess great ability to form reactive oxygen species (ROS) like hydroxyl radicals and superoxide anions when exposed to light radiation. These capabilities form the basis for efficient application in water treatment [3]. Furthermore, these MONCs can be engineered to improve their selectivity and photocatalytic efficiency, thereby improving their efficiency in water treatment. Highly reactive species, like ROS, are capable of mineralizing resistant organic molecules. And the capability of MONCs to generate such ROS, in situ, informs the cruciality of MONCs in the advanced oxidation processes (AOPs) necessary for the degradation of a wide array of organic pollutants. This versatility of MONCs necessitates their utilization in diverse water treatment processes, thereby offering a scalable solution to water purification challenges [4].

This chapter provides an overview of MONCs' potential in water treatment and purification, concerning their mechanisms and applications. The chapter also examines the roles of MONCs in degrading organic contaminants, eliminating heavy metals, and pharmaceutical residues, besides functioning in various AOPs.

11.2 Metal Oxide Catalysts for Degrading Organic Contaminants in Water

The use of MONCs, in water treatment, for the degradation of organic pollutants is an improvement, and indeed a breakthrough over the traditional methods. Many organic contaminants like endocrine-disrupting compounds (EDCs), phenols, synthetic dyes, and numerous industrial products are not easily biodegraded. Hence, traditional methods of water treatment are not effective in eliminating [5]. MONCs offer a robust solution by facilitating AOPs, where highly reactive species degrade these compounds. Among the various MONCs, titanium dioxide (TiO_2) has garnered the most attention consequent to its strong oxidizing potential, photochemical stability, cost-effectiveness, and environmental benignity. TiO_2, upon ultraviolet (UV) illumination, produces electron-hole pairs that catalyze oxidation-reduction reactions, ultimately yielding hydroxyl radicals (•OH), and these radicals are capable of non-selectively oxidizing a broad spectrum of organic molecules [6–8]. Zinc oxide (ZnO) exhibits similar photocatalytic behavior with a comparable bandgap (~3.3 eV) but is susceptible to photocorrosion during extended UV exposure. Nevertheless, ZnO is favored for its high quantum efficiency and ability to be synthesized into various morphologies, including nanorods and nanoparticles, which enhance surface reactivity [7]. Iron oxides, such as hematite (Fe_2O_3), supports Fenton and photo-Fenton reactions, which are advantageous for generating hydroxyl radicals.

Manganese dioxide (MnO_2) and cerium oxide (CeO_2) also exhibit catalytic properties suitable for oxidative degradation. MnO_2 demonstrates oxidizing ability in both

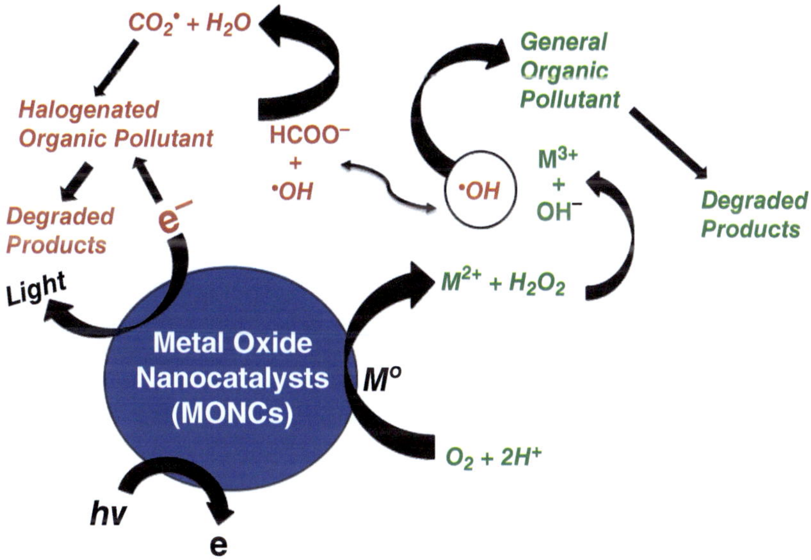

Figure 11.1 Mechanism of organic contaminants degradation using metal oxide catalysts.

acidic and neutral media, and it can activate oxidants such as ozone and persulfate, further boosting ROS production [9]. CeO_2 is particularly notable for its oxygen storage capacity and reversible redox cycle between Ce^{3+} and Ce^{4+}, enabling continuous catalytic activity [10]. These properties make it suitable for degrading EDCs, dyes, etc. The mechanism of organic contaminants degradation using metal oxide catalysts is represented in Figure 11.1.

To optimize the photocatalytic properties of MONCs, several strategies have been explored. One such strategy involves the addition of elements like nitrogen, sulfur, silver, or copper to the MONCs to improve their photocatalytic properties. Another is the coupling of the MONCs with other semiconductors to form heterojunctions with the aim of improving charge separation and light absorption. MONCs are also coated with carbon-rich materials like graphene oxide or carbon nanotubes with a view to increasing the MONCs' surface area for better/improved reactions [11].

Recent studies demonstrate the effectiveness of composite nanocatalysts. One of such composite nanocatalyst is TiO_2-graphene. This composite increases the free movement of electrons and prevents them from quickly reuniting with the "holes" left behind, thereby resulting in improved degradation of pollutants like methylene blue and bisphenol-A [12]. Similarly, ternary composites combining ZnO, g-C_3N_4, and Ag nanoparticles show synergistic effects, improving photocatalytic efficiency under solar irradiation [13].

MONCs therefore provide a transformative/innovative approach over traditional methods for the efficient degradation of organic contaminants in water. Their various unique properties make them ideal for AOP-based treatments, and the possibility of producing composite structures with them for the desired purpose makes them very promising in overcoming a series of limitations associated with traditional methods of water purification. Considering their large-scale applications, MONCs can be said to be the future for water purification.

11.3 Removal of Heavy Metals and Emerging Pollutants like Pharmaceuticals and Pesticides

Heavy metals are nonbiodegradable, and this results in their accumulation in the environment, including water bodies. Examples of these heavy metals are As, Cd, Cr, and Pb. Accumulation of these heavy metals in the environment can culminate in their bioaccumulation in living organisms, both terrestrial and aquatic, and this poses serious public health risks. Similarly, emerging pollutants like pharmaceuticals and pesticides are washed from land into water bodies, endangering aquatic organisms besides disrupting endocrine and compromising immune systems even at trace levels [14]. But MONCs have multiple mechanisms for the removal of this group of contaminants.

Some of the mechanisms employed by MONCs in eliminating heavy metals include adsorption, redox transformation, and precipitation. For example, iron oxide nanoparticles have a strong affinity for metal ions and magnetic properties. These properties aid in the wide use of iron oxide nanoparticles in water treatment, as the nanoparticles are easily recovered after use. In addition, the surface hydroxyl groups of iron oxide-based MONCs interact with metal ions through complexation, which promotes efficient adsorption [15].

Taking Arsenic (As) as an example, As(V) is more stable and less mobile than As(III); hence, As(V) is more readily removed from water. So, to eliminate As from water, Fe_3O_4 nanoparticles can either bind directly to As(V) or convert As(III) to As(V). Taking Cr, Cr(VI), which is carcinogenic, is more toxic than Cr(III). In eliminating Cr from water, Cr(VI) is reduced to Cr(III) by either MnO_2 or Fe_3O_4. The Cr(III) then forms a solid hydroxide, which is then precipitated out at neutral or alkaline pH [16]. This capability of MONCs to drive redox reactions is a unique merit, as they can transform toxic species into safer ones, and also trap them. This contributes to their effectiveness and efficiency in water treatment/purification [17].

There are various nonbiodegradable pharmaceuticals that have been detected in treated effluent and water bodies. Some of these include antibiotics like ciprofloxacin and tetracycline; nonsteroidal anti-inflammatory drugs (NSAIDs) like ibuprofen and diclofenac; and hormonal substances such as estradiol. To address this, researchers have successfully identified some MONCs that can degrade this class of pollutants under ultraviolet or visible light using photocatalytic reactions. Some of the identified MONCs in this category include TiO_2, ZnO, and $BiVO_4$. These catalysts generate ROS that attack the structure [16–18].

Pesticides such as atrazine, imidacloprid, and glyphosate are also resistant to natural degradation processes. Studies have shown that TiO_2 and ZnO can degrade these pesticides through photocatalytic oxidation. For example, atrazine is oxidized

Table 11.1 MONC applications for pollutant removal.

Pollutant category	Example contaminants	Effective MONCs	Primary mechanism	Removal efficiency (%)
Pharmaceuticals	Ciprofloxacin, carbamazepine	Cu-doped TiO_2, $BiVO_4$/rGO	Photocatalysis	85–98
Pesticides	Atrazine, glyphosate	TiO_2, ZnO	•OH radical oxidation	90–99
Heavy metals	As(III/V), Cr(VI), Pb(II)	Fe_3O_4, MnO_2	Adsorption/ redox	92–99
Endocrine disruptors	Bisphenol A, estradiol	CeO_2, TiO_2-graphene	Catalytic degradation	88–95
Industrial dyes	Methylene blue	N-doped TiO_2, ZnO/g-C_3N_4/Ag	Visible-light photocatalysis	

by •OH radicals generated on TiO_2 surfaces, leading to dechlorination and ring cleavage, eventually forming nontoxic end products like carbon dioxide and water [18, 19]. The degradation of efficiency depends heavily on catalyst properties (e.g., crystallinity, surface area). To further enhance removal performance, composite and doped MONCs are being developed. For instance, Cu-doped TiO_2 exhibits improved visible light absorption and enhanced degradation rates of fluoroquinolone antibiotics. Similarly, $BiVO_4$ combined with reduced graphene oxide improves charge carrier separation and increases photocatalytic degradation efficiency for pharmaceuticals such as carbamazepine [20]. Table 11.1 provides an overview of various pollutant categories, for example, pollutants and the MONCs employed for their removal, along with the primary removal mechanisms.

Overall, the versatility of MONCs in adsorbing, transforming, and degrading heavy metals and emerging contaminants makes them invaluable tools for advanced water treatment. Continuous research on improving their selectivity, stability, and environmental compatibility will be crucial for practical implementation.

11.4 Photocatalysis and AOPs for Water Treatment

Photocatalysis is a cornerstone of AOPs, wherein a semiconductor catalyst is activated by light to induce the generation of ROS that degrade organic contaminants. MONCs, especially those based on TiO_2, ZnO, and Fe_2O_3, serve as key components in these mechanisms because of their semiconductor characteristics, extensive surface area, and strong redox potential [21].

When exposed to ultraviolet or visible light with energy equal to or exceeding the photocatalyst's bandgap, electrons are promoted from the valence band to the conduction band, resulting in the formation of electron vacancies, or holes (h^+), in the valence band. These charge carriers initiate a cascade of redox reactions at the catalyst surface. The electrons generated upon light excitation reduce molecular oxygen to form superoxide radicals ($O_2^{\bullet-}$), whereas the corresponding holes oxidize water molecules or hydroxide ions to yield hydroxyl radicals (•OH). These ROS exhibit strong oxidative potential and are instrumental in breaking down organic contaminants into harmless end-products [22, 23].

TiO_2 is the archetypal photocatalyst, but its activation is restricted to UV light due to its wide bandgap. To overcome this limitation, doping with nonmetal elements (e.g., N, S) or coupling with narrow-bandgap semiconductors (e.g., CdS, g-C_3N_4) has been used to enable visible-light-driven photocatalysis [24]. For example, N-doped TiO_2 exhibits a redshift in light absorption, while TiO_2/CdS heterojunctions show improved charge separation and extended light response.

Iron oxide (Fe_2O_3), with a smaller bandgap (~2.1 eV), is suitable for visible light applications. The presence of hydrogen peroxide activates the catalyst to initiate Fenton and photo-Fenton processes, where Fe^{2+} reacts with H_2O_2 to produce •OH radicals. Photo-Fenton reactions are particularly attractive as they utilize solar light to regenerate Fe^{2+} from Fe^{3+}, enhancing the sustainability and efficiency of the process [25]. The applicability of photo-Fenton processes is largely influenced by pH, as acidic conditions (pH 2.5–3.5) are typically required for optimal performance.

Persulfate-based AOPs have also gained attention in recent years. In these systems, persulfate ($S_2O_8^{2-}$) is activated by MONCs to produce sulfate radicals ($SO_4^{\bullet-}$), which have a higher redox potential than •OH and are effective over a broader pH range. Oxides of some transition metals like Mn, Co, and Ce have been identified to have the capability to activate persulfate both under heat and light conditions [26]. An example is MnO_2 nanoparticle, which is used to decompose peroxymonosulfate (PMS) to yield sulphate radicals ($SO_4^{\bullet-}$). These sulphate radicals are then utilized in degrading contaminants that are otherwise difficult to degrade.

Ozone-assisted catalysis is another approach to degrading contaminants. Under this approach, MONCs like TiO_2 and MnO_2 interact with ozone layers to generate hydroxyl radicals in addition to other radicals. The generated radicals then degrade into a series of nonbiodegradable organic contaminants like pharmaceuticals, pesticides, and industrial dyes [27]. The use of MONCs in pollutant degradation can be upscaled. This is because the efficiency of MONCs in degrading pollutants can be increased, and it is achieved via nanostructure engineering aimed at improving AOP performance. Some strategies employed for this include creating mesoporous structures, core-shell architecture, and the incorporation of magnetic materials. These are meant for the purpose of increasing the surface area of MONCs for reaction, improving charge dynamics, and easy recovery of the MONCs. Surface modification with hydrophilic or hydrophobic groups also tailors the interaction between the catalyst and target pollutants, enhancing selectivity [28].

In summary, MONC-enabled photocatalysis and AOPs present versatile, efficient, and scalable solutions for the treatment of recalcitrant pollutants in water. Continuous innovation in catalyst design, light utilization, and integration with hybrid technologies will be vital in optimizing these systems for real-world deployment.

11.5 Challenges with Scaling Up Nanocatalysts for Widespread Water Filtration

While MONCs demonstrate remarkable performance in laboratory-scale studies, transitioning these materials into full-scale water treatment systems presents significant technical, economic, and environmental challenges. Addressing these issues is crucial to unlocking the full potential of MONCs for real-world applications.

11.5.1 Catalyst Recovery and Reusability

One of the most pressing challenges in scaling up MONC applications is the efficient recovery of the catalyst posttreatment. Most MONCs are synthesized in nanoparticulate form to maximize surface area and catalytic activity; however, their small size makes them difficult to separate from treated water. This not only adds to operational costs but also raises concerns over the secondary release of nanoparticles into the environment [29].

To mitigate this, researchers have explored immobilization techniques, such as anchoring MONCs onto inert substrates (e.g., glass beads, silica, ceramics, and membranes) or incorporating them into composite structures. Immobilized systems allow easy recovery and reuse but often suffer from reduced surface area and catalytic efficiency. Another promising strategy involves the synthesis of magnetically separable MONCs, such as Fe_3O_4-based composites, which can be retrieved using external magnetic fields [30].

11.5.2 Stability and Long-Term Performance

MONCs are prone to deactivation due to various factors, including agglomeration, fouling, and leaching of active components during extended use. For instance, ZnO nanoparticles can undergo photocorrosion, and iron-based catalysts may lose their Fenton activity in the presence of chelating agents or high salinity. Moreover, the performance of MONCs in real water matrices, which contain competing ions, natural organic matter, and fluctuating pH, is often inferior to results obtained in controlled laboratory settings [31].

Ensuring the long-term stability of MONCs requires careful material design. Approaches include doping to enhance structural integrity, coating particles with protective layers (e.g., silica or carbon), and engineering hierarchical structures that resist aggregation. Reusability tests, life cycle assessments (LCAs), and performance evaluations under realistic conditions are necessary to assess operational viability.

11.5.3 Cost of Synthesis and Scalability

The synthesis of MONCs can involve complex procedures, expensive precursors, and high-temperature treatments, which hinder cost-effective production at scale. For instance, methods like hydrothermal synthesis, sol-gel processing, and chemical vapor deposition yield high-purity nanomaterials but are resource-intensive. Additionally, scaling up laboratory protocols without compromising material quality remains a key obstacle [32].

To address this, researchers are investigating green and scalable synthesis routes. These include biosynthesis using plant extracts or microorganisms, low-temperature precipitation methods, and solvent-free techniques. Moreover, industry-focused research is shifting toward the use of earth-abundant and low-cost materials (e.g., iron oxides, manganese oxides) over rarer or expensive alternatives like $BiVO_4$ or CeO_2 [33].

11.5.4 Environmental and Health Concerns

As engineered nanomaterials, MONCs pose potential ecological and toxicological risks upon environmental release. The fate, transport, bioaccumulation, and transformation of these particles in aquatic systems are not yet fully understood. For example, there is evidence that nanoparticles can induce oxidative stress in aquatic organisms or interact with microbial communities in unintended ways [34].

Hence, the development of MONCs must be guided by environmental safety assessments. Strategies to minimize environmental risks include the use of immobilized or encapsulated systems, selecting biodegradable or inert support materials, and performing comprehensive ecotoxicological studies. Regulatory frameworks and standardized testing protocols will also be essential to ensure safe deployment.

11.5.5 Integration with Existing Water Treatment Infrastructure

Another practical barrier to MONC adoption is compatibility with existing water treatment plants. Integrating nanocatalysts into traditional systems requires modular designs that accommodate flow rates, pressure requirements, and maintenance constraints. Pilot-scale demonstrations and cost-benefit analyses are crucial to convince stakeholders of their feasibility and long-term advantages.

11.6 Prospects for MONC-Based Sustainable Water Treatment

Considering the unique properties of MONCs, there is no doubt that they will continuously remain relevant in water treatment and purification as long as their use remains sustainable, cost-effective, and environmentally safe. Several forward-looking strategies are being pursued to ensure that MONC technologies transition from promising laboratory results to impactful, real-world applications.

11.6.1 Development of Solar-Active and Visible-Light Photocatalysts

A major limitation of conventional photocatalysts such as TiO_2 and ZnO is their dependence on UV light, constituting only a narrow spectral band (~4%) of incoming solar energy. To maximize solar utilization and reduce energy costs, the development of visible-light-active MONCs is a top research priority. Doping with nonmetals (e.g., N, C, S), coupling with narrow-bandgap semiconductors (e.g., $BiVO_4$, g-C_3N_4), and forming heterojunctions have all shown promise in extending light absorption into the visible range [35].

Emerging materials such as black TiO_2, plasmonic-metal-decorated oxides (e.g., Ag, Au), and oxygen-deficient semiconductors are being explored for their enhanced light harvesting and charge separation capabilities. These advances support the design of photocatalysts that operate effectively under sunlight, opening the door to decentralized and off-grid treatment systems powered by renewable energy [36].

11.6.2 Hybrid and Multifunctional Treatment Systems

Integrating MONCs into hybrid systems that combine physical, chemical, and biological processes can significantly enhance treatment efficacy. For example, coupling photocatalysis with membrane filtration, adsorption, or biodegradation can result in synergistic effects that improve contaminant removal, reduce energy consumption, and extend the lifespan of each component [37].

Multifunctional MONCs that exhibit photocatalytic, adsorptive, antimicrobial, and redox properties simultaneously are particularly attractive. These materials can target a wide range of contaminants, not limited to pathogens, heavy metals, and organic micropollutants, thus supporting a more holistic water purification strategy.

11.6.3 Smart and Modular Reactor Designs

The successful deployment of MONCs in field settings will depend on the design of efficient and scalable reactor systems. Recent advances include the development of modular photoreactors that allow continuous monitoring and regulation of treatment parameters. These systems are designed to be compact, portable, and easy to maintain, which makes them well-suited for localized water treatment applications, notably within communities with limited access to centralized infrastructure [38].

Immobilized catalysts on optical fibers, membranes, or glass surfaces can be integrated into continuous flow systems, ensuring sustained contact between the catalyst and contaminants while minimizing catalyst loss. There is also the development of innovative reactors, like 3D-printed reactors and fluidized beds, which improve water movement through the system and access to light by the catalyst to aid photocatalysis.

11.6.4 Green and Scalable Synthesis Approaches

Sustainability is becoming a major concern of nanotechnology, and this is propelling research in eco-friendly/safe methods of synthesizing MONCs. These methods utilize the use of safe solvents, low energy, and natural materials like plant extracts, bacteria, and fungi to synthesize required nanoparticles. MONCs synthesized through this approach seem to possess unique surface features and are more biocompatible. With this, these MONCs are better in environmental applications since they are biodegradable [39]. Furthermore, to scale up this approach for large-scale production of MONCs, researchers are exploring the use of industrial waste and renewable materials as starting materials for nanoparticle synthesis. This will ensure a low cost of production and prudent use of resources.

11.6.5 Life Cycle Assessment and Environmental Impact Evaluation

The sustainability of MONC technologies is hinged on the outcome of the assessment of their environmental impacts, and this can be via LCA. Parameters to assess would include starting materials and energy requirement for synthesis; environmental impact of the process; and any associated risk during and after the process. Careful assessment of each of these parameters would culminate in the selection of the best starting materials and manufacturing methods that are most eco-friendly, as environmental risks and health hazards would be minimized optimally [40]. However, this would need an interdisciplinary approach requiring input from scientists, toxicologists, engineers, and policy makers to ensure that guidelines/protocols and standards are developed for the sustainable and safe use of MONCs in the treatment and purification of water [40, 41].

Considering the advances being witnessed in material science, reactor engineering, and green chemistry, it is safe to say the future of MONC-based water treatment and purification is bright. With the right investment, favorable government policies, and interdisciplinary collaborations, MONCs will no doubt play a crucial role in the global effort to ensure clean and safe water for all.

11.7 Conclusion

MONCs are indeed an innovative development in the treatment and purification of water. This is because water contaminants that were previously identified as difficult to degrade are now found to be degradable by a variety of MONCs. This is made possible by their unique properties, like high surface-to-volume ratio, redox activity, and ROS generation capability. These properties equipped MONCs with the capabilities to degrade a wide array of water contaminants like persistent organic compounds, heavy metals, pharmaceuticals, and pesticides. In this chapter, we have considered various roles that MONCs can play in the degradation of pollutants concerning water treatment and purification. They are effective in degrading complex contaminants using photocatalysis and AOPs simultaneously. MONCs like TiO_2, ZnO, Fe_2O_3, MnO_2, and CeO_2 can also be structurally engineered and doped with other elements to enhance their activity in visible light, improving their efficiencies. MONCs are capable of detoxifying emerging contaminants and heavy metals by trapping, transforming, and chemically reacting with them thus, making them safer for the environment. Innovations in composite and hybrid nanomaterials have enabled multifunctional systems capable of addressing a broad spectrum of contaminants within a single treatment cycle. Despite their high potential, MONCs face significant challenges in terms of real-world deployment. Issues such as catalyst recovery, long-term stability, environmental safety, and economic scalability must be addressed through smart design strategies and policy frameworks. The incorporation of MONCs into modular, decentralized, and solar-powered treatment units represents an exciting direction for future water treatment solutions, particularly in low-resource and off-grid communities.

Looking ahead, interdisciplinary efforts will be crucial. Material scientists must work closely with environmental engineers, toxicologists, and policymakers to ensure the safe, efficient, and sustainable application of MONCs. Emphasis on green synthesis, LCAs, and environmental risk evaluation will further solidify MONCs as environmentally responsible tools in the global quest for clean water. Conclusively, MONCs stand at the forefront of innovative water purification technologies. With continuous research, thoughtful engineering, and sustainable practices, these materials have the potential to address some of the most pressing challenges in environmental remediation and public health.

Acknowledgment

The Department of Chemical Sciences, Anchor University, Lagos, Nigeria, and the Research Directorate of the University of South Africa, Florida Campus, are gratefully acknowledged by the authors.

Conflict of Interest

The authors declare that they have no conflicts of interest regarding the publication of this chapter.

References

1 Olaleru SA, Molokwu MI, Mathew S, Ejidike IP, Oyebamiji OO. Enhanced photocatalytic degradation of methylene blue dye using TiO_2 nanoparticles obtained via chemical and green synthesis: a comparative analysis. *Pure Appl Chem*. 2025;97(5):541–53.
2 Qu X, Alvarez PJJ, Li Q. Applications of nanotechnology in water and wastewater treatment. *Water Res*. 2013;47(12):3931–46.
3 Schneider J, Matsuoka M, Takeuchi M. Understanding TiO_2 photocatalysis: mechanisms and materials. *Chem Rev*. 2014;114(19):9919–86.
4 Chong MN, Jin B, Chow CWK, Saint C. Recent developments in photocatalytic water treatment technology: a review. *Water Res*. 2010;44(10):2997–3027.

5 Brillas E, Martinez-Huitle CA. Decontamination of wastewaters containing synthetic organic dyes by electrochemical methods: a general review. *Appl Catal B Environ*. 2015;166–167:603–43.

6 Fujishima A, Zhang X, Tryk DA. TiO_2 photocatalysis and related surface phenomena. *Surf Sci Rep*. 2008;63(12):515–82.

7 Verma A, Haritash AK. Evaluation of TiO_2, ZnO, and Fe_2O_3 nanoparticles for photocatalytic degradation of chlorpyrifos in aqueous solution. *J Environ Chem Eng*. 2019;7(4):103222.

8 Umar M, Roddick F, Fan L. Recent advances in the application of TiO_2-based catalysts for degrading emerging contaminants in water: a review. *Chem Eng J*. 2015;266:121–38.

9 Wang S, Ang HM, Tade MO. Volatile organic compounds in indoor environment and photocatalytic oxidation: state of the art. *Environ Int*. 2007;33(5):694–705.

10 Wang Y, Wang Z, Li Y. Enhanced photocatalytic activity of CeO_2-based nanomaterials: a review. *Chem Eng J*. 2020;401:126074.

11 Dong F, Zhao W, Wu Z, Guo S. Band structure and visible light photocatalytic activity of multi-type nitrogen doped TiO_2 nanoparticles prepared by thermal decomposition. *J Hazard Mater*. 2009;162 (2–3):763–70.

12 Zhang H, Lv X, Li Y, Wang Y, Li J. P25–graphene composite as a high performance photocatalyst. *ACS Nano*. 2010;4(1):380–6.

13 Huang Y, Li X, Bai Y. Efficient photocatalytic degradation of antibiotics over ternary g-C_3N_4/ZnO/Ag composites under visible light. *Appl Catal B Environ*. 2018;237:219–28.

14 aus der Beek T, Weber FA, Bergmann A. Pharmaceuticals in the environment—global occurrences and perspectives. *Environ Toxicol Chem*. 2016;35(4):823–35.

15 Li L, Fan M, Brown RC. Synthesis, properties, and environmental applications of nanoscale iron-based materials: a review. *Crit Rev Environ Sci Technol*. 2006;36(5):405–31.

16 Wu T, Liu G, Zhao J, Hidaka H, Serpone N. Photoassisted degradation of dye pollutants. 7. Comparative photocatalytic efficiency of TiO_2, ZnO, and other oxides in degrading methylene blue. *J Phys Chem B*. 1998;102(30):5845–51.

17 Michael I, Rizzo L, McArdell CS. Urban wastewater treatment plants as hotspots for antibiotic-resistant bacteria and genes spread into the environment: a review. *Sci Total Environ*. 2013;447:345–60.

18 Lhomme L, Brosillon S, Wolbert D. Photocatalytic degradation of pesticides under simulated solar light: case study of diuron, isoproturon and atrazine. *Chemosphere*. 2008;70(2):381–86.

19 Kotlhao K, Mtunzi FM, Pakade V, Ejidike IP, Klink MJ. Synthesis, characterization and evaluation of Ag-TiO_2 nanocomposites for photo-catalytic degradation of selected chlorophenols. *Dig J Nanomater Biostruct*. 2018;13(3):835–84.

20 Zheng Y, Luo L, Lin H. Visible-light-driven degradation of antibiotics using $BiVO_4$/rGO nanocomposite. *J Photochem Photobiol A Chem*. 2016;321:200–8.

21 Lee J, von Gunten U, Kim JH. Persulfate-based advanced oxidation: critical assessment of opportunities and roadblocks. *Environ Sci Technol*. 2020;54(6):3064–81.

22 Chen X, Mao SS. Titanium dioxide nanomaterials: synthesis, properties, modifications and applications. *Chem Rev.* 2007;107(7):2891–959.

23 Dimo SN, Obidi OF, Nejo AO, Olaleru SA, Ejidike IP, Adetona AJ. Preparation of nanosized TiO_2 in the presence of bacteria *Bacillus paralicheniformis* and its photocatalytic properties in the process of decolorization of dyes. *Theor Exp Chem.* 2024;60(3):194–201.

24 Yu J, Low J, Xiao W, Zhou P, Jaroniec M. Enhanced photocatalytic CO_2-reduction activity of anatase TiO_2 by coexposed {001} and {101} facets. *J Am Chem Soc.* 2014;136(25):8839–42.

25 Neyens E, Baeyens J. A review of classic Fenton's peroxidation as an advanced oxidation technique. *J Hazard Mater.* 2003;98(1–3):33–50.

26 Anipsitakis GP, Dionysiou DD. Radical generation by the interaction of transition metals with common oxidants. *Environ Sci Technol.* 2004;38(13):3705–12.

27 Huang W, Brigante M, Wu F, Hanna K, Mailhot G. Development of a new hybrid $UV/O_3/TiO_2$ process for advanced oxidation of chlorinated phenols. *Water Res.* 2008;42(15):3511–20.

28 Li J, Xu Q, Yang P. Hierarchical nanostructured metal oxides for environmental applications. *Chem Soc Rev.* 2021;50(2):970–1038.

29 Kumar PS, Korving L, van Loosdrecht MCM, Witkamp GJ. Nanoparticles in water treatment: from laboratory-scale studies to full-scale applications. *Ind Eng Chem Res.* 2019;58(14):5467–85.

30 Choi J, Lee S, Park S. Magnetically separable Fe_3O_4-TiO_2 composite photocatalysts for wastewater treatment. *J Ind Eng Chem.* 2010;16(2):235–40.

31 Nasrollahzadeh M, Sajjadi M, Iravani S, Varma RS. Green-synthesized nanocatalysts and nanomaterials for water purification: a review. *RSC Adv.* 2019;9(44):24452–81.

32 Bhatkhande DS, Pangarkar VG, Beenackers AACM. Photocatalytic degradation for environmental applications—a review. *J Chem Technol Biotechnol.* 2002;77(1):102–16.

33 Roopa MC, Thirumala S, Kallimani S, Manohara BM, Basavarajaiah SM. Green synthesis of nanoparticles for enhanced wastewater treatment and other applications: A review and future perspectives. Next Mater. 2025;8:100664.

34 Bundschuh M, Filser J, Lüderwald S. Nanoparticles in the environment: where do we come from, where do we go to? *Environ Sci Eur.* 2018;30:6.

35 Li H, Yu H, Quan X, Chen S. Facet-dependent photo-reactivity of $BiVO_4$ and Ag loading for enhanced visible-light photocatalysis. *Appl Catal B Environ.* 2012; 123–124:174–81.

36 Mao C, Xiang Y, Liu E. Recent advances in plasmonic metal-semiconductor heterojunction photocatalysts for solar-driven water purification. *Coord Chem Rev.* 2021;432:213742.

37 Zhang Y, Zeng GM, Tang L. A review on activated carbon/graphene-based nanomaterials for treating antibiotics in water: mechanism, application, and challenges. *Chemosphere.* 2016;153:19–29.

38 Thakur A, Sharma S, Amini A, Agarwal S. Recent advances in photocatalytic reactor design and light engineering for water purification. *J Environ Chem Eng.* 2022;10(1):106934.

39 Ying S, Guan Z, Ofoegbu PC, Clubb P, Rico C, Feng He F, Hon J. Green synthesis of nanoparticles: Current developments and limitations. *Environ. Technol. Inno.* 2022;26(1):102336.

40 Keller AA, Lazareva A. Predicted releases of engineered nanomaterials: from global to regional to local. *Environ Sci Technol Lett.* 2014;1(1):65–70.

41 Dimo SN, Obidi OF, Nejo AO, Olaleru SA, Ejidike IP, Adetona AJ. Biofabrication, spectroscopic, and photocatalytic studies of titania nanoparticles mediated by *Proteus mirabilis* strain NG-ABK-32 for smart applications. *Smart Sci.* 2024;12(2):373–86.

Part 4

Advanced, Industrial, and Environmental Impact of Metal Oxide Nanocatalysts Systems

12

Metal-Organic Frameworks (MOFs) and Metal Oxide Nanocatalysts

Olayinka Oderinde[1], Saheed Abiola Raheem[2], Chiamaka Linda Mgbechidinma[3, 4], Joshua Iseoluwa Orege[5], Yuanfeng Wu[6], Adetola Abiola Ajayi[7, 8], Olakunle Ibrahim Oresegun[9], Emmanuel Ohifueme Alegbe[1], Dayo A Ayeni[8, 10], and Oluseyi Olaniyi Olaide[11]*

[1] Department of Chemistry, Nile University of Nigeria, Abuja, Nigeria
[2] Department of Applied and Environmental Chemistry, University of Szeged, Szeged, Hungary
[3] School of Life Sciences, Centre for Cell and Development Biology and State Key Laboratory of Agrobiotechnology, The Chinese University of Hong Kong, Shatin, New Territories, Hong Kong, China
[4] Department of Microbiology, University of Ibadan, Ibadan, Nigeria
[5] Chemical Sciences Division, Oak Ridge National Laboratory, Oak Ridge, TN, USA
[6] Henan Key Laboratory of Coal Green Conversion, International Joint Laboratory of Coal Clean Utilization, Collaborative Innovation Center of Coal Work Safety and Clean High Efficiency Utilization, College of Chemistry and Chemical Engineering, Henan Polytechnic University, Jiaozuo, Henan, China
[7] Department of Science Laboratory Technology, School of Pure and Applied Sciences, The Federal Polytechnic, Ilaro, Ilaro, Nigeria
[8] Department of Chemistry, Faculty of Natural and Applied Sciences, Lead City University, Ibadan, Nigeria
[9] School of Mechanical Engineering, Zhejiang University, Hangzhou, Zhejiang, China
[10] Department of Chemistry, Obafemi Awolowo University, Ile-Ife, Nigeria
[11] Department of Chemical and Pharmaceutical Chemistry, University of South Wales, Pontypridd, UK

*Corresponding author: yinkaoderinde@yahoo.com

12.1 Overview of Metal-Organic Frameworks (MOFs) and Their Combination with Metal Oxide Nanocatalysts (MONCs)

Functional composite materials like MOF-MONCs combine the distinct physicochemical characteristics of metal oxides, such as high thermal stability, electrical conductivity, and catalytic activity, with the structural tunability and high porosity of MOFs [1, 2]. The complementary properties of the additional materials lead to synergistically improved performance for a variety of applications. These hybrid systems take advantage of the advantages of both MOF structures and metal oxides, in contrast to pristine MOFs [3–5]. Recent research has been directed to the integration of MONCs with MOFs, with the aim to overcome the intrinsic limitations of

Metal Oxide Nanocatalysts for Sustainable Energy Production, First Edition. Edited by Abayomi Bamisaye et al.
Published 2026 by Wiley-VCH GmbH.

individual components while maximizing their synergistic potential. These hybrid materials are very effective and versatile for real-world applications because they do not only increase chemical durability and structural stability but also promote better charge or mass transfer and add new active sites [6, 7].

In this chapter, we explore the innovative integration of MOFs with MONCs, offering a comprehensive perspective on their synergistic behavior and potential in energy-related and environmental technologies.

12.1.1 Synergistic Catalytic Interaction Between MOFs and MONCs

The fundamental mechanism behind the synergistic catalytic behavior of MOF-MONC composites lies in the mutual enhancement of active site accessibility, electronic interactions, and structural stability [8]. MONCs embedded within or deposited on MOFs benefit from improved dispersion due to the spatial confinement and anchoring effect provided by the MOF matrix. This guarantees maximum active site exposure and inhibits nanoparticle agglomeration. Site-specific reactions with enhanced selectivity can be facilitated by the MOF framework's ability to simultaneously encourage reactant channeling and selective adsorption to these active centers [9]. An additional important factor in catalysis is electron transfer at the MOF-MONC interface. The interface frequently produces a distinct electronic environment where charge redistribution takes place as a result of the metal oxide and MOF's different Fermi levels. The lifetime of charge carriers can be increased and recombination suppressed in photocatalytic systems, for instance, by effectively separating and transferring photoexcited electrons produced in the semiconductor MOF or MONC across the interface [10, 11].

Oxygen vacancies and surface defects in MONCs like CuO and CeO_2 can greatly increase catalytic turnover in redox reactions like CO oxidation, water splitting, or CO_2 reduction when controlled by the nearby MOF environment [8, 12, 13]. The MOF can utilize the electronic coupling to stabilize these vacancies and even adjust the redox potential of the MONC [12]. Furthermore, the closeness of Lewis or Brønsted acid/base sites within the MOF structure can alter the local chemical environment even more, resulting in enhanced-site catalysis pathways that are impossible to accomplish in isolated components.

12.1.2 Applications in Carbon Dioxide (CO_2) Capture and Conversion

The constant rise in atmospheric CO_2 concentration due to anthropogenic activities has emerged as a critical threat to climate stability and global sustainability. Effective carbon capture, utilization, and storage (CCUS) technologies are key to mitigating CO_2 emissions, especially from large point sources such as fossil fuel power plants [14, 15]. For instance, amongst the various classes of materials investigated in CO_2 capture, MOFs have demonstrated considerable CO_2 uptake capacities under ambient or flue gas conditions owing to their high surface areas, tunable pore environments, and functional versatility. However, their practical application has been limited by issues such as hydrothermal instability, moderate selectivity, and limited scalability, and they are susceptible to structural degradation in humid or acidic environments. To address these challenges, hybrid materials based on MONC-integrated MOFs are being developed and are gaining increasing attention. These MOF-metal oxide hybrids exhibit synergistic behavior, offering improved structural stability, enhanced CO_2/H_2 separation, and selective adsorption under realistic conditions. Based on the aforementioned challenges, Zhou et al. [16] employed ZIF-67 membranes synthesized on α-Al_2O_3-plate-supported cobalt nanosheets with amine modification, achieving enhanced H_2/CO_2 permselectivity. The as-fabricated hybrid membrane demonstrated a high hydrogen permeance along with an ideal separation factor of 30.0, underscoring the potential of MONC-MOF hybrids in separation applications. Additionally, conductive metal oxides enhance electron transfer, which is advantageous in energy-related applications. Beyond capture, MOF-MONC hybrids also show significant promise in CO_2 conversion processes. These materials have been reported to prevent the loss of active sites caused by surface reorganization and nanoparticle growth. For example, An et al. [4] anchored ultrasmall Cu/ZnO_x nanoparticles within the cavities of UiO-bpy MOFs. The catalyst featured preassembled bpy and $Zr_6(\mu_3\text{-}O)_4(\mu_3\text{-}OH)_4$ sites that prevented the Cu NPs agglomeration and phase separation between Cu and ZnO_x. This design achieved 100% CO_2 hydrogenation to methanol at 100% selectivity over 100 h high activity, with a space-time yield of up to 2.59 gMeOH kgCu^{-1} h^{-1}. In another study, Crake and coworkers [17] developed TiO_2/NH_2-UiO-66 nanocomposite for enhanced CO_2 capture and photoconversion to CO under UV-Vis irradiation. The nanocomposite leveraged the high CO_2 adsorption capacity of the MOF and the photocatalytic properties of the preformed TiO_2 nanosheet, resulting in significant durability and efficiency compared to the performance of their single components. Table 12.1 shows previously reported works involving employed MOF-MONCs in CO_2 capture and conversion.

Table 12.1 Recent advances in MOFs-MONCs hybrid systems in CO_2 capture and conversion.

MOF-MONC hybrid system	Process	Product	Key findings	References
Cu/ZnO_x@ UiO-bpy MOF	CO_2 hydrogenation	CH_3OH	100% selectivity	An et al. [18]
TiO_2@ CoNi-MOF NTs	Photocatalysis	CH_4	99% selectivity	Liang et al. [19]
$Cu_3(BTC)_2/Cu_2O$ photelectrode	Photoelectrochemical reduction	N/A	Enhanced activity and durability in PEC CO_2 reduction	Deng et al. [20]
Core-shell ZnO/ Pt@ZIF-8 hybrid	Photocatalysis	CH_3OH	Achieved CH_3OH efficiency of 1.13 μmol g^{-1} h^{-1} being 16 and 1.8 times more efficient than the pristine ZnO and binary ZnO/Pt nanorods, respectively	Li et al. [21]

12.1.3 Applications of MOFs and MOF-MONCs in Water Treatment

In recent times, the research into the MOF-MONCs hybrid system for the light-assisted catalytic degradation of organic pollutants, pathogens, and heavy metals removal has been intensified. For example, Zhang and Park [22] using a facile two-step method to prepare a three-dimensional (3D) MoO_3/ZIF-8 core-shell nanorod composite for hexavalent chromium (Cr(VI)) ion removal from wastewater. Notably, 15% ZIF-8 in the MoO_3@ZIF-8 was reported to show the best photocatalytic activity, promoting 100% reduction of $15\,mg\,L^{-1}$ Cr(VI) within 45 min under visible light, compared to the pure ZIF-8 and MoO_3 nanowires. Also, Yang et al. [23] demonstrated the photocatalytic degradation of methyl orange (MO) using a TiO_2@UiO-66 composite. The composite with 5 wt% TiO_2 achieved 97.59% MO degradation under UV light, surpassing pure TiO_2 and UiO-66 due to improved electron-hole separation via a Type-II heterojunction. Optimizing conditions revealed that the degradation efficiency peaked at pH 2 and 0.2 g/L catalyst dosage had minimal pore blockage and optimal light penetration. Additionally, Hidayat and coworkers [24] demonstrated the Al_2O_3@UiO-66 superior adsorption capacity (268.82 mg/g) for methyl orange, MO through an exothermic and spontaneous process compared to pristine UiO-66 (233.65 mg/g). The team of Chen et al. [25] further revealed the generation of a new composite structure combining ultrasmall (sub-2 nm) hydrous ferric oxide nanoparticles (HFeO) with MOFs (Fe-based MIL-100) at room temperature. The size-exclusion effect of the uniform micropores originating from MOFs leads to the structure exhibiting high anti-interference ability in As(V) adsorption, maintaining more than 97% capacity even within abundant competitive ions or species. Other reported researches on the application of MOF-MONC hybrids in water treatment is indicated in Table 12.2.

12.1.4 Applications of MOF-MONCs Composites for Hydrogen Production

Growing concerns about the global energy crisis and the role of fossil fuel combustion in global warming have prompted research into efficient and clean energy technologies. The objective is to supplant traditional fuels with energy technologies that produce zero carbon emissions. In light of this, hydrogen has emerged as a promising energy carrier and substitute for fossil fuels due to its environmental sustainability and high energy density [29, 30]. In this context, electrocatalytic water splitting using wind energy or solar systems to generate hydrogen represents a highly sustainable approach [31]. However, the reaction suffers from sluggish kinetics and a large overpotential, making it reliant on electrocatalysts to improve the reaction kinetics and lower the overpotential. Although noble metal catalysts remain the most effective for hydrogen evolution reactions (HERs), their expensive cost and scarcity render them impractical for large-scale commercial use [32], hence the shift towards non-noble metal catalysts that exhibit appreciable catalytic performance. Compared to other non-noble catalysts, MOFs and MONCs offer a wide range of properties and considerable structural flexibility, making them specifically promising candidates for

Table 12.2 Recent water treatment applications using MOFs and MONCs.

MOF/MONC hybrid system	Pollutant removed	Efficiency/ performance	Major controlled conditions/factors	References
Cu-BTC@Alg/ Fe_3O_4	Rhodamine B (RhB) dye from wastewater	97% removal efficiency	Surface area	Hemdan et al. [26]
HFeO@MIL-100	Arsenic (As(V)) removal from contaminated water	Concentration decreased from 3900 μg/L to 5 μg/L	Adsorption capacity and uniform micropores	Chen et al. [25]
MAA-Zn$CrFeO_4$/ HKUST-1	Methylene blue (MB) removal from wastewater	High absorption speed (15 min), low-cost, high surface area, selectivity, and efficiency.	Magnetic characteristic and high porosity	Geranmayeh et al. [27]
ZnO_x-MOF@MnO_2	Reusable and sustainable strontium ion removal from aqueous solutions	Maximum adsorption capacity of 147.094 mg/g	pH, coexisting competing ions	Choi et al. [28]
TiO_2@UiO-66	Methyl orange (MO) degradation	97.59% degradation rate	pH 2 and 0.2 g/L catalyst dosage	Yang et al. [23]

replacing platinum-based catalysts [33, 34]. The fundamentals of HER, evaluation parameters and recent developments in MOFs and MONCs for hydrogen generation are discussed herein.

12.1.5 Fundamentals of Electrochemical HER

HER represents the cathodic part and half-reaction of the multi-electron/multi-proton water splitting. In an acidic electrolyte, an adsorbed hydrogen atom (H_{ad}) is generated when the electron transferred to the surface of the electrocatalyst interacts with the adsorbed proton. Studies have shown that the HER reaction mechanism consists of Volmer, Heyrovsky, and Tafel steps, respectively. Volmer is the first step that involves the release of electrons from the cathodic surface to form H_{ad} while the latter reacts with a second electron and proton to produce an hydrogen molecule (H_2) during the Heyrovsky step. In the final step (Tafel), two adjacent H_{ad} combine to form an H_2 in a process known as chemical desorption. The reaction steps in alkaline and neutral solutions are similar and involve water dissociation, a more energy-intensive process. The overall reaction is depicted in Figure 12.1, while the reaction steps are summarized in Table 12.3.

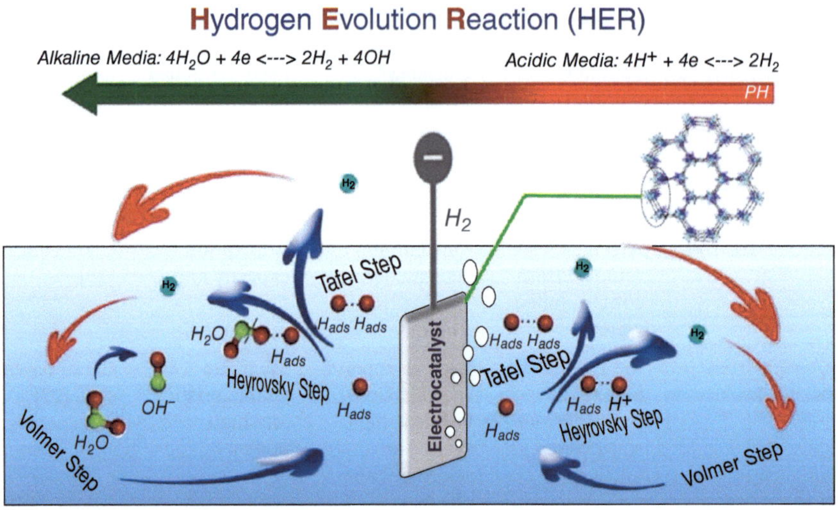

Figure 12.1 The schematic mechanism for HER.
Source: (36) / American Chemical Society / CC BY 4.0.

Table 12.3 Reaction mechanism of HER under different electrolyte solutions.

Steps	Acidic solutions	Basic and neutral solutions
Volmer step	$H_3O^+ + e + M \rightarrow M - H_{ad} + H_2O$	$H_2O + e + M \rightarrow M - H_{ad} + OH^-$
Heyrovsky step	$H_3O^+ + e + M - H_{ad} \rightarrow H_2 + H_2O + M$	$M - H_{ad} + H_2O \rightarrow H_2 + OH^-$
Tafel step	$2M - H_{ad} \rightarrow 2M + H_2$	$2M - H_{ad} \rightarrow 2M + H_2$
Overall reaction	$2H^+ + 2e^- \rightarrow H_2$	$2H_2O + 2e^- \rightarrow H_2 + 2OH^-$

M is the active site on the surface of the catalyst.

12.1.6 MOF-MONCs Catalyst for Hydrogen Production

Despite recent advances in the electrocatalytic activity of MOFs and MONCs towards HERs, concerns remain regarding their conductivity, stability, and overall performance. In light of this, Chu et al. [36] in an attempt to improve hydrogen production in an acidic medium, prepared carbon-coated nickel-nickel oxide composites (Ni/NiO@C/Gr) via carbonization of graphene oxide (GO) and Ni-MOF-74. When tested in 0.5 M H_2SO_4, the Ni/NiO@C/Gr composite showed an overpotential of 108 mV and good stability over 24 h with minimal decline in potential. The improved performance is primarily ascribed to the synergistic interaction between metallic Ni and NiO, which enhances water molecule adsorption on the catalyst surface and increasing electrical conductivity provided by the graphene. Also, Bhabal and co-researchers [37] prepared cobalt phospho-borate on Co-MOF on nickel foam

Table 12.4 HER performance of MOF-MONC composite catalysts under different conditions.

Catalysts	Electrolyte	Overpotential η_{10} (mV)	Tafel slope (mV dec^{-1})	References
$Co/Co_3O_4/C$	1.0 M KOH	169	60.7	Jiang et al. [38]
$MIL\text{-}53/Ti_3C_2@MnO_2$	0.5 M KOH	85.7	56.72	Albargi et al. [39]
$Co_3O_4\text{-}CoO@Co$	0.5 M H_2SO_4	28 @ η_{50}	30.9	Karuppasamy et al. [40]
CuO–NiO@N, P – PC	1.0 M KOH	80	42	Kumar et al. [41]
$CoFe_2O_4/CoFe@NC$	1.0 M KOH	86	99.5	Song et al. [42]

(CoPBO/Co-MOF@NF) via the wet chemical reduction method. When evaluated in an electrolyte solution (1.0 M KOH and 0.33 M urea), it exhibited a low potential of 1.26 V when compared to Co-MOF@NF (1.27V). The improved electrocatalytic performance of CoPBO/Co-MOF@NF arose from their larger ECSA, at 22.61 mF/cm^2, which is more than 40 times higher than Co-MOF@NF, at 0.55 mF/cm^2. This performance emphasizes the importance of composite catalyst fabrication for high-performance hydrogen production. Table 12.4 shows the HER performance of various MOF-MONC catalysts in different solutions.

12.1.7 Methods for Synthesizing Hybrid MOF-MONC Systems

MOFs and composites have been synthesized using a variety of techniques, and these include hydro-/solvothermal [43–46], microwave [47–49], surfactant-assisted [50–52], spray-drying [53, 54], ultrasonic spray pyrolysis [55], ultrasonication/sonochemical [56, 57], just to mention a few. These synthetic strategies significantly contribute to defining the material's morphology, sizes, tunability, crystallinity, flexibility, precise control of crystalline phase structure, specific surface areas as well as adsorption capacities, which ultimately, is crucial for optimizing their target pollutants' usage and efficiency. Some of the most utilized synthetic methods for MOF-MONC systems are discussed in the next section.

12.1.7.1 Solvo-/Hydrothermal Synthesis

This is one of the most straightforward and practical methods that does not require complicated preconditions or specialized equipment, where MOF-MONCs crystals with high levels of surface area, porosity, purity, and crystallinity are produced. Controlling the chemical composition and structure of the as-obtained final products relies on adjustable reaction parameters—including temperature, time, concentration of reactants, and particle size. The starting materials, such as metal salts or complexes, are treated under elevated temperature (70–260 °C) and pressure conditions for 24–48 h in a chemical reactor using either a solvent-based (solvothermal) or water-based (hydrothermal) solution [45, 58]. For example, Asadevi et al. [59] synthesized zinc oxide-zinc-based MOF, ZnO@Zn-MOF-5 nanocomposites solvothermally from pure zinc oxide nanoparticles (initially prepared greenly using *Annona muricata* leaf extract) and as-prepared MOF-5 in a dimethyl formamide, DMF, solvent inside a sealed autoclave reactor, after which it was subjected to a temperature of 150 °C for 3 h. The as-synthesized ZnO@Zn-MOF-5 exhibited polycrystallinity in its structure with a pore size of 0.23 nm and a surface area of 9.335 m^2/g. The composite was further utilized as an ultrasensitive probe for Cd^{2+} and Cu^{2+} ions detection in drinking water samples, with 99.46 and 99.85% quenching efficiencies observed, respectively. The team in another work [60] utilized ZnO@MOF-5 as a bifunctional catalyst for the adsorptive removal of Pb^{2+} ions and light-assisted degradation of methylene blue (MB) in industrial wastewater. The as-prepared exhibited a crystallite size of 32.23 nm, while its surface area was determined at 9.335 m^2/g. The MB degradation efficiency was found to be 97.47% in H_2O_2, while its Pb^{2+} removal was found to be 67.25% with 1 ppm of lead solution. Also, Abdi and coworkers [61] also solvothermally synthesized SnO_2 porous hollow nanofiber (PHNF)/UiO-66-NH_2 nanocomposite. The as-synthesized nanocomposite exhibited an enhanced surface area of 916 m^2/g compared to that obtained for the pure SnO_2 PHNF. The nanocatalyst was thereafter used for the ultrasonic irradiation-assisted degradation of tetracycline (TC), achieving an excellent degradation performance of 98% at optimum conditions, while 90% degradation performance was reported after six cycles, implying that it can be applied in actual circumstances.

12.1.7.2 Microwave-Assisted Synthesis

Microwave heating operates through two primary mechanisms: ionic conduction, which affects ions, and dipolar polarization, which influences dipolar molecules. This method allows for direct heating of the reaction mixture, resulting in rapid and uniform temperature elevation, as well as selective heating as a result of the components' dielectric properties. The combination of short reaction times and a consistent microwave field offers several benefits, including accelerated crystal nucleation, controlled crystal size growth, high product yield and purity, and enhanced phase selectivity [48, 62]. Wan et al. [63] facilely synthesized hollow tubular iron-doped iron-based metal-organic frameworks (Fe-MOFs-MW) composites using microwave-assisted method in addition to using sodium hydroxide as a modulator. The as-prepared Fe-MOFs were thereafter utilized to degrade sulfamethoxazole (SMX), of which 98% efficiency was reported when the catalyst dosage was 0.5 g/L at 120 mins and SMX concentration of 20 mg/L. When other pollutants like TC and ciprofloxacin (CIP) were utilized, the degradation efficiencies were still maintained at >90%. Also, Li and coworkers [64] prepared superparamagnetic Fe_3O_4@MIL-100(Fe) via microwave irradiation. The as-prepared catalyst was reported to possess a surface area of 1244.6 m^2/g and pore volume of 0.6608 cm^3/g, which was later used in the degradation of diclofenac sodium (DCF). In an H_2O_2-assisted visible light photocatalytic process, the as-prepared catalyst was able to remove DCF exhibiting more than 99.4% degradation efficiency, while 87.8% removal efficiency was reported for total organic carbon, TOC under optimal conditions.

12.1.7.3 Surfactant-Assisted Synthesis

Surfactants are amphiphilic organic molecules composed of both hydrophobic (nonpolar) tails and hydrophilic (polar) head groups, enabling them to adsorb readily at the interfaces between solid-liquid or liquid-liquid phases. This dual affinity imparts surfactants with the unique ability to function in both water-insoluble (lipophilic) and water-soluble (hydrophilic) environments. Their ability to reduce surface and interfacial tension underpins a wide range of physicochemical processes [65, 66]. Surfactants are typically categorized into six distinct types based on their functional roles. The categories are dispersing, structure-directing, capping, mild reducing, oxidative etching, and ion-exchange agents (Figure 12.2) [67].

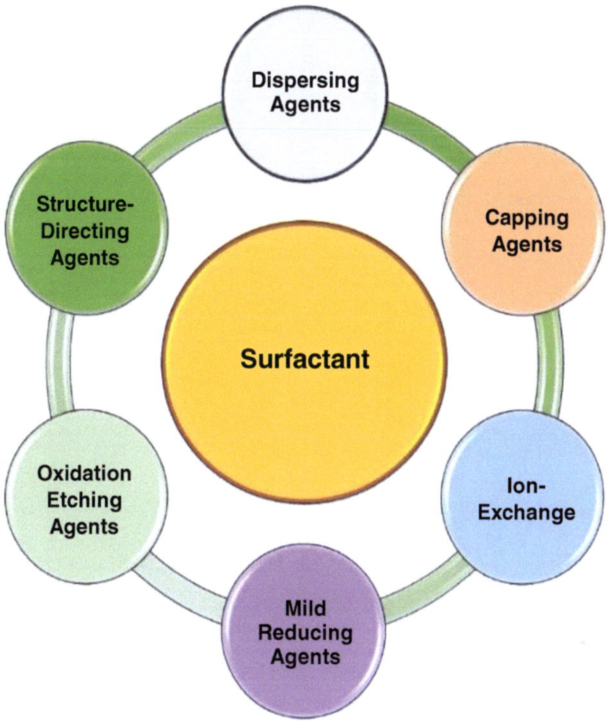

Figure 12.2 Distinct categories of surfactant.
Source: Adapted from [68, 69].

Bayram Hashemzadeh et al. [68] reported the surfactant-assisted solvothermal synthesis of α-Fe_2O_3@ZIF-67 core-shell nanostructured composites. In this approach, cetyltrimethylammonium bromide (CTAB) was employed as a structure-directing agent during the initial formation of α-Fe_2O_3 nanospheres, serving to modulate their morphology. Subsequently, polyvinylpyrrolidone (PVP) was utilized to facilitate the heterogeneous growth of the ZIF-67 shell on the surface of the α-Fe_2O_3 core nanoparticles. The resulting α-Fe_2O_3@ZIF-67 nanocomposite was then applied as a catalyst for the degradation of CIP antibiotics in aqueous solution. Also, Tang and colleagues [69] employed polyglycerol polyricinoleate (PGPR 90) as a surfactant in the synthesis of BMA@Cu/Ni@MOF-TiO_2 microspheres, which were subsequently utilized for the degradation of MB. An increase in the surfactant concentration to 14% w/v was reported to have enhanced the formation of pores on the microsphere surface. These pores were identified as the primary channels facilitating the degradation of organic pollutants. Other reported MOF-MONCs publications based on synthesis methods are highlighted in Table 12.5.

Table 12.5 Synthesis methods for selected MOF-MONC hybrid composites and their efficiencies in applied pollutant removal.

MOF-MONCs hybrid catalyst	Synthetic method	Surface area ($m^2\ g^{-1}$)/pore volume ($cm^3\ g^{-1}$)	Targeted environmental usage	Efficiency (%)	References
MIL-53 (Al)@SiO_2	Solvothermal	667.80/0.47	Bisphenol A (BPA) removal	85.04	Chatterjee et al. [70]
$CoFe_2O_4$/ZIF-8	Hydrothermal	1723.57/0.82	Methylene blue (MB) degradation	97.6 (60 min)	Zhang et al. [71]
$Zn_4O(BDC)_3$-5/gC_3N_4/Fe_3O_4	Solvothermal	17.8893/0.080865	Rhodamine blue (RhB)/Methylene violet (MV)/Coriacide black MR	95/53/31 (90 min)	Mohammadipour-Nodoushan et al. [44]
SnO_{2-x}/CD-MOF	Solvothermal		Rhodamine blue (RhB)/Tetracycline (TC)	93.5 (45 min)/83.1 (60 min)	Zhang et al. [72]
Bi_2O_3@Fe-MOF	Hydro-/solvothermal		Rhodamine blue (RhB) and triclopyr pesticide	91.79 and 88.42 (120 min)	Sharma et al. [73]
$Fe_{0.75}Cu_{0.25}$-BDC@Alg-MoO_3/GO	Solvothermal		Cr(VI)	>98	Mahmoud et al. [74]
Fe_3O_4/Fe_2O_3-MIL-101	Solvothermal/Microwave irradiation		Cd(II)/Ni (II)	41.17/16.8	Gusain et al. [75]
nFe_3O_4@MIL-88A(Fe)/APTMS	Microwave	62.21/9.45	Cd(II)/Pb(II)/Cr(VI)	98.49/96.88/99.69	Mahmoud et al. [76]
Fe_3O_4-WDS-biochar@ZIF-67	Microwave	51.16/0.64	Tartrazine (TA)/Sunset yellow (SY)	99.06/99.54	Haghdoust et al. [77]

12.2 Conclusion

MOF-MONC composites represent a highly versatile and promising class of functional hybrid materials with significant potentials in addressing emerging environmental and energy-related challenges. Their synergistic integration combines the high surface area, tunable porosity, and functional versatility of MOFs with the chemical stability, redox activity, and catalytic robustness of metal oxides. This unique combination enables enhanced performance in CO_2 capture and conversion processes, particularly by promoting selective adsorption and facilitating catalytic transformation into value-added chemicals and fuels. Moreover, MOF-MONC composites demonstrate excellent activity in photocatalytic and electrocatalytic hydrogen production, owing to improved charge separation and light-harvesting properties. In the fields of water treatment, these composites exhibit high efficiency in the removal of a broad range of organic and inorganic pollutants through adsorption and catalytic degradation mechanisms. Their multifunctionality, structural tunability, and reusability underscore their practicality for real-world deployment. Continued advancements in synthesis techniques, particularly those enabling scalable, cost-effective, and environmentally benign fabrication, are essential for expanding their commercial viability. Overall, MOF-MONC composites hold significant promise as next-generation catalysts for sustainable environmental remediation, renewable energy production, and climate change mitigation, supporting global efforts toward a cleaner and more resilient future.

12.3 Future Outlook for MOF-MONC Composites in Climate-Related Technologies

MOFs and metal oxide-based composites have drawn significant attention as advanced functional materials for addressing inherent climate and environmental challenges. These materials have demonstrated considerable potential in applications such as CO_2 conversion, selective gas separation, photocatalytic degradation of organic contaminants, and water purification. A promising research trajectory involves the development of MOF-metal oxide hybrid composites, which integrate the high porosity and molecular sieving capabilities of MOFs with the structural robustness, thermal stability, and catalytic efficiency of metal oxides. Recent advances in computational materials science, particularly in machine learning-assisted screening and high-throughput simulation techniques, are accelerating the rational design of MOF-based composites with targeted physicochemical properties. The translation of these materials into practical environmental technologies is expected to expand significantly in the near future. However, in order to enable widespread implementation, significant obstacles need to be addressed. The main drawbacks are high production costs, limited recyclability, complexity in scalable synthesis routes, and structural instability under real-world operating conditions. Innovations in scalable synthesis methods and cost reduction will be vital for commercialization. Additionally, in order to guarantee responsible deployment, enhanced research on the ecological effects and environmental-benignness of these materials is necessary. Finally, continued interdisciplinary collaboration and investment in overcoming current limitations will determine their success in mitigating environmental challenges and climate impacts.

References

1 Huang X, Yan S, Deng D, Zhang L, Liu R, Lv Y. Novel strategy for engineering the metal-oxide@MOF core@shell architecture and its applications in cataluminescence sensing. *ACS Appl Mater Interfaces.* 2021 Jan 20;13(2):3471–80.

2 Chong X, Bao W, Wang Y, Sun Z, Chen L, Chen Q, et al. Preparation of magnetic Fe_3O_4 composite Cu-MOFs and their removal of perrhenate ions. *J Solid State Chem.* 2025 Aug;348:125390.

3 Li R, Li W, Jin C, He Q, Wang Y. Fabrication of ZIF-8@TiO_2 micron composite via hydrothermal method with enhanced absorption and photocatalytic activities in tetracycline degradation. *J Alloys Compd.* 2020 Jun;825:154008.

4 An Y, Lv X, Jiang W, Wang L, Shi Y, Hang X, et al. The stability of MOFs in aqueous solutions—research progress and prospects. *Green Chem Eng.* 2024 Jun;5(2):187–204.

5 Li C, Chandresh A, Zhang Z, Moulai S, Heinke L. Stability and degradation of metal–organic-framework films under ambient air explored by uptake and diffusion experiments. *Adv Mater Interfaces.* 2022 Jan 4;9(3):2101947.

6 Yoon J, Kim J, Kim C, Jang HW, Lee J. MOF-based hybrids for solar fuel production. *Adv Energy Mater.* 2021 Jul 15;11(27): 2003052.

7 Nurbayti S, Adawiah A, Bale UF, Fadhilla R, Ramadhan FN, Zulys A, et al. Sonochemical assisted synthesis of Cr-PTC metal organic framework, ZnO, and Fe_3O_4 composite and their photocatalytic activity in methylene blue degradation. *Bull Chem React Eng Catal.* 2024 Aug 1;19(2):318–26.

8 Vali SA, Moral-Vico J, Font X, Sánchez A. Cu/ZnO/CeO_2 supported on MOF-5 as a novel catalyst for the CO_2 hydrogenation to methanol: a mechanistic study on the effect of CeO_2 and MOF-5 on active sites. *Catal Letters.* 2024 Jul 8;154(7):3157–73.

9 Li J, Xu X, Liu X, Qin W, Wang M, Pan L. Metal-organic frameworks derived cake-like anatase/rutile mixed phase TiO2 for highly efficient photocatalysis. *J Alloys Compd.* 2017 Jan;690:640–6.

10 Hussain I, Zhang K. MOF-derived scaffolds as electrode materials: a mini-review. *Nanoscale.* 2024;16(33):15515–28.

11 Huang Z, Dong P, Zhang Y, Nie X, Wang X, Zhang X. A ZIF-8 decorated TiO_2 grid-like film with high CO_2 adsorption for CO_2 photoreduction. *J CO_2 Util.* 2018 Mar;24:369–75.

12 Zhao T, Zhong D, Hao G, Zhao Q. Vacancy engineering and hydrophilic construction of CoFe-MOF for boosting water splitting by in situ phytic acid treatment. *Appl Surf Sci.* 2023 Jan;607:155079.

13 Ullah A, Dong G, Javed Y, Anwer A, Geng D. Boosted CO_2 electroreduction performance by enhanced oxygen vacancies in Ag-MOF/CuO nanosheets heterostructures for selective C_2H_4 production. *Fuel* 2025 Nov;400:135795.

14 Cardoso JC, Stulp S, de Brito JF, Flor JBS, Frem RCG, Zanoni MVB. MOFs based on ZIF-8 deposited on TiO_2 nanotubes increase the surface adsorption of CO_2 and its photoelectrocatalytic reduction to alcohols in aqueous media. *Appl Catal B.* 2018 Jun;225:563–73.

15 Juma AK, Merican ZMA, Haruna A, Ayodele BV, Atta MR, Shaharun MS. Advances in methanol synthesis via MOF-based catalysts: a comprehensive review and critical analysis. *Inorg Chem Commun.* 2025 Jun;176:114266.

16 Zhou A, Dou Y, Zhao C, Zhou J, Wu XQ, Li JR. A leaf-branch TiO_2/carbon@MOF composite for selective CO_2 photoreduction. *Appl Catal B.* 2020 May;264:118519.

17 Crake A, Christoforidis KC, Kafizas A, Zafeiratos S, Petit C. CO_2 capture and photocatalytic reduction using bifunctional TiO_2/MOF nanocomposites under UV–vis irradiation. *Appl Catal B.* 2017 Aug;210:131–40.

18 An B, Zhang J, Cheng K, Ji P, Wang C, Lin W. Confinement of ultrasmall Cu/ZnO_x nanoparticles in metal–organic frameworks for selective methanol synthesis from catalytic hydrogenation of CO_2. *J Am Chem Soc.* 2017 Mar 15;139(10):3834–40.

19 Liang S, Chen Y, Han W, Jiao Y, Li W, Tian G. Hierarchical S-scheme titanium dioxide@cobalt-nickel based metal–organic framework nanotube photocatalyst for selective carbon dioxide photoreduction to methane. *J Colloid Interface Sci.* 2023 Jan;630:11–22.

20 Deng X, Li R, Wu S, Wang L, Hu J, Ma J, et al. Metal–organic framework coating enhances the performance of Cu_2O in photoelectrochemical CO_2 reduction. *J Am Chem Soc.* 2019 Jul 10;141(27):10924–9.

21 Li X, He W, Li C, Song B, Liu S. Synergetic surface modulation of ZnO/Pt@ZIF-8 hybrid nanorods for enhanced photocatalytic CO_2 valorization. *Appl Catal B.* 2021 Jun;287:119934.

22 Zhang Y, Park SJ. Facile construction of MoO_3@ZIF-8 core-shell nanorods for efficient photoreduction of aqueous Cr (VI). *Appl Catal B.* 2019 Jan 1;240:92–101.

23 Yang J, Chang X, Wei F, Lv Z, Liu H, Li Z, et al. High performance photocatalyst TiO_2@UiO-66 applied to degradation of methyl orange. *Discover Nano.* 2023 Sep 11;18(1):112.

24 Hidayat ARP, Sulistiono DO, Murwani IK, Endrawati BF, Fansuri H, Zulfa LL, et al. Linear and nonlinear isotherm, kinetic and thermodynamic behavior of methyl orange adsorption using modulated Al_2O_3@UiO-66 via acetic acid. *J Environ Chem Eng.* 2021 Dec 1;9(6):106675.

25 Chen T, Ji M, Wen L, Guo T, Pan S, Cheng S, et al. In-situ forming Sub-2 nm hydrous iron oxide particles in MOFs for deep-treatment and high anti-interference in arsenic removal. *Chem Eng J.* 2022 Mar 1;431:133813.

26 Hemdan M, Ragab AH, Gumaah NF, Mubarak MF. Sodium alginate-encapsulated nano-iron oxide coupled with copper-based MOFs (Cu-BTC@Alg/Fe_3O_4): versatile composites for eco-friendly and effective elimination of Rhodamine B dye in wastewater purification. *Int J Biol Macromol.* 2024 Aug 1;274:133498.

27 Geranmayeh S, Mastali S, Mohammadnejad M. Functionalized iron Oxide/MOF nanocomposite exhibiting excellent performance for methylene blue adsorption. *Inorg Chem Commun.* 2024 Apr 1;162:112222.

28 Choi JW, Park YJ, Choi SJ. Synthesis of metal-organic framework ZnO x-MOF@ MnO_2 composites for selective removal of strontium ions from aqueous solutions. *ACS Omega*. 2020 Apr 21;5(15):8721–9.

29 Raheem SA, Shen H, Thomas T, Yang M. Integrating trace amounts of Pd nanoparticles into Mo_3N_2 nanobelts for an improved hydrogen evolution reaction. *Phys Chem Phys*. 2022;24(2):771–7.

30 Abiola Raheem S, Shen H, Saad A, Guo H, Thomas T, Yang M. Mo_3N_2/VO_2 composite as electrocatalysts for hydrogen evolution reaction. *Inorg Chem Commun*. 2022 Aug 1;142:109614.

31 Zhou KL, Wang Z, Han CB, Ke X, Wang C, Jin Y, et al. Platinum single-atom catalyst coupled with transition metal/metal oxide heterostructure for accelerating alkaline hydrogen evolution reaction. *Nat Commun*. 2021 Dec 1;12(1):3783.

32 Zhang L, Zhao H, Xu S, Liu Q, Li T, Luo Y, et al. Recent advances in 1D electrospun nanocatalysts for electrochemical water splitting. *Small Structures*. 2021;2(2):2000048.

33 Liu P, Sun C, Liu G, Jiang Z, Zhao H. Ultra-small-sized multi-element metal oxide nanofibers: an efficient electrocatalyst for hydrogen evolution. *Nanoscale Adv*. 2022 Feb 23;4(7):1758–69.

34 Zheng W, Lee LYS. Metal-organic frameworks for electrocatalysis: catalyst or precatalyst? *ACS Energy Lett*. 2021;6(6):2838–43.

35 Kazemi A, Manteghi F, Tehrani Z. Metal electrocatalysts for hydrogen production in water splitting. *ACS Omega*. 2024 Jan 29;9(7):7310–35.

36 Chu M, Wang L, Li X, Hou M, Li N, Dong Y, et al. Carbon coated nickel – nickel oxide composites as a highly efficient catalyst for hydrogen evolution reaction in acid medium. *Electrochim Acta*. 2018 Feb 20;264:284–91.

37 Bhabal R, Gupta S, Fernandes R, Gupta M, Patel N. Bifunctional CoPBO/Co-MOF composite electrocatalyst for energy-efficient hydrogen evolution by urea-assisted water splitting. *Int J Hydrogen Energy*. 2025 Apr 4;116:299–311.

38 Jiang J, Dou H, Cao M, Du L, Yan Y, Wang X, et al. Amorphous nickel oxide electrodes with high-current-density electrocatalytic performance for hydrogen evolution. *Int J Hydrogen Energy*. 2024 Jan 2;51:887–94.

39 Albargi HB, Zeeshan M, Iqbal MW, Nermin A, Sunny MA, Ahmad J, et al. Advanced hierarchical MIL-53/Ti_3C_2@MnO_2 composites for superior supercapattery performance and efficient hydrogen evolution reaction. *Diam Relat Mater*. 2025 Jun 1;156:112409.

40 Karuppasamy L, Gurusamy L, Anandan S, Barton SC, Liu CH, Wu JJ. Nonprecious and shape-controlled Co_3O_4-CoO@Co electrocatalysts with defect-rich and spin-state altering properties to accelerate hydrogen evolution reaction at large current densities over a wide pH range. *Chem Eng J*. 2024 Sep 1;495:153442.

41 Kumar RD, Nivetha R, Asrami MR, Naderi A, Jourshabani M, Nguyen DV, et al. Cluster-level tuning CuO—NiO implanted on porous carbon for efficient electrocatalytic oxygen evolution and hydrogen evolution reactions. *J Alloys Compd*. 2025 May 5; 1026:180333.

42 Song X, Zhang J, Feng X, Qi Y, Cui J, Xiong Y. $CoFe_2O_4$/CoFe loaded 3D ordered hierarchical porous N-doped carbon for efficient oxygen reduction in Zn-air battery and hydrogen evolution. *J Energy Chem*. 2025 Jul 1;106:220–30.

43 Biju JR, Palanichamy B, Vasantha Kumar SSB, Subramanian S. Visible-light-activated Zn-MOF-5/g-C_3N_4/Fe_3O_4 nanocomposites: synthesis, structure and application in photodegradation of three dyes. *New J Chem.* 2025;49(18):7560–75.

44 Mohammadipour-Nodoushan R, Shekarriz S, Shariatinia Z, Heydari A, Montazer M. Improved cotton fabrics properties using zinc oxide-based nanomaterials: a review. *Int J Biol Macromol.* 2023;242:124916.

45 Naseer M, Siyal SH, Najam T, Afzal S, Iqbal R, Ismail MA, et al. Engineering of metal oxide integrated metal organic frameworks (MO@ MOF) composites for energy and environment sector. *Mater Sci Eng B.* 2025 Mar 1;313:117909.

46 Ejeromedoghene O, Oderinde O, Kang M, Agbedor S, Faruwa AR, Olukowi OM, et al. Multifunctional metal-organic frameworks in oil spills and associated organic pollutant remediation. *Environ Sci Pollut Res.* 2020 Dec 29;27(34):42346–68.

47 Głowniak S, Szczęśniak B, Choma J, Jaroniec M. Advances in microwave synthesis of nanoporous materials. *Adv Mater.* 2021;33(48):e2103477.

48 Mujawar N, Mohammad K, Rashmi W, Arshid N, Editors. *Contemporary Nanomaterials in Material Engineering Applications* [Internet]. Mubarak NM, Khalid M, Walvekar R, Numan A, editors. Cham: Springer International Publishing; 2021. (Engineering Materials). Available from: https://link.springer.com/10.1007/978-3-030-62761-4

49 Meshram AA, Sontakke SM. Rapid degradation of metamitron and highly complex mixture of pollutants using MIL-53(Al) integrated combustion synthesized TiO_2. *Adv Powder Technol.* 2021 Aug 1;32(8):3125–35.

50 El-Hankari S, Aguilera-Sigalat J, Bradshaw D. Surfactant-assisted ZnO processing as a versatile route to ZIF composites and hollow architectures with enhanced dye adsorption. *J Mater Chem A Mater.* 2016;4(35):13509–18.

51 Wu Y, Li X, Zhao H, Yao F, Cao J, Chen Z, et al. Core-shell structured Cu_2O@HKUST-1 heterojunction photocatalyst with robust stability for highly efficient tetracycline hydrochloride degradation under visible light. *Chem Eng J.* 2021 Dec 15;426:131255.

52 Gao Q, Kang H, Cai Y, Xue D, Yu F, Fang J, et al. Improving photoelectrochemical response of ZnO nanowire arrays by coating with p-type ZnO-resembling metal-organic framework. *Dalton Trans.* 2019;48(25):9310–6.

53 Carné-Sánchez A, Imaz I, Cano-Sarabia M, Maspoch D. A spray-drying strategy for synthesis of nanoscale metal-organic frameworks and their assembly into hollow superstructures. *Nat Chem.* 2013 Mar;5(3):203–11.

54 Mitsuka Y, Ogiwara N, Mukoyoshi M, Kitagawa H, Yamamoto T, Toriyama T, et al. Fabrication of integrated copper-based nanoparticles/amorphous metal–organic framework by a facile spray-drying method: highly enhanced CO_2 hydrogenation activity for methanol synthesis. *Angew Chem - Int Ed.* 2021 Oct 4;60(41):22283–8.

55 Kubo M, Kawamoto A, Shimada M. Preparation of N-doped nanoporous carbon from ZIF-8 metal-organic framework via ultrasonic spray pyrolysis. *Adv Powder Technol.* 2024 Dec;35(12):104727.

56 Ahmad W, Tahir M, Bibi B, Gong F, He L, Tang H, et al. Advances in ZIF-8 variants: synthesis, design, and applications in gas sensing. *Coord Chem Rev.* 2025 Sep;539:216733.

57 Sundarraj NK, Pandiyan JQS, Souwaileh AAl, Wu JJ, Anandan S. Sonochemical synthesis of Ni-MOF using 2-methylimidazole as an organic linker: pushing the boundaries of energy storage. *Electrochim Acta*. 2025 May;523:145975.

58 Annamalai J, Murugan P, Ganapathy D, Nallaswamy D, Atchudan R, Arya S, et al. Synthesis of various dimensional metal organic frameworks (MOFs) and their hybrid composites for emerging applications – a review. *Chemosphere*. 2022 Jul;298:134184. Available from: https://linkinghub.elsevier.com/retrieve/pii/S0045653522006774

59 Asadevi H, Prasannakumaran Nair Chandrika Kumari P, Padmavati Amma R, Khadar SA, Charivumvasathu Sasi S, Raghunandan R. ZnO@MOF-5 as a fluorescence "Turn-Off" sensor for ultrasensitive detection as well as probing of copper(II) ions. *ACS Omega*. 2022 Apr 19;7(15):13031–41.

60 Harisankar A, Preethi PC, Sreeja TG, Rejani P, Murali M, Raghunandan R. Zinc oxide functionalized MOF-5 for the adsorptive removal of Pb(II) metal ions and photocatalytic degradation of methylene blue dye in aqueous medium. *Ionics*. 2024 Apr 1;30(4):2313–31.

61 Abdi J, Mazloom G, Hayati B. Sonocatalytic degradation of tetracycline hydrochloride using SnO_2 hollow-nanofiber decorated with UiO-66-NH_2. *J Environ Manage*. 2024 Nov 1;370:122572.

62 Imtiyaz A, Singh A. A study of the fabrication of different-dimensional metal–organic frameworks and their hybrid composites for novel applications. *J Inorg Organomet Polym Mater*. 2023;33:3027–48.

63 Wan Y, Wan J, Zhao JR, Wang Y, Luo T, Yang S, et al. Facile preparation of iron oxide doped Fe-MOFs-MW as robust peroxydisulfate catalyst for emerging pollutants degradation. *Chemosphere*. 2020 Sep 1;254:126798.

64 Li S, Cui J, Wu X, Zhang X, Hu Q, Hou X. Rapid in situ microwave synthesis of Fe_3O_4@MIL-100(Fe) for aqueous diclofenac sodium removal through integrated adsorption and photodegradation. *J Hazard Mater*. 2019 Jul 5;373:408–16.

65 Wang Q, Xia W, Wang Z, Li J, Song Q, Du X, et al. A review on photosensitive surfactants: mechanism, synthesis, and application. *Chem Afr*. 2025;8(4):1257–70.

66 Shenoy P, Kedimar N, Rao SA. A comprehensive review on anticorrosive behaviour of surfactants across diverse metals using multiple techniques: current insights and future horizons. *Chem Eng J Adv*. 2024;20:100645.

67 Song T, Gao F, Guo S, Zhang Y, Li S, You H, et al. A review of the role and mechanism of surfactants in the morphology control of metal nanoparticles. *Nanoscale*. 2021;13:3895–910.

68 Hashemzadeh B, Alamgholiloo H, Noroozi Pesyan N, Asgari E, Sheikhmohammadi A, Yeganeh J, et al. Degradation of ciprofloxacin using hematite/MOF nanocomposite as a heterogeneous Fenton-like catalyst: a comparison of composite and core—shell structures. *Chemosphere*. 2021 Oct 1;281:130970.

69 Tang ZY, Mou CL, Li YD, Chen SH, Yang XM, Ding X, et al. BMA porous microspheres loaded with silane-modified Cu/Ni@MOF-TiO_2 for the photocatalytic degradation of methylene blue. *New J Chem*. 2025 Apr 14;49:8315–27.

70 Chatterjee A, Jana AK, Basu JK. A novel synthesis of MIL-53(Al)@SiO_2: an integrated photocatalyst adsorbent to remove bisphenol a from wastewater. *New J Chem.* 2020 Nov 21;44(43):18892–905.

71 Zhang K, Sun D, Ma C, Wang G, Dong X, Zhang X. Activation of peroxymonosulfate by $CoFe_2O_4$ loaded on metal-organic framework for the degradation of organic dye. *Chemosphere.* 2020 Feb 1;241:125021.

72 Zhang L, Zhang Y, Jiao Y, Yu J, Wang L, Chen J. SnO_2-x/CD-MOF S-Scheme heterostructure photocatalysts synergized by oxygen vacancies and highly porous for rhodamine B and tetracycline degradation: process, and mechanism analysis. *Colloids Surf A Physicochem Eng Asp.* 2025 Jan 20;705:135625.

73 Sharma A, Tahir M, Ahamad T, Kumar N, Sharma S, Kumari M, et al. Improved charge transfer and enhanced visible light photocatalytic activity of Bi_2O_3@ Fe-MOF for degradation of Rhodamine B and Triclopyr. *J King Saud Univ Sci.* 2023 Dec 1;35(10):102922.

74 Mahmoud ME, Amira MF, Azab MMHM, Abdelfattah AM. Inclusion of bimetallic $Fe_{0.75}Cu_{0.25}$-BDC MOFs into alginate-MoO_3/GO as a novel nanohybrid for adsorptive removal of hexavalent chromium from water. *Sci Rep.* 2022 Dec 1;12(1):19108.

75 Gusain D, Awolusi OO, Bux F. Synthesis and characterization of iron oxide/ MIL-101 composite via microwave solvothermal treatment. *Surf Sci.* 2022 Feb 1;716:121952.

76 Mahmoud ME, Amira MF, Seleim SM, Mohamed AK. Amino-decorated magnetic metal-organic framework as a potential novel platform for selective removal of chromium (Vl), cadmium (II) and lead (II). *J Hazard Mater.* 2020 Jan 5;381:120979.

77 Haghdoust S, Arabkhani P, Ghaderi S, Ghaedi M, Asfaram A. Rapid microwave-assisted synthesis of a magnetic biochar@ZIF-67: an efficient nanocomposite-based adsorbent for the dye-contaminated water cleanup. *New J Chem.* 2023 Apr 27;47(19):9257–70.

13

Bimetallic and Multimetallic Metal Oxide Nanocatalysts

*Shaima Hameed**

Department of Chemistry, University of Kashmir, Srinagar, Jammu and Kashmir, India

*Corresponding author: write2sheikhshyma07@gmail.com

13.1 Introduction

The field of catalysis plays an imperative role in modern science and technology. It facilitates many vital processes in various fields such as in the field of production of energy (e.g., fuel cells, batteries, hydrogen formation) [1]; in the manufacturing sector of chemical industries (e.g., pharmaceutical and polymer manufacturing industries, in Haber-Bosch processes, and in petroleum refining) [2]; in environmental management sectors [3] (e.g., industrial waste treatment sectors and the sector where reduction of harmful emissions like NO_x, CO, and volatile organic compounds (VOCs) is being done [4]. Catalysis works through the use of catalysts, which provides reaction pathway of lower activation energy that enables chemical transformations to occur more rapidly under mild conditions. The catalytic reaction improves selectivity in the product formation by reducing the undesirable side-product generation and hence improves energy efficiency [5], thus highlighting the indispensable contribution of the catalyst to both economic and sustainable development. Catalysts are classified as homogenous [6] or heterogeneous [7] depending on their phase relative to the reactants. However, heterogeneous catalysts are generally preferred in industrial applications [8] due to their stability, ease of separation, and reusability under harsh conditions [9]. In recent years, advances in nanoscience have revolutionized catalysis by enabling the development and synthesis of materials at the nanoscale range with large surface areas [10] and excellent surface properties [11] that significantly improve catalytic performance [12]. Among these nanocatalysts, metal oxides have attracted attention due to their structural diversity [13], flexible physicochemical properties, and wide application [14, 15]. But the metal oxides with single metals face limitations in catalytic activity, selectivity, and durability [16].

Metal Oxide Nanocatalysts for Sustainable Energy Production, First Edition. Edited by Abayomi Bamisaye et al.
Published 2026 by Wiley-VCH GmbH.

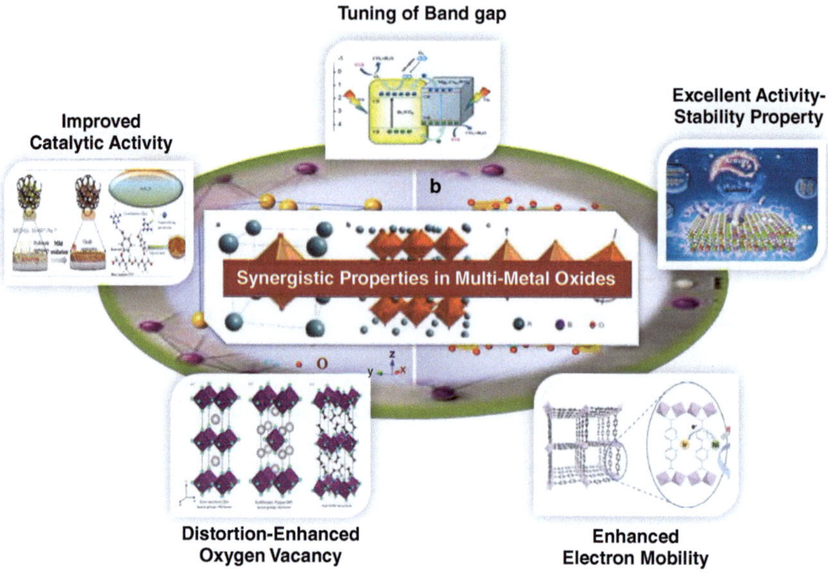

Figure 13.1 Synergistic properties of adding more than one metal to the metal oxide.

To overcome these challenges bimetallic and multimetallic metal oxides, collectively called mixed metal oxides (MMOs) [17], have emerged as a promising alternative. The incorporation of multiple metal cations into a single metal oxide framework [18] induces synergistic effects between the metals (Figure 13.1), leading to enhanced overall performance [19], like improved electrical conductivity [20], catalytic efficiency [21], physicochemical properties [22], and thermal and chemical stability [23] compared to its monometallic counterparts.

The integration of metals with different characteristics into MMOs offers the advantage of fine-tuning their properties. This inclusion provides a means to control their electronic structures, tune their band gap energies, and enhance defect engineering to generate higher density of active sites in the form of defects [24]. With the result, such materials possess improved photocatalytic performance, charge transport, and surface reactivity [24]. MMOs have a versatile structural configuration, such as perovskite [25], spinel [26], fluorite [27], wurtzite [28], metal-organic framework (MOF) and a complex structure of layered MMOs [29]. Their versatility in compositional and structural modifications gives a means to alter their catalytic properties. For this reason, MMOs can be used in next-generation batteries, in fuel cells [30], in water splitting devices [31], or in pollution degradation processes [32]. They can be designed to meet the demands of advanced catalytic system by employing metals with variable oxidation states or strategically choosing the different combinations of metals [33] which can synergize with each other both electronically and chemically.

MMOs have proven useful not only as active catalysts but also as robust catalysts, which supports in a variety of chemical reactions [34]. Metal oxides integrated with elements like aluminum, magnesium, and zirconium are well known for their high thermal and chemical stability. For example, magnesium aluminate spinel ($MgAl_2O_4$)

retains its catalytic activity even at elevated temperatures due to its excellent thermal stability [35]. MMOs are used in electronic materials such as capacitors, sensors, and memory devices (e.g., $BaTiO_3$, $LiNbO_3$) as well as in high-temperature supercapacitors like $YBa_2Cu_5O_{7.x}$ due to their functional properties, including dielectric [36], ferroelectric [37], and pyroelectric behavior [38]. Similarly, bimetallic spinel oxides like $NiCo_2O_4$, $ZnFe_2O_4$, and $CuMn_2O_4$ have been extensively used in lithium-ion batteries and supercapacitors due to their high theoretical capacities, excellent rate performance, and long-term stability [39]. Contrary to this, multimetallic perovskite and spinel oxides have shown outstanding catalytic activity in various electrochemical reactions such as oxygen reduction reaction (ORR) [40], oxygen evolution reaction (OER) [40], and hydrogen evolution reaction (HER) [41], making them critical components in fuel cells and water-splitting technologies. Their application further extends to environmental fields, like in wastewater remediation, gas sensing, and carbon dioxide (CO_2) reduction, highlighting their importance in sustainable and green technologies. Despite these advantages, several challenges remain in MMOs that needs to be solved, such as difficulty in developing a cost-effective, scalable synthesis method for industrial use, ensuring long-term structural stability, and achieving a deeper mechanistic understanding.

13.2 Structure and Morphology of MMOs

MMOs (bimetallic or multimetallic oxides) are categorized based on their crystalline structure, as discussed below:

13.2.1 Perovskites (ABO_3)

Among various oxide-based catalysts, perovskite oxides have emerged as a central focus due to their remarkable advancements in pollutant degradation [42], energy storage, and optoelectronic applications [43]. Perovskite oxides are multimetallic oxide materials that have been widely used in various photoelectric and catalytic devices like light-emitting diode (LED) [44], solar cells [45], lasers, photodetectors [46], sensors, supercapacitors, and batteries [47]. These widespread applications are attributed to their tunable crystal structure, combined electronic and ionic conductivity, and excellent electronic stability. The perovskite oxide mineral $CaTiO_3$ was first seen in the Ural Mountains in 1839 by Gustav Rose, a German mineralogist, whereas it was named after a Russian mineralogist Count Lev Alekseevich Perovski. In 1926, Victor Goldschmidt introduced the concept of tolerance factor to provide a detailed crystal structure of it, while in 1945, Helen Dick Megaw further refined its structural model using X-ray diffraction data [48]. Perovskites are a versatile class of crystalline materials with the chemical formula ABO_3, where A represents larger-sized cations and B represents small-sized and both these cations are coordinated with oxide anions (O^{2-}) [49]. The ideal structure of perovskite crystallizes is a primitive cubic lattice as shown in Figure 13.2 with the space lattice Pm3m, in which A cations occupy the center of cubic unit cells and are 12-fold coordinated by surrounding oxygen atoms.

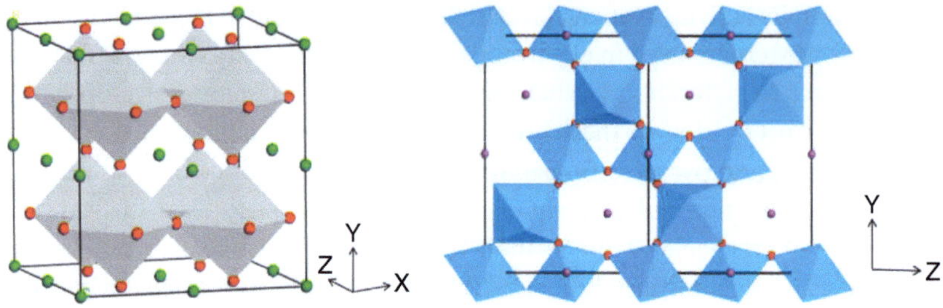

Figure 13.2 Crystal structure of perovskites (a) $BaTiO_3$ and (b) $Na_2Ta_2O_6$ (red represents oxygen; green and purple represent A-site cation; and grey and blue represent BO_6 octahedra).
Source: [50] / MDPI / CC BY 4.0.

The B cations occupy the corners of the cube and are octahedrally coordinated (sixfold) by oxygen atoms. The oxides occupy the face-center of the cube, forming a three-dimensional (3D) network with the corner-sharing BO_6 octahedra. Oxidation states of the A cation and B cation are such that the overall charge is balanced with the three oxide ions [49]. Common cationic combinations include $A^{1+}B^{5+}$, $A^{2+}B^{4+}$, $A^{3+}B^{3+}$, which allow for considerable compositional flexibility. Many technologically important oxides, such as $SrTiO_3$, $CaTiO_3$, $BaTiO_3$ [50], $Na_2Ta_2O_6$ [50], $SrZnO_3$, $SrHfO_3$, $SrSnO_3$, and $BaSrO_3$, adopt this structure or one of its slightly distorted variants. Perovskites exhibit an ideal cubic structure only at elevated temperatures. At room temperature or lower temperature, they undergo distortions due to size mismatches between A and B cations, which prevent them from fitting in the crystal framework neatly. These distortions results in the tilting and rotation of the BO_6 octahedra [51] leading to lower-symmetry structures such as tetragonal, monoclinic, orthorhombic, or rhombohedral [52]. To assess the stability of these distorted structures, Goldschmidt tolerance factor [53] is used, which relates the ionic radii of the A-site cation(r_A) (r_A), the B-site cation (r_B), and the oxygen anion (r_O) as shown in equation (1) [54]:

$$t = \frac{\sqrt{2}\,(r_A + r_O)}{2\,(r_B + r_O)} \tag{1}$$

A wide range of tolerance factor values indicates that perovskites can show a wide range of structural adaptability, making them highly versatile for various applications, like solid oxide fuel cells (SOFCs), dielectric capacitors, catalysts, thermoelectrics, superconductors, and next-generation solar cells.

13.2.1.1 Factors Influencing Stability and Structure of Perovskite

The stability of the perovskite structure plays an important role in determining their functional capabilities. Due to differences in size of A and B cations, or the location of these cations in the oxide framework, distortion can occur in the structure of perovskites [55]. Even though distorted perovskites have reduced symmetry, but it becomes important in materials which show magnetic and electric properties. These distortions are caused by the three main factors: size incompatibility, Jahn-Teller Distortion (JTD), and compositional variation [49].

1. **Size Incompatibility:** The structural stability of perovskites is inherently tied to the size compatibility of their constituent cations [56]. Deviations from their ideal cationic sizes lead to distortions resulting in reduced symmetry and coordination numbers [57]. For example, when the A cation is smaller than its ideal size, tilting of the BO_6 octahedra occurs [58], lowering its coordination number from 12 to 8 [59] as shown in Table 13.1. On the other hand, an undersized B-cation can move off-center within its octahedral site, creating an electric dipole that enables ferroelectricity—a property observed in materials $BaTiO_3$ [60].

   For predicting a stable crystal structure of perovskite material, an empirical index, Goldschmidt tolerance factor (*t*) [53], is used which reflects how closely the ionic size matches the geometric requirements of a cubic perovskite [61]. It is calculated using the radius of cations and anions involved in the perovskite structure (ABO_3). In the ideal cubic crystal, the cell axis (a) is related geometrically to the ionic radii (r_A, r_B, and r_O) as:

$$a = \sqrt{2}(r_A + r_O) = 2(r_B + r_O) \qquad (2)$$

Table 13.1 Perovskite tilt system permutations based on Glazer's classification.

Tilt system	Crystal structure	Space group symmetry	Space group number	Example perovskites
O-Tilt	Cubic	*Pm3m*	221	$CaTiO_3$, $SrTiO_3$
1-Tilt	Tetragonal	*P4/mbm*	127	$NaNbO_3$
2-Tilt	Monoclinic	*C2/m*	12	$PtAlO_3$
	Orthorhombic	*Cmcm*	63	$NaNbO_3$
	Orthorhombic	*Immm*	71	$Sr_3CoSb_2O_9$, Sr_2CoSbO_6
	Orthorhombic	*Imma*	74	$PrAlO_3$
3-Tilt	Triclinic	*PI*	2	$CaSiO_3$
	Monoclinic	*P21/m*	11	$Sr_2Ta_2O_7$
	Monoclinic	*B2/b*	15	Pb_2MnTeO_6
	Orthorhombic	*Pmmm*	59	$NaNbO_3$
	Orthorhombic	*Pnma*	62	$SmAlO_3$, $BaCeO_3$, $GdFeO_3$
	Orthorhombic	*Immm*	71	La_2LiHO_3
	Trigonal	*R3c*	167	$LaCoO_3$, $BiFeO_3$, $LiNbO_3$
	Cubic	*Im3*	204	$Na_{0.69}WO_3$

Source: [64] / with permission of Royal Society of Chemistry.

The ratio of two expressions for the cell length is called Goldschmidt tolerance factor (*t*) [53]. Equation (1) represents the tolerance factor equation. A tolerance factor (*t*) value of 0.8 to 1 is favorable for an ideal perovskite cubic structure (e.g., in $SrTiO_3$) and a value below 0.8 or above 1 tolerance factor usually results in distorted perovskite structure. When the A cation is smaller, *t* becomes less than 1, causing the BO_6 octahedra to tilt in order to fill the available space, leading to the ilmenite structure, as in the case of $GdFeO_3$ whose tolerance factor is 0.81. Various structures that gets formed depends on the level of tilt of MO_6 unit as shown in Table 13.2. On the other hand, if the A is very large or the B is very small, the tolerance factor becomes greater than 1, and the structure shifts to a hexagonal shape, as seen in $BaTiO_3$, where *t* is 1.13. Therefore, by fine-tuning the effective tolerance factor, perovskites can achieve stability in lower-symmetry forms of their crystal lattice, which will open ways to optimize functional properties of advanced materials.

2. **JTD:** Structural distortion in various perovskites can also occur due to J-T active ions located at the B-site of the lattice [62]. For instance, in perovskites such as $LnMnO_3$ (where Ln represents La, Nb, or Pr), the Mn^{3+} ions play a key role in its structural distortion. The ion Mn^{3+} has a $3d^4$ electronic configuration, where three electrons occupy t_{2g} orbital and one electron occupies the e_g orbital. This uneven distribution of electrons causes tetragonal elongation in its MnO_6 octahedra [49], making perovskite a distorted one. Similarly, J-T distortion occurs in $LaCoO_3$ due to the presence of Co^{3+} ions [63], contributing to structural deviations from the ideal perovskite lattice.
3. **Defect Engineering:** Some perovskites have defects in it and these materials are called defect perovskites [64]. They are a class of materials with structural deviations due to ordered oxygen vacancies. These vacancies influence the coordination environment of ions, often leading to mixed oxidation states. For example, in compounds like in $SrFeO_x$ [49] oxygen content is in between 2.5 and 3, resulting in Fe ions adopting oxidation states of +3 and +4. This creates unique geometric arrangements, such as FeO_5 square pyramids. The controlled ordering of vacancies gives rise to properties like enhanced catalytic activity, electronic conductivity, and stability under different environmental conditions [49]. Other examples showing oxygen defects are $Ca_2Mn_2O_4$ and $La_2Ni_2O_5$.

13.2.2 Spinels (AB_2O_4)

Another type of MMOs is spinel-oxides with the general formula AB_2O_4. In their crystal structure, oxygen anion occupies an FCC close-packed structure wherein the divalent A cations and trivalent B cations occupy either tetrahedral (Td) voids or octahedral (Oh) voids [65]. They are broadly classified into three main categories: normal spinel [66], inverse spinel [66], and mixed spinel [67], which are classified based on occupancy of divalent (A^{2+}) and trivalent (B^{3+}) cations within the two principal voids of the cubic lattice, that is, tetrahedral (Td) and octahedral (Oh) voids. In a normal spinel as shown in Figure 13.3, the divalent (A^{2+}) cation occupy one-eighth of the tetrahedral sites and its trivalent (B^{3+}) cations fill two-fourth of the octahedral sites.

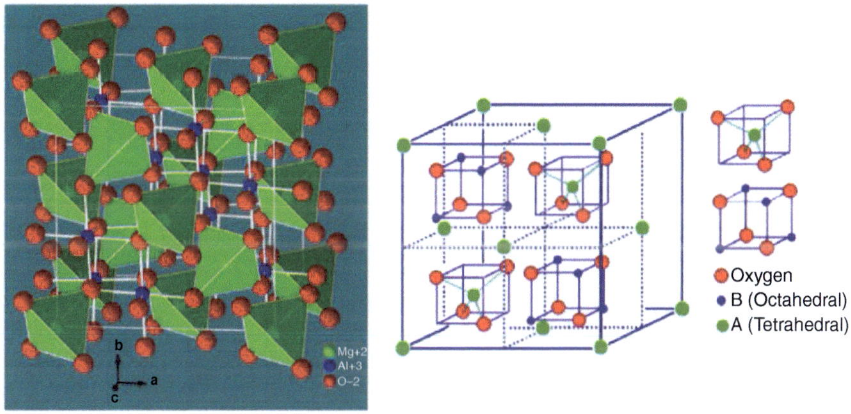

(a) (b)

Figure 13.3 Crystal Structure of a normal spinel: (a) Structural representation of $MgAl_2O_4$ where Mg^{2+} ions (light green) occupy tetrahedral sites surrounded by oxygen atoms (green), while Al^{3+} ions (blue) are situated in the octahedral coordination. (b) Spatial arrangement of tetrahedral (A-site) and octahedral (B-site) sites within the spinel lattice. Each cubic unit cell comprises eight formula units.
Source: [68] / with permission of Elsevier.

This configuration gives rise to $A^{2+}_{(Td)}B^{3+}_{2(oh)}O_4$ type of spinels [68], for example, $MgAl_2O_4$. An inverse spinel, on the other hand, has a different cation distribution in which the A^{2+} fill one-half of the octahedral voids while as B^{3+} occupy both one-eighth Td void and one-half octahedral void making the spinel structure as $A^{2+}_{(Oh)}B^{3+}_{(Td)}B^{3+}_{(Oh)}O_4$ [69], for example, Fe_3O_4. When the A^{2+} and B^{3+} cations occupy both tetrahedral and octahedral sites simultaneously in varying proportions, a mixed spinel or partially inverse spinel is created, for example, in $NiFe_2O_4$ [69]. Like the perovskites, the properties of spinels can be tuned too [70]. It allows unique structural flexibility through the incorporation of different metals at the sites of A, B, and O that enables the accommodation of intrinsic defects and vacancies. Such tunability is important because the optical, structural, and magnetic properties of spinels are strongly affected by the distribution of cations. This distribution plays an important role in determining the magnetic behavior of spinel material. Many mixed-metal spinels exhibit ferrimagnetic properties, making them useful as magnetic materials. In these structures, the magnetic moment of cations within the same sublattice (either Td or Oh) tends to align parallel due to strong super-exchange interactions [71]. The magnetic property and strength of the net moment can be altered by changing the metal cation distribution within the lattice, which offers a versatile platform for designing materials with desired magnetic responses. A well-known example is an inverse spinel magnetite (Fe_3O_4), where the cationic arrangement can be represented as $(Fe^{3+})_{Td}[Fe^{2+}Fe^{3+}]_{Oh}O_4$. This configuration is responsible for its strong magnetic properties. Another example is mixed spinels such as $NiFe_2O_4$ and $CoFe_2O_4$, which display magnetic and electronic properties. Hence, spinel oxide is other class of MMOs whose cation distribution can greatly influence its magnetic and catalytic properties, making them useful for various materials.

13.2.3 Layered Structure Multimetal Oxides

Layered MMOs are a versatile family of 2D materials that possess two or more layers of metal oxides with different metals held together by weak interlayer interactions and hence possess wide interlayer spacing [72, 73] as shown in Figure 13.4. The presence of a broad interlayer space enables numerous.

Guest molecules like cations, anions, and polymers intercalate [74] into layered oxide materials. This makes materials particularly attractive for applications in energy storage, catalysis, and environmental remediation. These materials are represented as $A_xM_yO_z$ [75] where A are alkali metal ions (e.g., Li^+, Na^+, etc.) and M are transition metal ions (e.g., Co, Ni, or Mn), but combinations involving post-transition metals like Sn, Pb, or Bi have also been reported. The M metal ions exhibit mixed oxidation states or a mixture of tetravalent oxidation states, such as Al^{3+} and Ti^{4+}, or divalent and pentavalent cations [76]. These mixed valence states contribute to charge ordering within the lattice and can induce local structural distortions. The value of x usually ranges from $0.5 \leq x \leq 1$. These lattices are built by sheets of edge-sharing MO_6 octahedra inside which the alkali metal ions are placed in tetrahedral, octahedral, or trigonal prismatic environment [77]. Based on the concentration of alkali metals (value of x), the type of environment an alkali metal occupy and the number of different oxide layers present, layered

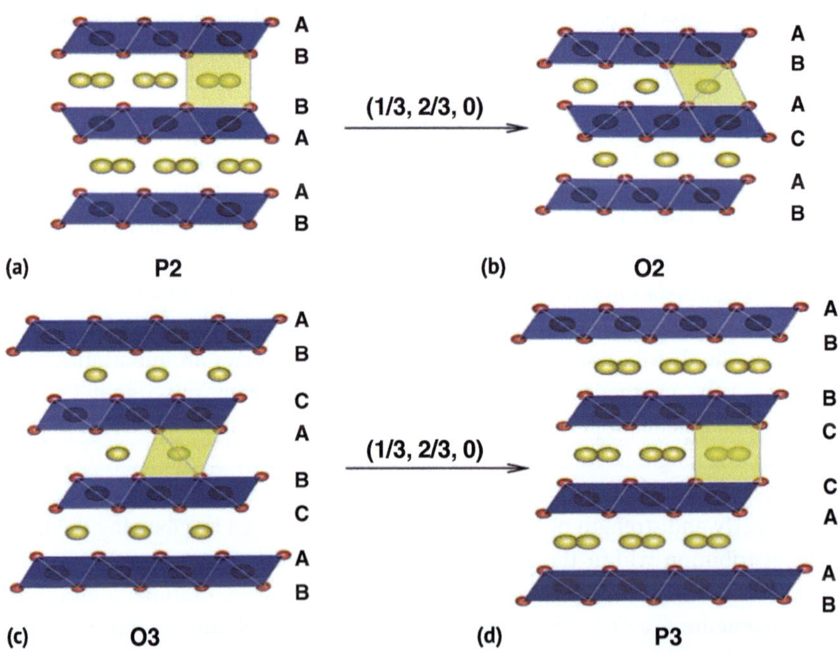

Figure 13.4 Four main types of crystal structure for layered oxides (NaxTMO2). (a) P2-type, (b) O2-type, (c) O3-type, and (d) P3-type stacking. The blue atoms and yellow atoms are transition metals and Na ions in the O-type frameworks respectively. *Source*: [73] / John Wiley & Sons / CC BY 4.0.

metal oxides are of five different types: O_3-type (Alkali metal occupies octahedral sites, $0.7 < x < 1$ in the general formula), P_2-type (Alkali metal occupy prismatic site, $0.3 < x < 0.7$), O_2-type, P_3-type, and T_1-type of layered metal oxides. In these structures, the oxide ion adopt a cubic close-packed (CCP) arrangement while the large-sized alkali metal ion (A^+), compared to the smaller trivalent metal ions (M^{3+}), resides in the interstitial sites that are adjacent to MO_6 octahedron. This arrangement leads to alternating layers of edge-sharing AO_6 and MO_6 layers that share edges. The layered structure formed through this configuration enables various stacking sequences within the lattice. Common stacking sequences such as ABCABC (O_3-type), ABACAB (O_2-type), and ABAB (T_1-type) result in octahedral (O) and tetrahedral (T) environments respectively. These packing patterns not only influence the site symmetry and coordination but also impact the stoichiometry [1, 76].

More complex arrangements arise as the alkali metal ions occupy trigonal prismatic sites. These occur in structures like P_2-type in which the stacking sequence is ABBA and P_3-type, in which the stacking sequence is ABBCCA. Within intralayer, robust bonding between transition metal and oxygen atoms arises, and because of the transition metal-oxygen interaction, bonding tends to exhibit significant covalent character, which supports structural flexibility to the lattice [78]. This covalency stabilizes face-sharing configurations and reflects the structural adaptability of these materials, especially in chemically nonideal or vacancy-rich environments. The structural adjustments in interlayer spacing and substitutional doping at the metal sites allow tailored control over redox behavior, electronic conductivity, and ion transport pathways. These layered frameworks are of particular interest for energy storage and for energy conversion applications, notably in sodium- and lithium-ion batteries [79]. As ongoing research continues to refine synthetic methods and explore doped or mixed-metal variants, layered metal oxides remain a promising platform for the development of next-generation functional materials.

13.2.4 MOF

MOFs are a unique class of crystalline but porous materials that are composed of metal nodes coordinated with multidentate organic linkers. This forms a highly ordered 3D network. The metal nodes like Fe^{2+} and Cu^{2+}, are connected with the organic linkers like carboxylates and imidazolates, resulting in a porous framework which bears larger surface area. These features make MOFs excellent precursors for MMOs, especially upon thermal and chemical treatment. The metal nodes can reorganize into stable oxide frameworks with high dispersion, controlled morphology, and abundant active sites through the calcination process. This transformation enables the formation of MOF with tailored compositions and oxidation states, offering a versatile platform for application in catalysis, energy storage, etc. For example, MOF Cu-BDC consists of CuO clusters as nodes and benzene-1,4-dicarboxylate (BDC) as organic linkers forming a cubic lattice with large pores.

13.3 Synthesis of MMOs

Various synthetic methods are commonly employed for MMOs, each offering an advantage in terms of particle size, morphology, and crystallinity. Some of these methods are as follows:

13.3.1 Sol-Gel Methods

It is a liquid-phase technique used for MMOs synthesis. In this method, both composition and morphology of metal oxide can be controlled. This approach involves the conversion of a liquid sol phase to a solid gel phase using metal precursors such as alkoxides or metal salts. For example, sol-gel synthesis of $ZnFe_2O_4$.

13.3.2 Coprecipitation Method

It is the simplest method of MMOs synthesis. It involves the dissolution of metal oxides in water, followed by their simultaneous precipitation using base to form mixed metal hydroxides or carbonates. The resulting hydroxides or carbonates are calcinated to form MMOs, for example, coprecipitation of Ni^{2+} and Co^{2+} ions to yield $NiCo_2O_4$ after calcination.

13.3.3 Combustion Synthesis

It is also called self-propagating high-temperature synthesis. It utilizes an exothermic reaction between metal nitrates and a fuel like urea to combust rapidly, producing MMOs with high surface area. For example, synthesis of $LaCoO_3$ using glycine as a fuel.

13.3.4 Spray Pyrolysis

In this method, solution of metal salts is aerosolized and passed through a high-temperature furnace where droplets rapidly decompose to form fine MMO particles. This method offers good control over particle size and uniformity. For example, formation of ZnO-SnO_2 metal oxide.

13.3.5 Microwave-Assisted Synthesis

Microwave heating provides rapid and uniform energy, enhancing reaction rates and leading to small particle size and high purity as microwave synthesis in case of $CuFe_2O_4$.

Therefore, the synthesis of MMOs is an important and versatile step in tailoring materials for various applications. Each synthetic method offers a unique advantage in controlling particle size, morphology, crystallinity, and surface area. Thus, the selection of an appropriate method depends mainly on the desired structure and function of the final product.

13.4 Role of MMOS in Catalysis

13.4.1 Green Hydrogen Production Reaction

An electrochemical reaction plays an important role in the clean energy production field using either water electrolyzer process or fuel cells (Figure 13.5). In water electrolysis, the splitting of water into hydrogen and oxygen occurs through two steps: the OER at the anode and the HER at the cathode. In fuel cells, the reverse reaction occurs, that is, HOR at the anode and ORR at the cathode. These reactions are important for the generation of green hydrogen. But both ORR and OER needs high overpotential to get initiated, which ultimately shows sluggish reaction kinetics and leads energy losses to drive the reaction. To address this limitation, MMO nanocatalysts have emerged as powerful catalysts to enhance OER and ORR activity by lowering the energy requirements and improving reaction kinetics. By reducing the energy barrier, hydrogen can be produced in a way that does not depend on the usage of fossil fuels, thereby contributing to the reduction of carbon emissions.

A range of monometallic oxides have been used for the catalytic potential in OER like rutile-type oxides of 4d and 5d transition metals, specifically that of iridium oxide (IrO_2) and ruthenium oxide (RuO_2) have been used as electrocatalysts, but both of them show poor stability by getting dissolved at the same potential at which OER occurs. Similarly, platinum and platinum-based metals have been found to be an active catalysts for ORR, but it forms an oxide layer over the surface at higher overpotential. As a result, the ongoing research aimed at improving the stability, cost-effectiveness, and efficiency of these by incorporating any other metal along with these monometallic oxides into their composition. Various metals have been explored as a potential

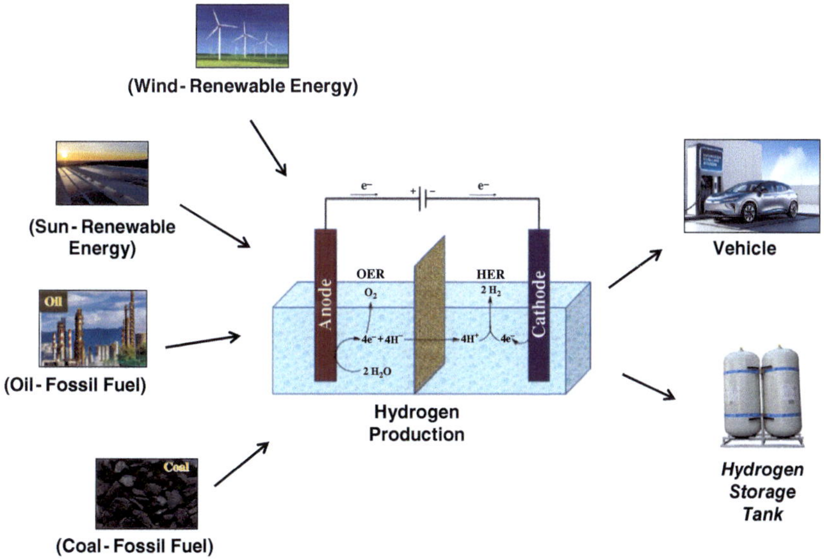

Figure 13.5 Schematic representation of industrial hydrogen production. *Source*: Ardian / Adobe Stock Photos.

additives to improve the performance and stability of these monometallic catalysts, for example, Cr, Ni, Co, Zn, and Sn. For instance, adding Sn to IrO_2 has shown to maintain its high catalytic performance for OER by reducing the susceptibility of IrO_2 towards degradation as the amount of iridium required gets reduced, which makes the catalyst more feasible and stable for large-scale hydrogen production.

Various MMOs have been used till date for OER and ORR reactions to produce green hydrogen like manganese-based spinel oxides $MnCo_2O_4$ and $CoMn_2O_4$, lanthanide-based perovskite oxides [80], that is, $LnMO_3$ (where Ln = La and M = a transition metal, i.e., Cr, Mn, Fe, Co, and Ni). Their electrochemical activity was evaluated using cyclic voltammetry (CV), linear sweep voltammetry (LSV), and Tafel slope analysis, which showed that among the series, $LaMnO_3$ had an excellent catalytic activity towards ORR operating at lower overpotentials and exhibiting a Tafel slope of 53 mV/dec, which indicated fast kinetics. The Tafel slope graphs are shown in the Figure 13.6, which depicts that $LaCoO_3$ and $LaNiO_3$ had an excellent catalytic activity for OER with Tafel slopes around 58–60 mV/dec. While $LaFeO_3$ shows moderate performance activity for both OER and ORR, but did not surpass the efficiency of its cobalt and nickel-based metal oxides. Most importantly, they observed that $LaCoO_3$ shows superior electrochemical durability for ORR reaction by retaining about 72% of its initial current for over 3 h and showing minimal degradation even after 2000 electrochemical cycles. The OER catalytic performance across the series followed an order of $LaCoO_3 > LaNiO_3 > LaCrO_3 > LaMnO_3 > LaFeO_3$ demonstrating $LaCoO_3$ as the most active OER catalyst. Conversely, the catalytic efficiency of ORR followed an order of $LaMnO_3 > LaCoO_3 > LaNiO_3 > LaFeO_3 > LaCrO_3$ indicating that $LaMnO_3$ as the most effective catalyst for ORR. Similarly, UiO-66(Zr) MOF, MIL-101(Cr), etc. have been studied for hydrogen production too.

Hence, MMOs can significantly enhance the efficiency for OER and ORR, making them an ideal candidate for green hydrogen production.

13.4.2 CO_2 Reduction

Due to the continuous burning of fossil fuels like oil, coal, and natural gas, the CO_2 level has increased in the atmosphere, which contributes to the climate change causing global warming as it traps heat in the earth's atmosphere. To combat this

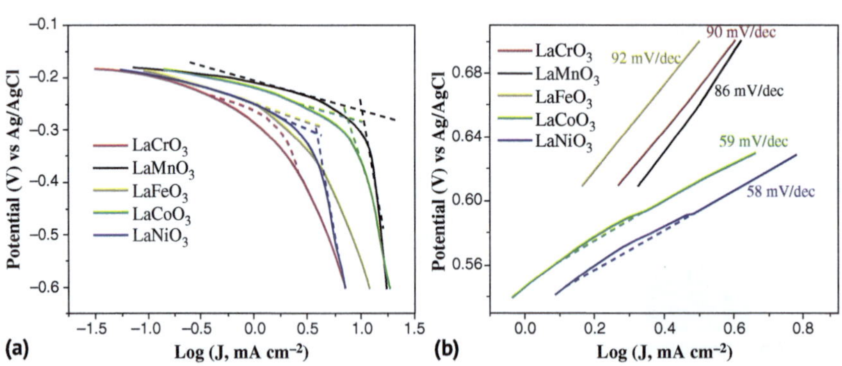

Figure 13.6 Tafel slope of various lanthanide perovskite oxides for (a) ORR and (b) OER using 1M KOH as electrolyte at 25°C.
Source: [80] / with permission of Elsevier.

problem, various CO_2 capture technologies have been developed, which could manage to capture CO_2 from air and industrial processes. However, it has managed to capture about 35 million tonnes of CO_2, whereas the global emissions from fossil fuels reach nearly 35 gigatonnes annually [81]. This shows that capturing CO_2 alone is not enough. We also need smart and efficient ways to reuse or recycle CO_2 into useful products. Various methods have been used in the transformation of CO_2 into fuels and chemicals. Among them, MMOs catalysts have been widely used for their ability to transform CO_2 into compounds such as hydrogen, methanol, methane, and ammonia. Showing tunable composition, structural versatility, and improved catalytic efficiency, MMOs enhance both the selectivity and activity of CO_2 conversion [82], offering an efficient path towards sustainable chemical production. The $NiCo_2O_4$, Ni-Al@Al_2O_3, $La_{2-x}Ce_xNiO_4$, Ni/CeO_2/Al_2O_3, and NiO/Co_3O_4 have been used for their ability to convert CO_2 to methane. Their conversion rate to methane was found to be equal to 56%, 96%, 79%, 68%, and 55% respectively. Another promising approach for CO_2 conversion is the reverse water-gas shift (RWGS) reaction, where CO_2 is converted to water gas, that is, carbon monoxide (CO) and water [83]. The metals such as Pt, Pd, Rh, and Cu combined with metal oxide like CeO_2 convert CO_2 to CO [84]. Similarly, perovskite-type oxides like $LaMnO_3$ perovskites doped with Co and Fe show isothermal CO production capability, for example, $La_{0.75}Sr_{0.25}FeO_{3-\delta}$, $La_{0.75}Sr_{0.25}FeO_{3-\delta}/SiO_2$, CeO_2@$La_{0.75}Sr_{0.25}FeO_{3-\delta}$, NiO@ $La_{0.75}Sr_{0.25}FeO_{3-\delta}$, and Co_3O_4@ $La_{0.75}Sr_{0.25}FeO_{3-\delta}$ all were used for the CO_2 splitting to CO. The reduction of CO_2 to methanol has also been studied. Some of the complex MMOs and their conversion of CO_2 to suitable product have been tabulated in Table 13.2. Therefore, MMOs have proved the best candidate for CO_2 reduction, offering a sustainable route beyond CO_2 capture.

13.4.3 Pollutant Degradation

MMOs have been used to combat with environmental pollutants like organic contaminants such as synthetic dyes, pesticides, and herbicides; VOCs like benzene, toluene, and formaldehyde; inorganic pollutants such as heavy metals like mercury, arsenic,

Table 13.2 Showing various MMOs that have been used for the CO_2 conversion.

MMOs	CO_2 conversion (%)	Temperature (K)	Product	References
$NiGa_2O_x$/$CuGa_2O_x$/ $CoGa_2O_x$	99	973	CO + H_2O	Kang et al. [85]
Cu/$CaTiO_3$	—	420–660	Methanol	Huš et al. [86]
Ni/Mg/$Ce_{0.6}Zr_{0.4}O_2$/Al_2O_3	61	1133	Methane	Zhao et al. [87]
Ni-Al_2O_3	90	1023	Methane	Luna and Iriarte [88]
CuO-Fe_2O_3/Al_2O_3	90	1073	Methane	Siriwardane et al. [89]

and lead; and gaseous pollutants like NO_x, CO, and H_2S. These pollutants collectively threaten environmental sustainability, biodiversity, and human health, emphasizing the need for efficient remediation technologies for their degradation and detoxification [90]. Various spinel-based metal oxides like $CuCr_2O_4$, Co_3O_4, Fe_3O_4, Fe_4O_5, and Fe_5O_6 are known for their high catalytic performance in CO oxidation. Researchers also find out that supporting these oxides on materials like γ-Al_2O_3 or TiO_2, like Co_3O_4/Al_2O_3 and Fe_2O_3/Al_2O_3 improves their thermal stability and enhances their overall activity. Similarly, gold-based catalysts such as Au-ZnO/Al_2O_3 and Au/Al_2O_3 demonstrate excellent efficiency in boosting CO oxidation. G. Chandra Dhal et al. synthesized a series of $LaMnO_3$ catalysts for checking the reduction activity of NO_x gas, whose percentage conversion is tabulated in Table 13.3.

Similarly using light source, the cubic phase of $KTaO_3$ degrades toluene into CO_2, $(Ag_{0.75}Sr_{0.25})(Nb_{0.75}Ti_{0.25})O_3$ is used to degrade acetaldehyde (CH_3CHO) into CO_2, Bi_2WO_6 has been used to remove rhodamine B. Other photocatalytic activities of MMOs and the pollutants on which they show power to degrade them have been outlined in Table 13.4. Therefore, MMOs exhibit strong catalytic and photocatalytic behavior for the degradation of a wide range of environmental pollutants.

Table 13.3 Showing different catalytic activity of MMOs with and without doping.

Catalyst	T_{may}-NO (°C)	NO conversion (%)
$LaMnO_3$	386	45.3
$La_{0.65}Ag_{0.35}MnO_3$	301	53.2
$La_{0.65}Ag_{0.1}K_{0.25}MnO_3$	368	64.4

Source: Adapted from [91].

Table 13.4 Showing various MMOs that have been used for the degradation of various organic pollutants.

Photocatalyst	Absorbance (nm)	Pollutant degraded	Degradation (%)
$Ag/BiFeO_3$	496	RhB (10 mg L^{-1})	32%, 7 h
Pt/Fe-$Bi_2Ti_2O_7$	517	MO (10 μM)	88%, 2 h
$LaFeO_3/Ag_3PO_4$	760	Phenol (20 mg L^{-1})	92%, 2 h
$Ca_2Fe_2O_5/ZnFe_2O_4$	653	SA (10 mg L^{-1})	50%, 3 h
Bi_2WO_6-rGO	450	BPA (10 mg L^{-1})	90%, 80 min
Bi_2MoO_6/TiO_2	477	ARS (5 mg L^{-1})	85%, 4 h
$BiOI/Bi_2WO_6$	599	DCP (30 mg L^{-1})	98%, 3 h

RhB: rhodamine B; MO: methyl orange; DCP: 2,4-dichlorophenol; BPA: bisphenol A; ARS: alizarin red.
Source: Adapted from [61].

13.5 Mechanism of MMOs Catalytic Action

The mechanism of catalytic action of MMOs is primarily driven by the combined redox activity of the metal constituents present in them. Unlike single metal catalysts, where the reactive species (S) interacts with only one type of the metal center (either M^1 or M^2), each possessing specific properties only, mixed or bimetallic oxides enable the reactive species (S) to interact simultaneously with both M^1 and M^2 centers, which possess complementary properties, as illustrated in the scheme below. This combined interaction leads to synergistic effects (shown in Figure 13.7) that increase surface adsorption, improve charge transfer efficiency, and hence increase the catalytic oxidation process.

An excellent example of such synergistic behavior is observed in the CO oxidation. It was studied that the $CuMnO_x$ and its supported forms $CuMnO_x/\gamma\text{-}Al_2O_3$ show high catalytic efficiency for CO oxidation. However, Cu or Mn oxides individually show very low activity compared to mixed-oxide counterparts. Another MMOs Co_3O_4 spinels have been also used for CO oxidation. Its mechanism is clearly shown in Figure 13.8, which illustrates how Co_3O_4 facilitates the CO oxidation process through a multistep process.

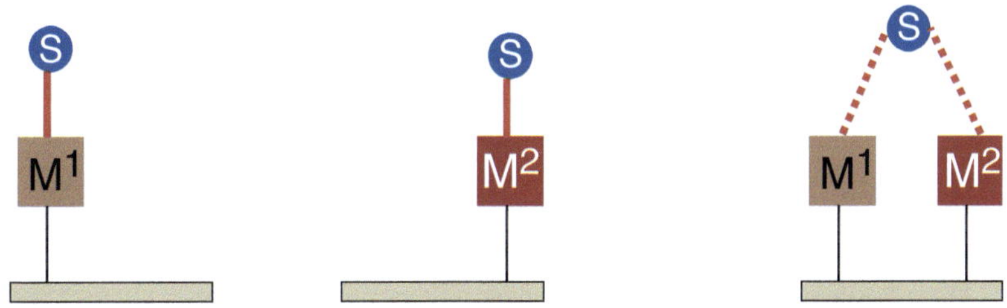

Figure 13.7 Schematic representation of substrate binding to single metals M^1 and M^2 and to the bimetallic catalysts.

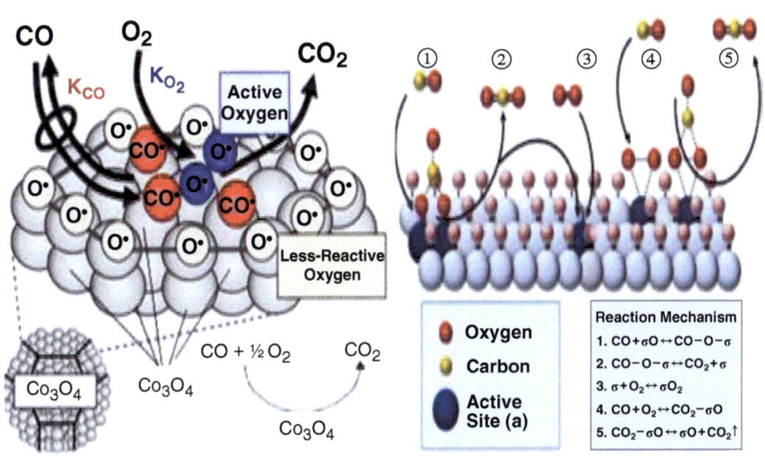

Figure 13.8 Depicts mechanism of oxidation of CO to CO_2 over Co_3O_4 catalyst. *Source*: Reproduced from [92], with permission of Springer Nature.

Therefore, the superior performance of $CuMnO_x$, $CuMnO_x/\gamma\text{-}Al_2O_3$, and Co_3O_4 MMOs for CO oxidation validates the role of metal-metal cooperation in enhancing catalytic efficiency.

13.6 Conclusions

The aim of this chapter to provide insight into MMOs and their superior catalytic ability compared to single metal oxides. These nanocatalysts exhibit synergistic interaction between multiple cations, thus enhancing their stability and catalytic activity. They can adopt various crystal structures, such as perovskites, spinels, and layered types, through synthetic and compositional modifications. Owing to these features, MMOs have been applied in green hydrogen production, CO_2 reduction, and for pollutant degradation. Ongoing research is focusing on improving synthetic methods and nanostructuring strategies to better control their morphology and electronic properties, thereby improving their efficiency and expanding their applicability in addressing global challenges.

References

1 Wang PF, You Y, Yin YX, Guo YG. Layered oxide cathodes for sodium-ion batteries: phase transition, air stability, and performance. *Adv Energy Mater.* 2018;8(8):1701912.
2 Zybert M. Applied catalysis in chemical industry: synthesis, catalyst design, and evaluation. *MDPI.* 2023;13:607.
3 Othman SMA. Catalysts of the preparation and industrial importance of catalysis and catalyst deactivation. *Int J Adv Chem Res.* 2020;1:23–7.
4 Vinodkumar T, Reddy BM. *Catalytic Combustion over Cheaper Metal Oxides.* New York: Nova Science Publishers; 2011;3. p. 36–7.
5 Zheng R, Liu Z, Wang Y, Xie Z. Industrial catalysis: strategies to enhance selectivity. *Chin J Catal.* 2020;41(7):1032–8.
6 Mahmudov KT, Pombeiro AJL. Control of selectivity in homogeneous catalysis through noncovalent interactions. *Chem Eur J.* 2023;29(26):e202203861.
7 Gross E, Liu JH-C, Toste FD, Somorjai GA. Control of selectivity in heterogeneous catalysis by tuning nanoparticle properties and reactor residence time. *Nat Chem.* 2012;4(11):947–52.
8 Schmal M. *Heterogeneous Catalysis and its Industrial Applications.* Springer; 2016.
9 Njovu G, Lukashuk L. Exploration of single atom alloy catalysis for industrial applications: part I: preparation and characterisation. *Johnson Matthey Technol Rev.* 2025;69(2):247–59.
10 Kim H, Kim M-C, Kim S-b, Kim Y-S, Choi J-H, Park K-W. Porous SnO_2 nanostructure with a high specific surface area for improved electrochemical performance. *RSC Adv.* 2020;10(18):10519–25.
11 Pozzi M, Jonak Dutta S, Kuntze M, Bading J, Rüßbült JS, Fabig C, et al. Visualization of the high surface-to-volume ratio of nanomaterials and its consequences. *J Chem Educ.* 2024;101(8):3146–55.

12 Khalil M, Kadja GTM, Ilmi MM. Advanced nanomaterials for catalysis: current progress in fine chemical synthesis, hydrocarbon processing, and renewable energy. *J Ind Eng Chem.* 2021;93:78–100.

13 Das D. Structural diversity in heterobimetallic complexes of alkaline earth metals and molybdenum, and selective dye adsorption properties of mixed metal oxides synthesized from the heterobimetallic complexes. *ChemistrySelect.* 2019;4(4):1428–36.

14 Chavali MS, Nikolova MP. Metal oxide nanoparticles and their applications in nanotechnology. *SN Appl Sci.* 2019;1(6):607.

15 Grilli M. Metal oxides. *Metals* 2020;10:820.

16 Chen Y, Miao Y, Hu X, Kim D, Zhu Y, Su Y, et al. Evaluating the functions of the key dopant elements in multi-metal oxide electrocatalysts for high-performance Li-O2 batteries. *Energy Storage Mater.* 2023;63:102989.

17 Gawande MB, Pandey RK, Jayaram RV. Role of mixed metal oxides in catalysis science—versatile applications in organic synthesis. *Catal Sci Technol.* 2012;2(6):1113–25.

18 Koyale PA, Panda DK, Delekar SD. Metal oxide engineering. In: Delekar SD, editors. *Advances in Metal Oxides and Their Composites for Emerging Applications.* Elsevier; 2022. p. 3–56.

19 Bo X, Zan L, Jia R, Dastafkan K, Zhao C. The nature of synergistic effects in transition metal oxides/in-situ intermediate-hydroxides for enhanced oxygen evolution reaction. *Curr Opin Electrochem.* 2022;34:100987.

20 Saha R, Gupta K, Gómez García CJ. Strategies to improve electrical conductivity in metal–organic frameworks: a comparative study. *Cryst Growth Des.* 2024;24(5):2235–65.

21 Hooriabad Saboor F, Shahsavari S, Zandjou M, Asgari M. From structure to catalysis: advances in metal-organic frameworks-based shape-selective reactions. *ChemNanoMat.* 2024;10(7):e202400049.

22 He Y, Yin Z, Wang Z, Wang H, Xiong W, Song B, et al. Metal-organic frameworks as a good platform for the fabrication of multi-metal nanomaterials: design strategies, electrocatalytic applications and prospective. *Adv Colloid Interface Sci.* 2022;304:102668.

23 Yeom J, Giacomo L, Lea G, Claudia C, Janczak-Rusch J. The thermal stability and degradation mechanism of Cu/Mo nanomultilayers. *Sci Technol Adv Mater.* 2024;25(1):2357536.

24 Wang X, Hanson JC, Liu G, Rodriguez JA, Iglesias-Juez A, Fernández-García M. The behavior of mixed-metal oxides: physical and chemical properties of bulk $Ce_{1-x}Tb_xO_2$ and nanoparticles of $Ce_{1-x}Tb_x O_y$. *J Chem Phys.* 2004;121(11):5434–44.

25 Humayun M, Li Z, Israr M, Khan A, Luo W, Wang C, et al. Perovskite type ABO_3 oxides in photocatalysis, electrocatalysis, and solid oxide fuel cells: state of the art and future prospects. *Chem Rev.* 2025;125(6):3165–241.

26 Liu Y, Kishimoto F, Lu X, Li J, Takanabe K. CO_2 hydrogenation over Fe-Mn-Zn spinel oxide nanohybrids precatalysts. *Appl Catal B.* 2025;361:124675.

27 Zhou J, Guo L, Xu J, Ouyang R, Yu X, Fang X, et al. Fluorite phase La–M–O (M= Zr and Ce) composite oxides for oxidative dehydrogenation of ethane at low or high temperatures: redox sites vs. lattice oxygen. *Catal Sci Technol.* 2025;15(9):2829–37.

28 Adamski NL, Dreyer CE, Van de Walle CG. Polarization properties at rocksalt/wurtzite oxide interfaces. *Phys Rev B.* 2020;102(20):201301.

29 Wei F, Zhang Q, Zhang P, Tian W, Dai K, Zhang L, et al. Research progress on layered transition metal oxide cathode materials for sodium ion batteries. *J Electrochem Soc.* 2021;168(5):050524.

30 Saranya D, Selvaraj V. Double metal oxide based nickel hybrid nanocatalyst for electrooxidation and alkaline fuel cell device fabrication. *Int J Hydrogen Energy.* 2018;43(29):13450–61.

31 Vikraman D, Hussain S, Hailiang L, Karuppasamy K, Sivakumar P, Santhoshkumar P, et al. Spinel-structured metal oxide-embedded MXene nanocomposites for efficient water splitting reactions. *Inorg Chem Front.* 2022;9(22):5903–16.

32 Patil SM, Vanalakar SA, Delekar SD. Mixed metal oxide nanocomposites for environmental remediation. In: Delekar SD, editor. *Advances in Metal Oxides and their Composites for Emerging Applications.* Elsevier; 2022. p. 425–77.

33 Ahmed MG, Tay YF, Zhang M, Chiam SY, Wong LH. Tailoring surface electronic structure of spinel Co_3O_4 oxide via Fe and Cu substitution for enhanced oxygen evolution reaction. *ACS Mater Lett.* 2024;6(10):4756–64.

34 Song J, Wang S, Xu Y, Liu Q, Zhao Y. LDH derived $MgAl_2O_4$ spinel supported Pd catalyst for the low-temperature methane combustion: roles of interaction between spinel and PdO. *Appl Catal A: Gen.* 2021;621:118211.

35 Jiang X, Fan J, Xiang S, Mou J, Yao P, Jiao Y, et al. Superior catalytic activity and high thermal durability of $MgAl_2O_4$ modified $Pt/Ce_{0.5}Zr_{0.5}O_2$ TWC. *Appl Surf Sci.* 2022;578:151915.

36 Okeke IS, Obodo RM, Eze CN, Orji MP, Agwu KK, Ubachukwu AA, et al. Effect of surface modification on the dielectric properties of mixed transition metal oxides ceramics composites. In: Jose R, Ezema F, editors. *Surface Modification and Functionalization of Ceramic Composites.* Elsevier; 2023. p. 59–80.

37 Zhang BH, Liu XQ, Chen XM. Review of experimental progress of hybrid improper ferroelectricity in layered perovskite oxides. *J Phys D Appl Phys.* 2021;55(11):113001.

38 Sefadi JS, Mochane MJ, Gumede TP, Malebo NJ, Mokhena TC. Graphene-metal oxide nanoparticles on piezoelectric and pyroelectric effect of polyvinylidene fluoride (PVDF). In: Hussain CM, Thomas S, editors. *Handbook of Polymer and Ceramic Nanotechnology.* Cham: Springer; 2021. p. 1123–46.

39 Zhao X, Mao L, Cheng Q, Li J, Liao F, Yang G, et al. Two-dimensional spinel structured co-based materials for high performance supercapacitors: a critical review. *Chem Eng J.* 2020;387:124081.

40 Mladenović D, Mladenović A, Santos DMF, Yurtcan AB, Miljanić Š, Mentus S, et al. Transition metal oxides for bifunctional ORR/OER electrocatalysis in unitized regenerative fuel cells. *J Electroanal Chem.* 2023;946:117709.

41 Obayi CS, Nnamchi PS. Mixed transition metal oxides for photoelectrochemical hydrogen production. In: Ezema FI, Lokhande CD, Jose R, editors. *Chemically*

Deposited Nanocrystalline Metal Oxide Thin Films: Synthesis, Characterizations, and Applications. Springer; 2021. p. 279–92.

42 Mahmoudi F, Saravanakumar K, Maheskumar V, Njaramba LK, Yoon Y, Park CM. Application of perovskite oxides and their composites for degrading organic pollutants from wastewater using advanced oxidation processes: review of the recent progress. *J Hazard Mater*. 2022;436:129074.

43 Kostopoulou A, Kymakis E, Stratakis E. Perovskite nanostructures for photovoltaic and energy storage devices. *J Mater Chem A*. 2018;6(21):9765–98.

44 Quan LN, García de Arquer FP, Sabatini RP, Sargent EH. Perovskites for light emission. *Adv Mater*. 2018;30(45):1801996.

45 Shi Z, Jayatissa AH. Perovskites-based solar cells: a review of recent progress, materials and processing methods. *Materials*. 2018;11(5):729.

46 Xie C, Liu CK, Loi HL, Yan F. Perovskite-based phototransistors and hybrid photodetectors. *Adv Funct Mater*. 2020;30(20):1903907.

47 Cao Y, Liang J, Li X, Yue L, Liu Q, Lu S, et al. Recent advances in perovskite oxides as electrode materials for supercapacitors. *Chem Commun*. 2021;57(19):2343–55.

48 Ortega-San-Martin L. Introduction to perovskites: a historical perspective. In: Arul NS, Nithya VD, editors. *Revolution of Perovskite: Synthesis, Properties and Applications*. Singapore: Springer; 2020. p. 1–41.

49 Johnsson M, Lemmens P. Crystallography and chemistry of perovskites. Handbook of Magnetism and Advanced Magnetic Materials. Edited by Helmut Kronmüller and Stuart Parkin. Volume 4: Novel Materials. 2007 John Wiley & Sons, Ltd. ISBN: 978-0-470-02217-7.

50 Kanhere P, Chen Z. A review on visible light active perovskite-based photocatalysts. *Molecules* 2014;19:19995–20022.

51 Balachandran PV, Rondinelli JM. Interplay of octahedral rotations and breathing distortions in charge-ordering perovskite oxides. *Phys Rev B—Condens Matter Mat Phys*. 2013;88(5):054101.

52 Barnes PW. *Exploring Structural Changes and Distortions in Quaternary Perovskites and Defect Pyrochlores using Powder Diffraction Techniques*. The Ohio State University; 2003.

53 Bartel CJ, Sutton C, Goldsmith BR, Ouyang R, Musgrave CB, Ghiringhelli LM, et al. New tolerance factor to predict the stability of perovskite oxides and halides. *Sci Adv*. 2019;5(2):eaav0693.

54 Gzyl AS, Addo PK, Sanchez BM. A new ionic radii scale derived from perovskite oxides. *J Solid State Chem*. 2025;346:125254.

55 Vasala S, Karppinen M. A2B′ B″O6 perovskites: a review. *Prog Solid State Chem*. 2015;43 (1–2):1–36.

56 Rodriguez-Martinez LM, Attfield JP. Cation disorder and size effects in magnetoresistive manganese oxide perovskites. *Phys Rev B*. 1996;54(22):R15622.

57 Talapatra A, Uberuaga BP, Stanek CR, Pilania G. A machine learning approach for the prediction of formability and thermodynamic stability of single and double perovskite oxides. *Chem Mater*. 2021;33(3):845–58.

58 Xiang HJ, Guennou M, Íñiguez J, Kreisel J, Bellaiche L. Rules and mechanisms governing octahedral tilts in perovskites under pressure. *Phys Rev B*. 2017;96(5):054102.

59 Li X, Zhao H, Liang J, Luo Y, Chen G, Shi X, et al. A-site perovskite oxides: an emerging functional material for electrocatalysis and photocatalysis. *J Mater Chem A*. 2021;9(11):6650–70.

60 Anderson MT, Greenwood KB, Taylor GA, Poeppelmeier KR. B-cation arrangements in double perovskites. *Prog Solid State Chem*. 1993;22(3):197–233.

61 Kong J, Yang T, Rui Z, Ji H. Perovskite-based photocatalysts for organic contaminants removal: current status and future perspectives. *Catal Today*. 2019;327:47–63.

62 Herlihy A, Chen W-T, Ritter C, Chuang Y-C, Senn MS. Interplay between Jahn–Teller distortions and structural phase transitions in Ruddlesden–Poppers. *J Am Chem Soc*. 2025;147(9):7209–13.

63 Jia T, Hao Y, Hao H. The spin-state effect on the redox performance of $LaCoO_3$ perovskites. *Results Phys*. 2024;61:107766.

64 Arandiyan H, Mofarah SS, Sorrell CC, Doustkhah E, Sajjadi B, Hao D, et al. Defect engineering of oxide perovskites for catalysis and energy storage: synthesis of chemistry and materials science. *Chem Soc Rev*. 2021;50(18):10116–211.

65 Zhu H, Xie H, Zhao Y, Dai S, Li M, Wang X. Structure and magnetic properties of a class of spinel high-entropy oxides. *J Magn Mater*. 2021;535:168063.

66 Szotek Z, Temmerman WM, Ködderitzsch D, Svane A, Petit L, Winter H. Electronic structures of normal and inverse spinel ferrites from first principles. *Phys Rev B*. 2006;74(17):174431.

67 Chinnasamy C, Narayanasamy A, Nagamony P, Shinoda K, Jeyadevan B, Tohji K, et al. Mixed spinel structure in nanocrystalline $NiFe_2O_4$. *Phys Rev B*. 2001;63:184108.

68 Tsurkan V, Nidda H-A, Deisenhofer J, Lunkenheimer P, Loidl A. On the complexity of spinels: magnetic, electronic, and polar ground states. *Phys Rep*. 2021;926:1–86.

69 Mo S-D, Ching WY. Electronic structure of normal, inverse, and partially inverse spinels in the $MgAl_2O_4$ system. *Phys Rev B*. 1996;54(23):16555–61.

70 Norman JE, Dupuyb AD, Schoenung JM. Crystal lattice engineering of spinel phases in Mn-oxide doped (Co,Cu,Mg,Ni,Zn)O. *Mater Res Lett*. 2025; 13:1–10.

71 Wang X, Musicó BL, Kons C, Metz PC, Keppens V, Gilbert DA, et al. Local cation order and ferrimagnetism in compositionally complex spinel ferrites. *APL Mater*. 2022;10(12):121102.

72 Zhang A, Yin X, Zhang X, Ba J, Li J, Wei Y, et al. 2D Layered MnO_2 with an ultralarge interlayer spacing for aqueous zinc ion batteries. *ACS Appl Energy Mater*. 2024;7(3):1298–305.

73 Sun L, Zeng J, Wan X, Peng C, Wang J, Lin C, et al. Recent progress of interface modification of layered oxide cathode material for sodium-ion batteries. *Electron*. 2024;2:e31.

74 Stolz RM, Mahdavi-Shakib A, Frederick BG, Mirica KA. Host–guest interactions and redox activity in layered conductive metal–organic frameworks. *Chem Mater*. 2020;32(18):7639–52.

75 Ganguly P, Rao CNR. Crystal chemistry and magnetic properties of layered metal oxides possessing the K_2NiF_4 or related structures. *J Solid State Chem.* 1984;53(2):193–216.

76 Delmas C, Fouassier C, Hagenmuller P. Structural classification and properties of the layered oxides. *Physica B+C*. 1980;99 (1–4):81–5.

77 Augustyn V. Tuning the interlayer of transition metal oxides for electrochemical energy storage. *J Mater Res*. 2017;32(1):2–15.

78 Zhang Z, Xu X, Ma P, Asakura Y, Wang Z, Yamauchi Y, et al. Pushing the limit of layered transition metal oxides with heterolattice oxygen-mediated redox for capacitive deionization. *Nat Comm*. 2025;16(1):3652.

79 Xu Z, Song K, Chang X, Li L, Zhang W, Xue Y, et al. Layered oxide cathodes: a comprehensive review of characteristics, research, and development in lithium and sodium ion batteries. *Carbon Neutraliz*. 2024;3(5):832–56.

80 Ashok A, Kumar A, Bhosale RR, Almomani F, Malik SS, Suslov S, et al. Combustion synthesis of bifunctional $LaMO_3$ (M= Cr, Mn, Fe, Co, Ni) perovskites for oxygen reduction and oxygen evolution reaction in alkaline media. *J Electroanal Chem*. 2018;809:22–30.

81 Tabish A, Varghese AM, Wahab MA, Karanikolos GN. Perovskites in the energy grid and CO2 conversion: current context and future directions. *Catalysts* 2020;10(1):95.

82 Chitturi HS, Lavanya Y, Varun Y, Ramesh A, Pamu SH, Sreedhar I, et al. Unraveling the CO2 methanation and capture ability of NiO@metal oxides. *J Mater Chem A*. 2025;13(10):7422–44.

83 Kang K, Sampei H, Sekine Y. CO_2 conversion to CO by reverse water gas shift and dry reforming using chemical looping. *RSC Sustainability*. 2025;3(4):1598–628.

84 Zhang Q, Bown M, Pastor-Pérez L, Duyar MS, Reina TR. CO_2 Conversion via reverse water gas shift reaction using fully selective Mo–P multicomponent catalysts. *Ind Eng Chem Res*. 2022;61(34):12857–65.

85 Kang K, Kakihara S, Higo T, Sampei H, Saegusa K, Sekine Y. Equilibrium unconstrained low-temperature CO_2 conversion on doped gallium oxides by chemical looping. *Chem Commun*. 2023;59(74):11061–4.

86 Huš M, Kopač D, Likozar B. Catalytic hydrogenation of carbon dioxide to methanol: synergistic effect of bifunctional Cu/perovskite catalysts. *ACS Catal*. 2019;9(1):105–16.

87 Zhao X, Ngo HT, Walker DM, Weber D, Maiti D, Cimenler U, et al. Tri-reforming of surrogate biogas over Ni/Mg/ceria–zirconia/alumina pellet catalysts. *Chem Eng Comm*. 2018;205(8):1129–42.

88 Luna AEC, Iriarte ME. Carbon dioxide reforming of methane over a metal modified Ni-Al_2O_3 catalyst. *Appl Catal A Gen*. 2008;343 (1–2):10–5.

89 Siriwardane R, Tian H, Miller D, Richards G. Fluidized bed testing of commercially prepared MgO-promoted hematite and CuO–Fe_2O_3 mixed metal oxide oxygen carriers for methane and coal chemical looping combustion. *Appl Energy*. 2015;157:348–57.

90 Tahseen T, Farooq N, Iman W, Hareem A, Irshad R, Ibrahim TK, et al. G-C_3N_4-based multi-metal oxide nanocomposites: a new frontier in catalytic performance. *Solid State Sci.* 2025;163:107890.

91 Dhal GC, Dey S, Mohan D, Prasad R. Simultaneous control of NOx-soot by substitutions of Ag and K on perovskite ($LaMnO_3$) catalyst. *Bull Chem React Eng Catal.* 2018;13(1):144–54.

92 Dey S, Dhal G, Mohan D, Prasad R. Advances in transition metal oxide catalysts for carbon monoxide oxidation: a review. *Adv Compos Hybrid Mater.* 2019;2:178–98.

14

Composite and Hybrid Metal Oxide Nanocatalysts

Glory Valentine Umoh[1], Nelson Oshogwue Etafo[2,3,4], and Muyideen Olaitan Bamidele[4]*

[1] *El Centro de Investigación en Materiales Avanzados, S.C. (CIMAV), Chihuahua, Mexico*
[2] *Universidad Carolina, Saltillo, Coahuila, Mexico*
[3] *Department of Metallurgy and Materials Science, Institute of Technology, Saltillo, Coahuila, Mexico*
[4] *Facultad de Ciencias Químicas, Universidad Autónoma de Coahuila, Saltillo, Coahuila, Mexico*

*Corresponding author: gloryvals44@gmail.com

14.1 Introduction

14.1.1 Historical Context and Fundamental Concepts

The evolution of nanocatalysts has fundamentally renovated modern industrial processes, environmental remediation, and energy conversion technologies. Metal oxides have occupied a central role in this evolution due to their inherent catalytic properties, structural diversity, and thermal stability. The concept of composite and hybrid nanocatalysts emerged from the recognition that single-component systems often suffer from limitations, including rapid deactivation, limited selectivity, and insufficient activity under mild conditions. By integrating metal oxides with complementary materials, researchers have created synergistic systems where the whole exceeds the sum of its parts, ushering in a new paradigm in catalytic science [1, 2].

The foundation for modern hybrid nanocatalysts was laid with the 1992 discovery of the M41S mesoporous silica family by Mobil researchers. This breakthrough demonstrated that ordered nanoporous architectures could be engineered with tunable pore sizes, sparking intensive research into non-siliceous mesoporous materials, including transition metal oxides and their composites. According to IUPAC classification, these porous structures are categorized as microporous (<2 nm), mesoporous (2–50 nm), or macroporous (>50 nm), each offering distinct advantages for catalytic applications. The early 21st century witnessed accelerated development of metal oxide composites, where combinations like Fe_3O_4-SiO_2, CeO_2-SiO_2, and NiO-ZrO_2, see Table 14.1 demonstrated unprecedented performance in catalysis, biosensing, and drug delivery [1].

Metal Oxide Nanocatalysts for Sustainable Energy Production, First Edition. Edited by Abayomi Bamisaye et al.
Published 2026 by Wiley-VCH GmbH.

Table 14.1 Classification framework for hybrid and composite nanocatalysts.

Category	Main constituents	Representative examples	Primary advantages	Primary disadvantages	References
Oxide-oxide	Multiple metal oxides	NiO-ZrO_2, CeO_2-ZrO_2	Enhanced thermal stability and bifunctionality	• Phase segregation under thermal stress • Reduced surface area due to sintering • Limited redox flexibility in reducing environments	[1, 6]
Oxide-carbon	Metal oxides + carbon nanostructures	TiO_2/graphene, Fe_3O_4/CNTs	Improved conductivity and surface area	• Weak interfacial bonding causing delamination • Toxicity concerns from carbon components (e.g., CNTs) • Complex recovery in aqueous systems	[7–11]
Oxide-polymer	Inorganic oxides + organic polymers	PANI/ZnO, PPy/Fe_3O_4	Flexibility and processability	• Thermal instability (>200 °C) • Swelling/degradation in solvents • Microbial biofilm formation	[9–11]
Perovskite-based	ABO_3-type structures	$LaCoO_3$, $BiMnO_3$	Tunable redox properties	• Complex synthesis requiring precise stoichiometry • Leaching of B-site cations (e.g., Mn, Co) • Limited stability under acidic conditions	[3, 9, 10, 12]
Magnetic hybrids	Ferrites + functional coatings	Fe_3O_4@SiO_2, $CoFe_2O_4$/PVA	Magnetic recoverability	• Aggregation during magnetic separation • Oxide core corrosion in acidic media • ROS-induced toxicity in biological systems	[8, 7, 13]

14.1.2 Definitions and Classifications

Precise terminology is essential for navigating this complex field:

- **Composite Nanocatalysts:** Materials comprising two or more distinct constituents (e.g., metal-metal oxide, oxide-oxide) where phases remain physically separable but collectively enhance functionality. Examples include CuO/ZrO_2 systems where X-ray diffraction reveals distinct crystalline phases of both components [1].
- **Hybrid Nanocatalysts:** Materials where organic and inorganic components interact at the molecular level, creating new functionalities that are not present in individual materials. This category includes metal-organic frameworks (MOFs) and polymer-metal oxide conjugates [3, 4].
- **Doped vs. Composite Systems:** Crucially distinct, doping involves minimal incorporation of guest ions into a host lattice (typically <5%), causing subtle electronic modifications, while composites involve substantial integration where each component retains its fundamental properties [1]. For instance, H. Wu et al. [5] reported on the key structural characteristics of yolk–shell architectures, with emphasis on recent advances in their fabrication, including tunable internal void spaces and the integration of mono- or multimetallic composites. The influence of their customizable structures and functionalities on diverse electrochemical processes was also examined (as seen in Figure 14.1).

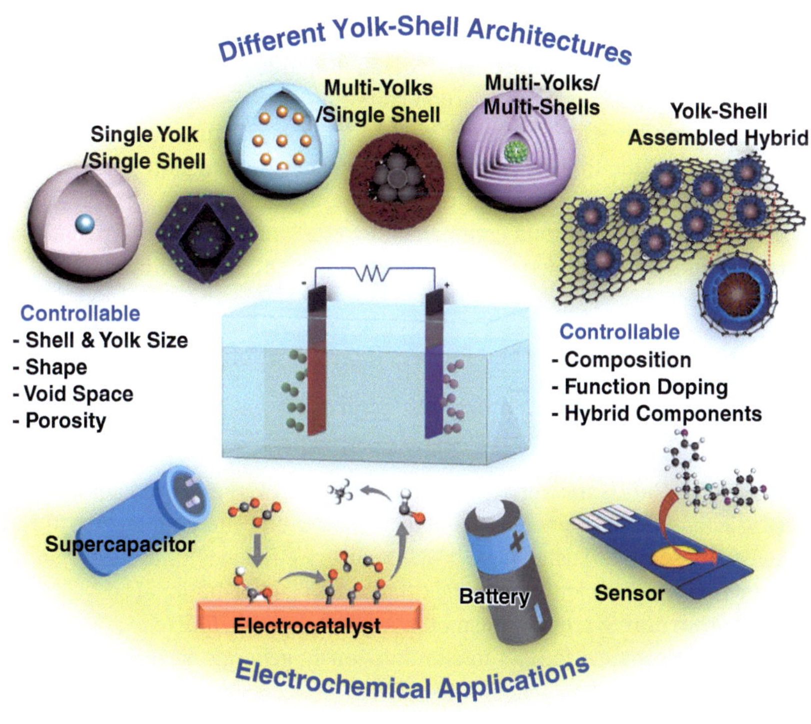

Figure 14.1 Evolution of hybrid nanocatalyst architectures.
Source: [5] / Taylor & Francis.

14.2 Significance of Composites and Hybrids in Catalysis

14.2.1 Overcoming Single-Component Limitations

The strategic importance of composite and hybrid nanocatalysts stems outlined in Table 14.1 from their ability to address intrinsic limitations of unitary metal oxides:

- **Stability Enhancement:** Hybridization significantly improves thermal stability and resistance to deactivation. For instance, silica-coated magnetite (Fe_3O_4@SiO_2) nanoparticles maintain catalytic activity across repeated thermal cycles by preventing magnetic core aggregation, a common failure mechanism in bare nanoparticles. Similarly, zirconia-toughened ceria composites exhibit exceptional resilience against phase transition and sintering at temperatures exceeding 800 °C, enabling operation in demanding environments like automotive exhaust treatment [14, 15].
- **Selectivity Modulation:** The electronic interactions at composite interfaces create tailored active sites for specific reaction pathways. In perovskite-type catalysts (e.g., $LaMnO_3$), B-site substitution with transition metals (Fe, Co) fine-tunes oxygen binding energy, directing selectivity in oxidative coupling reactions. Polymer-metal oxide hybrids demonstrate remarkable molecular recognition capabilities; molecularly imprinted polypyrrole-TiO_2 composites achieve enantioselectivity >95% in pharmaceutical synthesis, performance unattainable with either component alone [2, 16].
- **Reactivity Enhancement:** Hybridization induces synergistic effects that dramatically boost activity. Carbon nanotube (CNT)-supported Cu_2O nanoparticles exhibit 8-fold higher CO oxidation rates compared to unsupported counterparts due to interfacial electron transfer that facilitates oxygen activation. Similarly, graphene quantum dot-TiO_2 hybrids degrade pharmaceutical pollutants 2.3× faster than commercial P25 TiO_2 under visible light, attributed to quantum dot-mediated photon harvesting and charge separation [4, 9].

14.2.2 Multi-Functionality and Process Intensification

Beyond enhancing individual properties, composites enable function integration unachievable with single materials:

- **Magnetic Catalysts:** Ferrite-based hybrids (e.g., $CoFe_2O_4$@C) combine catalytic activity with magnetic separability, eliminating energy-intensive filtration steps. These materials achieve >98% recovery efficiency under external magnetic fields, significantly reducing catalyst loss and operational costs [15].
- **Photothermal Systems:** Plasmonic Au-TiO_2 hybrids utilize localized surface plasmon resonance (LSPR) to simultaneously generate heat and charge carriers, driving tandem reaction sequences under solar irradiation. This dual functionality increases solar-to-chemical efficiency by 40% compared to photocatalysis alone [4].

- **Self-Monitoring Catalysts:** Smart composites incorporating stimuli-responsive polymers (e.g., PNIPAM-ZnO) autonomously regulate activity based on environmental conditions. These systems reduce reaction temperature requirements by 30–50 °C while providing optical feedback about conversion progress [3].

14.3 Advantages of Composite Nanocatalyst

This section provides a comprehensive analysis of composite and hybrid metal oxide nanocatalysts with emphasis on structural diversity, performance advantages, application frontiers, design methodologies, and emerging trends. The scope encompasses both fundamental mechanisms and practical implementations while highlighting knowledge gaps and future research priorities.

14.3.1 Carbon-Metal Oxide Hybrids

Carbonaceous materials provide unparalleled advantages as catalyst supports due to their high surface area, electrical conductivity, and chemical tunability. The synergy at the hybrid interface often leads to substantial upgrades in catalytic performance, as detailed in Table 14.2, which measures the superior activity, selectivity, or stability of key carbon-metal oxide systems as opposed to their single-component counterparts. Recent advances demonstrate that hybridization fundamentally alters reaction mechanisms:

- **Graphene Oxide (GO) Composites:** GO's oxygenated functional groups anchor metal oxide nanoparticles while facilitating interfacial charge transfer. A breakthrough study demonstrated that TiO_2 nanosheets grown on reduced GO exhibit 94.1% tetracycline degradation under visible light as quantified in Table 14.2, 2.3× higher than pristine TiO_2. The enhancement mechanism involves electron injection from photoexcited TiO_2 into GO, suppressing recombination while generating superoxide radicals at hybrid interfaces [4].
- **Carbon Quantum Dot (CQD) Hybrids:** Sub-10 nm CQDs act as "electron reservoirs" and "photon antennas." When coupled with CuO_2, CQDs extend light absorption to 650 nm while prolonging charge separation lifetime to 9.8 ns (versus 2.3 ns in bare CuO_2). This enables 92.5% antibiotic removal efficiency with quantum yields approaching 0.43, performance metrics competitive with noble-metal catalysts [4, 7].
- **CNT Hybrids:** The curved graphene surfaces of CNTs induce strain in deposited metal oxides, creating defect-rich active sites. MnO_x/CNT composites achieve complete CO oxidation at 150 °C, 70 °C lower than MnO_x alone, attributed to accelerated Mars-van Krevelen cycles where CNTs facilitate oxygen vacancy refilling. CNT conductivity also enables electrocatalytic functionality; CNT-ZnO hybrids serve dual roles as propofol sensors and degradation catalysts in electrochemical systems [7, 9].

Table 14.2 Performance comparison of carbon-metal oxide hybrid nanocatalysts.

Hybrid system	Synthesis method	Application	Performance metrics	Enhancement vs. single component	References
TiO_2/graphene	Solvothermal	Tetracycline degradation	94.1% removal in 60 min (visible light)	2.3 × rate increase	[7, 17]
CuO_2/carbon dots	Microwave-assisted	Tetracycline degradation	92.5% removal in 45 min	Quantum yield 0.43 vs. 0.18	[18, 19]
Fe_3O_4/CNT	Coprecipitation	Heavy metal adsorption	Pb^{2+} capacity: 1457 mg/g	3.1 × capacity increase	[20, 21]

14.3.2 Polymer-Metal Oxide Hybrids

Polymer integration transforms brittle metal oxides into flexible, processable, and functionally adaptive systems with applications spanning environmental remediation to biomedical sensing:

- **Conducting Polymer Hybrids:** Polyaniline (PANI), polypyrrole (PPy), and polythiophene derivatives form charge-transfer complexes with metal oxides. PANI/NiO core-shell structures exhibit exceptional pseudocapacitive behavior (623 F/g at 1 A/g) while maintaining 89.4% capacitance after 5000 cycles, stability attributed to polymer-mediated strain accommodation during redox cycling. Similarly, $NiMoO_4$-PANI supercapacitors achieve 285 F/g with 92% retention, demonstrating hybrid advantages for energy storage [9, 14].
- **Molecularly Imprinted Polymers (MIPs):** MIPs on metal oxide surfaces create selective recognition cavities. A breakthrough creatinine sensor using Nb_2O_5-imprinted polythiophene achieves 4.614 μA cm^{-2} nM^{-1} sensitivity with 34 pM detection limit, sufficient for early-stage kidney disease diagnosis from saliva. The hybrid leverages Nb_2O_5's high isoelectric point to attract creatinine while polythiophene transduces binding events into electrical signals [3].
- **Stimuli-Responsive Systems:** Temperature/pH-sensitive polymers (e.g., PNIPAM, PAA) grafted onto oxides enable "smart" catalysts. PNIPAM-ZnO hybrids expand below their lower critical solution temperature (LCST), exposing active sites, while contracting above LCST to protect against poisoning. This self-regulating behavior maintains >80% activity in fouling environments where conventional catalysts fail completely [3].

14.3.3 Mixed Metal Oxide Systems

Combining multiple metal oxides creates bifunctional sites and tailored acid-base properties essential for complex reaction sequences:

- **Perovskite-Type Catalysts (ABO_3):** Lanthanum-based perovskites ($LaMO_3$; M = Co, Mn, Fe) demonstrate exceptional activity in transfer hydrogenation, a critical reaction for pharmaceutical synthesis. Their tunability enables substitution at A and B sites; partial substitution with Sr^{2+} creates oxygen vacancies that accelerate hydride transfer while maintaining structural stability at 600 °C. Bismuth-containing perovskites like $BiMnO_3$ exhibit visible-light photocatalysis due to Bi^{3+} 6s orbital hybridization with O 2p states, narrowing bandgaps to <2.1 eV [2]. The perovskite-type catalysts are promising in solar cells due to minimal thermal quenching [22].

- **Mesoporous Mixed Oxides:** Ordered SiO_2-ZrO_2 composites with 5 nm pores provide both acidic (Zr^{4+}) and basic (SiO^-) sites for cascade reactions. These materials convert glycerol to acrylic acid in one pot via dehydration-oxidation sequences with 74% yield, which is impossible on single oxides. Similarly, CeO_2-ZrO_2 oxygen storage materials undergo 1000 redox cycles without deactivation by balancing Ce^{4+}/Ce^{3+} redox potentials with Zr^{4+} structural stabilization [23].
- **Magnetic Composites:** Core-shell architectures like $Fe_3O_4@SiO_2@TiO_2$ combine magnetic recovery (Fe_3O_4), chemical stability (SiO_2), and photocatalytic activity (TiO_2). These trifunctional materials achieve 98% dye degradation with >95% magnetic recovery, solving the persistent catalyst separation challenge in slurry systems [24].

14.3.4 Emerging Hybrid Architectures

Recent innovations push beyond traditional composites:

- **MOF-Derived Hybrids:** Pyrolyzed ZIF-8(Fe) forms nitrogen-doped carbon with dispersed Fe nanoparticles, exhibiting both adsorption (1457 mg/g Cu^{2+} capacity) and Fenton-like catalysis for pollutant degradation. The preserved organic linkers provide additional coordination sites that enhance metal ion binding [25, 26].
- **Layered Double Hydroxides (LDHs):** ZnTi-LDHs exfoliated onto cellulose scaffolds create high-surface-area catalysts (>300 m^2/g) for solar water splitting. The intimate contact between components enables Z-scheme charge transfer that spatially separates redox reactions, minimizing back reactions [4].
- **Biohybrids:** Enzyme-metal oxide conjugates like CYP2B6/CNT/ZnO enable electrochemical propofol sensing with 0.5 μg/mL detection limits in serum. The enzyme selectively oxidizes propofol to quinones while CNTs enhance electron transfer and ZnO stabilizes enzyme conformation [7].

14.4 Major Application Sectors

In this section, we explored the main applications in terms of energy conversion technologies, environmental remediation, and biomedical and sensing as shown in Figure 14.2.

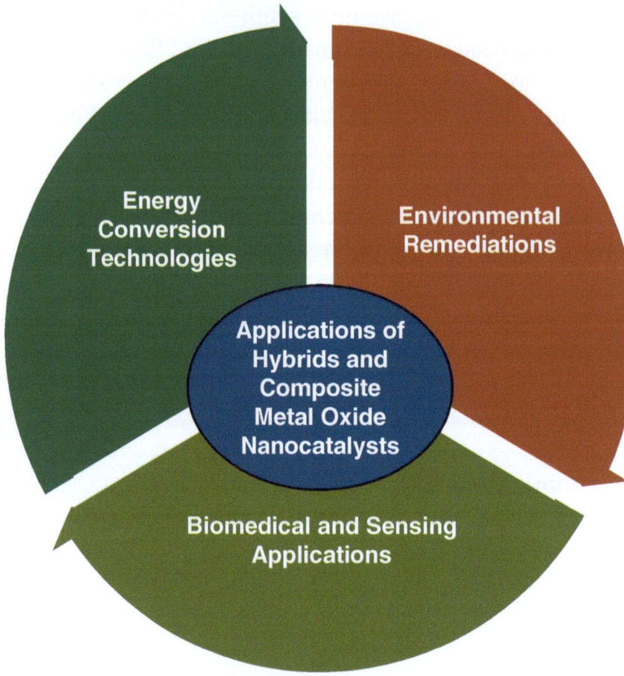

Figure 14.2 The applications of hybrids and composite metal oxide nanocatalysts.

14.4.1 Energy Conversion Technologies

- **Biodiesel Production:** Metal oxide nanocatalysts address key challenges in transesterification. MgO nanocrystals (20–50 nm) achieve 98% FAME yield at 65 °C, 20 °C lower than conventional catalysts, due to abundant edge sites and moisture tolerance. CaO/TiO_2 heterostructures prevent soap formation by neutralizing free fatty acids via Lewis acid sites while catalyzing triglyceride transesterification at basic sites. These advances reduce reaction times from hours to minutes while enabling waste cooking oil utilization [27, 28].
- **Water Splitting:** Perovskite hybrids like $BiMnO_3/g\text{-}C_3N_4$ achieve 12.8% solar-to-hydrogen efficiency by forming a Z-scheme heterojunction that preserves high redox potentials. Under visible light, the composite generates both hydrogen (HER: 48 μmol/h) and oxygen (OER: 24 μmol/h) without sacrificial agents, a rare feat for non-precious catalysts [2, 4, 29].
- **Fuel Cells:** $La_{0.8}Sr_{0.2}MnO_3$/CNT cathodes demonstrate 850 mW/cm^2 power density in solid oxide fuel cells at 700 °C (vs. 620 mW/cm^2 for platinum). The hybrid leverages perovskite stability under oxidizing conditions while CNTs enhance oxygen reduction kinetics through spillover effects [2, 25].

14.4.2 Environmental Remediation

- **Water Purification:** Magnetic nanohybrids enable simultaneous adsorption and degradation. Fe_3O_4@ZIF-8 removes 98% of Pb^{2+} (1457 mg/g capacity) while photocatalytic components decompose organic pollutants. These materials achieve ppb-level contaminant removal within minutes and are magnetically recovered for reuse over 10 cycles with <5% efficiency loss [8, 24]. Etafo et al. [30] reviewed the Ag_3PO_4/NiO heterostructure as a stable and efficient visible-light-driven photocatalyst. Charges accumulate in Ag_3PO_4, while NiO suppresses electron-hole recombination, facilitating redox reactions on the surface. The conduction band (CB) of one material overlaps with the valence band (VB) of the other, forming a straddling gap. This band alignment enhances light absorption and promotes electron transfer to a lower CB, enabling efficient charge separation. As a result, the composite is highly suitable for photocatalysis across a broad range of light wavelengths.

- **Air Pollution Control:** Mesoporous Co_3O_4-CeO_2 composites exhibit complete CO oxidation at −40 °C, unprecedented for non-noble catalysts. The hybrid utilizes Co^{3+} sites for CO adsorption while CeO_2 provides mobile oxygen species via Ce^{4+}/Ce^{3+} cycling. For NO_x abatement, MnO_x-TiO_2 nanotubes achieve 95% conversion at 200 °C by combining TiO_2's photocatalysis with MnO_x's selective catalytic reduction activity [9, 14].
- **Soil Remediation:** Hybrid nanoparticles (e.g., ZIF-8(Fe)) immobilize heavy metals through coordination and ion exchange. Their leaf-like morphology introduces additional imino groups that bind Cu^{2+} at 1457°mg/g capacity while preventing plant uptake. Field trials show 80% reduction in metal transfer to crops, enabling safe agriculture on contaminated lands [25].

14.4.3 Biomedical and Sensing Applications

- **Biosensors:** NiO@PANI core-shell sensors detect glucose at 0.1 μM concentrations by amplifying H_2O_2 oxidation signals. The hybrid's 3D porous structure immobilizes glucose oxidase while PANI's conductivity enables direct electron transfer without mediators, addressing a key limitation of enzymatic biosensors [3, 7]. Also, enhancing the photocatalytic antimicrobial efficacy of materials like TiO_2 and ZnO by combining them with biological entities or polymers, while ensuring biocompatibility with tissue [31].
- **Drug Delivery:** Mesoporous Fe_3O_4-SiO_2 carriers achieve pH-triggered doxorubicin release with >90% loading capacity. Magnetic guidance enables tumor-specific accumulation while SiO_2 pores protect drugs from premature degradation, reducing systemic toxicity by 60% in murine models [7]. According to Ogungbesan et al. [10], the surface modification can help retain the target photocatalytic effect while minimizing adverse side effects, such as ROS activity, in physiological environments, thereby addressing the issue of antibiotic-resistant bacteria.
- **Continuous Monitoring:** CYP2B6/CNT/GO/CuTi oxide electrodes measure propofol at 0.5 μg/mL in serum with 4.58 nA/μg/mL/mm^2 sensitivity. The hybrid avoids electrode fouling by enzymatically converting propofol to electroactive quinones while metal oxides enhance signal transduction [7].

14.5 Design and Analysis Methodologies

14.5.1 Synthesis Strategies

- **Templating Approaches:** Structure-directing agents create ordered architectures. Soft templates like Pluronic F127 produce mesoporous Co_3O_4 with 12 nm pores and 800 m^2/g surface area, ideal for gas diffusion-limited reactions. Hard templates like KCC-1 silica generate dendritic ZrO_2 with radial mesopores that shorten diffusion paths, doubling reaction rates compared to conventional catalysts [1].
- **Green Synthesis:** Plant-mediated methods offer eco-friendly alternatives. Bay leaf extract-synthesized CuTi oxide nanoparticles exhibit narrower size distributions (8 ± 2 nm) than chemical methods while retaining catalytic activity. The natural reducing agents (polyphenols) also functionalize surfaces, enhancing biocompatibility for medical applications [7].
- **Advanced Deposition:** Atomic layer deposition (ALD) creates precise oxide overlayers. Al_2O_3 ALD on Ni nanoparticles (3 cycles) prevents sintering at 600 °C while maintaining accessibility to active sites, enabling stable methane reforming where uncoated catalysts deactivate within hours [2].

14.5.2 Characterization Techniques

- **Surface Analysis:** XPS identifies Ce^{3+}/Ce^{4+} ratios in CeO_2-ZrO_2 to quantify oxygen vacancies, critical for Mars-van Krevelen mechanisms. In situ DRIFTS tracks CO adsorption configurations on Pd-TiO_2, revealing bridge-bonded CO as the active species for oxidation [2, 9].
- **Operando Methods:** Synchrotron XRD maps lattice expansion in Cu/ZnO during CO_2 hydrogenation, correlating strain with methanol yield. Environmental TEM visualizes pore collapse in mesoporous catalysts during reactions, guiding stability improvements [27].
- **Advanced Spectroscopy:** EPR spin trapping quantifies •OH radical generation rates in TiO_2/g-C_3N_4 hybrids, confirming Z-scheme charge transfer. Transient absorption spectroscopy measures 9.8ns charge lifetime in CQD-CuO_2, explaining its superior photocatalysis [4].

14.6 Emerging Trends and Future Perspectives

This section will focus on the emerging and next-generation innovations.

14.6.1 Multi-Functional Catalyst Systems

Next-generation hybrids integrate >2 functions:

- **Photothermal-Catalytic Hybrids:** Au-ZnO-CeO_2 composites utilize plasmonic heating ($\Delta T = 80\,°C$ under light) to accelerate CeO_2-catalyzed soot oxidation while ZnO captures intermediate CO, enabling complete particulate removal in diesel exhaust [4, 14].

- **Self-Healing Catalysts:** Polymers with dynamic bonds (e.g., Diels-Alder adducts) repair damage in hybrid coatings. After scratch tests, these systems recover 95% of original activity via thermally triggered rebonding, extending catalyst lifetimes in abrasive environments [3].
- **Biomimetic Systems:** Hemin-MnO_2 mimics peroxidase enzymes, achieving 10^7 turnover numbers in pollutant degradation. The hybrid leverages hemin's Fe-porphyrin site for H_2O_2 activation while MnO_2 stabilizes radical intermediates, surpassing natural enzymes in stability [14].

14.6.2 Sustainability and Scalability Challenges

Despite the promise, barriers remain:

- **Lifecycle Assessment:** Studies reveal nanoparticle synthesis contributes >60% of hybrid catalysts' environmental footprint. Continuous flow reactors with solvent recycling can reduce impacts by 40%, prioritizing green synthesis and energy-efficient processing [24, 25].
- **Stability Under Real Conditions:** Industrial testing shows ZIF-8 frameworks collapse in acidic waste streams. Developing pH-resistant coatings (e.g., polydopamine) maintains performance while enabling reuse, essential for economic viability [4, 15, 25].
- **Toxicity Concerns:** Long-term ecosystem impacts require evaluation. Surface passivation (e.g., silica shells) prevents metal leaching while maintaining activity, balancing efficacy and safety [8, 14].

14.6.3 Disruptive Innovations on the Horizon

- **Machine Learning-Guided Design:** Neural networks predict optimal $BiMnO_3$ doping for specific reactions, reducing experimental screening by 90%. Recent models accurately forecast light absorption edges (±0.1 eV) and turnover frequencies (±15%) using only composition inputs [2, 25, 29, 32].
- **Modular Nanofactories:** 3D-printed reactors with embedded catalysts customize chemical production. Proof-of-concept units produce pharmaceuticals on-demand using cartridge-swappable hybrid catalysts, potentially revolutionizing distributed manufacturing [9].
- **Quantum Dot Sensitization:** $CsPbBr_3$ quantum dots on TiO_2 nanotubes achieve 16.7% solar water-splitting efficiency, rivaling photovoltaic-electrolyzer systems. The dots' hot carrier injection enables multiple exciton generation, breaking traditional efficiency limits [27].

14.7 Conclusion

Composite and hybrid metal oxide nanocatalysts have evolved from scientific interests to indispensable tools addressing global sustainability challenges. By strategically combining metal oxides with carbonaceous materials, polymers, or secondary oxides, these advanced materials overcome fundamental limitations of single-component systems, achieving unprecedented activity, stability, and selectivity. The synergistic effects at hybrid interfaces, such as electron transfer in TiO_2/graphene or strain-induced activation in MnO_x/CNTs, create catalytic properties unattainable through conventional approaches. Applications spanning biodiesel production, pharmaceutical pollutant degradation, and biomedical sensing demonstrate the versatility of these materials. Perovskite-based hybrids like $BiMnO_3$ exhibit particular promise for multifunctional systems due to their tunable electronic structures. However, realizing the full potential of hybrid nanocatalysts requires addressing scalability challenges, environmental footprint concerns, and long-term stability issues under real-world conditions.

Future advancements will likely emerge from machine learning-guided design, multi-functional architectures, and closed-loop lifecycle management. As characterization techniques reveal ever-more-detailed structure-activity relationships and synthesis methods improve in precision and sustainability, hybrid nanocatalysts are poised to play pivotal roles in enabling circular economies and decarbonizing industrial processes, transforming catalysis from a solution to pollution into a cornerstone of sustainable technology.

References

1 Pal N. Nanoporous metal oxide composite materials: a journey from the past, present to future. *Adv Colloid Interface Sci.* 2020 Jun 1 [cited 2025 Aug 6];280:102156. Available from: https://www.sciencedirect.com/science/article/abs/pii/S0001868619304087

2 Pervaiz S, Javed M, Shah A, Latif A, Nasir S, Shah I. Environmental applications of magnetic nanohybrid materials. *RSC Adv.* 2025 Jun 11 [cited 2025 Aug 6];15(25):19899–936. Available from: https://www.researchgate.net/publication/392592006_Environmental_applications_of_magnetic_nanohybrid_materials

3 Umair M, Huma Zafar S, Cheema M, Usman M. New insights into the environmental application of hybrid nanoparticles in metal contaminated agroecosystem: a review. *J Environ Manage.* 2024 Jan 1 [cited 2025 Aug 6];349. Available from: https://pubmed.ncbi.nlm.nih.gov/37976639/

4 Wang D, Liu T, Yin J, Liu S. Stimuli-responsive fluorescent poly(N-isopropylacrylamide) microgels labeled with phenylboronic acid moieties as multifunctional ratiometric probes for glucose and temperatures. *Macromolecules* 2011 Apr 12 [cited 2025 Aug 7];44(7):2282–90. Available from: https://pubs.acs.org/doi/abs/10.1021/ma200053a

5 Wu H, Li J, Ji Q, Ariga K. Nanoarchitectonics for structural tailoring of yolk-shell architectures for electrochemical applications. *Sci Technol Adv Mater.* 2024 Dec 31;25(1):2420664.

6 Wang Y, Shen Z, Zhang L, Xu B, Wang K, He Z, et al. Synthesis of ZrO_2/CeO_2 composite coating with self-cleaning property for oil-water separation. *Mater*

Chem Phys. 2025 Mar 1 [cited 2025 Aug 8];333:130312. Available from: https://www.sciencedirect.com/science/article/abs/pii/S0254058424014421

7 Ansari AS, Azzahra G, Nugroho FG, Mujtaba MM, Ahmed ATA. Oxides and metal oxide/carbon hybrid materials for efficient photocatalytic organic pollutant removal. *Catalysts* 2025 Feb 1 [cited 2025 Aug 6];15(2):134. Available from: https://www.mdpi.com/2073-4344/15/2/134/htm

8 Fan Z, Gao F, Zhang J, Yi L, Luo N, Yi H, et al. Advances in CO catalytic oxidation on typical metal oxide catalysts: performance, mechanism, and optimization. *J Mater Chem A Mater*. 2025 [cited 2025 Aug 6];13(32). Available from: https://pubs.rsc.org/en/content/articlehtml/2025/ta/d5ta03067f

9 Akhtar M, Shahzadi S, Arshad M, Akhtar T, Saeed Ashraf Janjua MR. Metal oxide-polymer hybrid composites: a comprehensive review on synthesis and multifunctional applications. *RSC Adv*. 2025 Jun 2 [cited 2025 Aug 8];15(23):18173–208. Available from: https://pubs.rsc.org/en/content/articlehtml/2025/ra/d5ra01821h

10 Ogungbesan SO, Etafo NO, Anselm OH, Ejeromedoghene O, Kalulu M, Abdullah M, et al. Transition metal oxide nanohybrid materials: a review of their structures, properties, and applications. *J Mol Struct*. 2025 Aug 15 [cited 2025 Aug 8]; 1337:142209. Available from: https://riull.ull.es/xmlui/handle/915/42295

11 Nguyen TH, Vu MT, Nguyen NS, Nguyen TH, Vu MT, Nguyen NS. Hybrid magnetic-semiconductor oxides nanomaterial: green synthesis and environmental catalytic. In: Awwad NS, Alarfaji SS, Alomary A, editors. *Photocatalysts - New Perspectives* [Internet]. IntechOpen; 2022 [cited 2025 Aug 8]; Available from: https://www.intechopen.com/chapters/84050

12 Sahoo S, Wickramathilaka KY, Njeri E, Silva D, Suib SL. A review on transition metal oxides in catalysis. *Front Chem*. 2024 May 17;12:1374878.

13 Deng H, Lin L, Sun Y, Pang C, Zhuang J, Ouyang P, et al. Activity and stability of perovskite-type oxide $LaCoO_3$ catalyst in lignin catalytic wet oxidation to aromatic aldehydes process. *Energy Fuels*. 2009 Jan 22 [cited 2025 Aug 8];23(1):19–24. Available from: https://pubs.acs.org/doi/abs/10.1021/ef8005349

14 Garnier V, Gonçalves MA, de Almeida la Porta F, Candido da Silva A, Ramalho TC, Francisco de Aquino S. Molecularly imprinted polymer-supported ceramic catalysts for environmental applications: a comprehensive review. *Ceramics* 2025 May 10 [cited 2025 Aug 6];8(2):53. Available from: https://www.mdpi.com/2571-6131/8/2/53/htm

15 Li K, Miwornunyuie N, Chen L, Jingyu H, Amaniampong PS, Koomson DA, et al. Sustainable application of ZIF-8 for heavy-metal removal in aqueous solutions. *Sustainability*. 2021 Jan 19 [cited 2025 Aug 7];13(2):984. Available from: https://www.mdpi.com/2071-1050/13/2/984/htm

16 Ghosh SK, Rahaman H. Noble Metal–manganese oxide hybrid nanocatalysts. In: Mohapatra S, Nguyen TA, Nguyen-Tri P, editors. *Noble Metal-Metal Oxide Hybrid Nanoparticles: Fundamentals and Applications* [Internet]. Woodhead Publishing; 2019 [cited 2025 Jul 6]. p. 313–40. Available from: https://www.sciencedirect.com/science/article/abs/pii/B9780128141342000097

17 Choi N, Tang C, Park Y, Du A, Ayoko GA, Hwang Y, et al. Visible-light-driven photocatalytic degradation of tetracycline using citric acid and lemon juice-derived

carbon quantum dots incorporated TiO_2 nanocomposites. *Sep Purif Technol.* 2024 Dec 18 [cited 2025 Aug 8];350:127836. Available from: https://www.sciencedirect.com/science/article/abs/pii/S1383586624015752

18 Bahramian H, Fattah-alhosseini A, Karbasi M, Nikoomanzari E, Giannakis S. Synergy of Cu^{2+}-$Cu(OH)_2$-CuO with TiO_2 coatings, fabricated via plasma electrolytic oxidation: insights into the multifaceted mechanism governing visible light-driven photodegradation of tetracycline. *Chem Eng J.* 2023 Nov 15 [cited 2025 Aug 8];476:146588. Available from: https://www.sciencedirect.com/science/article/abs/pii/S1385894723053196

19 Malitha MD, Molla MTH, Bashar MA, Chandra D, Ahsan MS. Fabrication of a reusable carbon quantum dots (CQDs) modified nanocomposite with enhanced visible light photocatalytic activity. *Sci Rep.* 2024 Dec 1 [cited 2025 Aug 8];14(1):1–18. Available from: https://www.nature.com/articles/s41598-024-66046-5

20 Alsangur R, Doganay S, Ates I, Turgut A, Cetin L, Rebay M. Magnetic field dependent thermal conductivity investigation of water based Fe_3O_4/CNT and Fe_3O_4/graphene magnetic hybrid nanofluids using a Helmholtz coil system setup. *Diam Relat Mater.* 2024 Jan 1 [cited 2025 Aug 8];141:110716. Available from: https://www.sciencedirect.com/science/article/abs/pii/S0925963523010415?via%3Dihub

21 Borode A, Tshephe T, Olubambi P. A critical review of the thermophysical properties and applications of carbon-based hybrid nanofluids in solar thermal systems. *Front Energy Res.* 2024 Jan 10;12:1509437.

22 Etafo NO. Advancements of lanthanide-doped phosphors in solid-state lighting applications. *Curr Phys.* 2024 May 30; 1(1):e220124225904.

23 Liang XL, Dong X, Lin GD, Zhang HB. Carbon nanotube-supported Pd-ZnO catalyst for hydrogenation of CO_2 to methanol. *Appl Catal B.* 2009 May 20;88(3–4):315–22.

24 Mobarak MB, Sikder MF, Muntaha KS, Islam S, Rabbi SMF, Chowdhury F. Plant extract-mediated green-synthesized CuO nanoparticles for environmental and microbial remediation: a review covering basic understandings to mechanistic study. *Nanoscale Adv.* 2025 Apr 29 [cited 2025 Aug 7];7(9):2418–45. Available from: https://pubs.rsc.org/en/content/articlehtml/2025/na/d5na00035a

25 Veisi H, Pirhayati M, Mohammadi P, Tamoradi T, Hemmati S, Karmakar B. Recent advances in the application of magnetic nanocatalysts in multicomponent reactions. *RSC Adv.* 2023 Jul 7 [cited 2025 Aug 7];13(30):20530–56. Available from: https://pubs.rsc.org/en/content/articlehtml/2023/ra/d3ra01208e

26 Dodekatos G, Tüysüz H. Plasmonic Au/TiO_2 nanostructures for glycerol oxidation. *Catal Sci Technol.* 2016 Sep 26 [cited 2025 Aug 7];6(19):7307–15. Available from: https://pubs.rsc.org/en/content/articlehtml/2016/cy/c6cy01192f

27 Olaleru SA, Palaniyandy N, Mamba BB, Mwakikunga BW. Hybrid perovskite solar cells: a disruptive technology for hydrogen production through photocatalytic water splitting. *ChemistryOpen.* 2025 [cited 2025 Aug 7];14(8):e202500181. Available from: https://www.researchgate.net/publication/392071222_Hybrid_Perovskite_Solar_Cells_A_Disruptive_Technology_for_Hydrogen_Production_through_Photocatalytic_Water_Splitting

28 Modupe Abati S, Bamisaye A, Abidemi Adaramaja A, Rapheal Ige A, Adegoke KA, Olurotimi Ogunbiyi E, et al. Biodiesel production from spent vegetable oil with Al_2O_3 and Fe_2O_3-biobased heterogenous nanocatalysts: comparative and optimization studies. *Fuel* 2024 May 15;364:130847.

29 Xia W, Liu Y, Wang L, Gu D, Hao J. $BiMO_3$ (M = Co, Fe, Mn) perovskite activating peroxymonosulfate-based advanced oxidation process in imidacloprid degradation. *J Mater Sci Mater Electron.* 2024 Oct 1 [cited 2025 Aug 7];35(30):1953. Available from: https://link.springer.com/article/10.1007/s10854-024-13652-8

30 Etafo NO, Bamidele MO, Bamisaye A, Alli YA. Revolutionizing photocatalysis: unveiling efficient alternatives to titanium (IV) oxide and zinc oxide for comprehensive environmental remediation. *J Water Process Eng.* 2024 May;62:105369. Available from: https://linkinghub.elsevier.com/retrieve/pii/S2214714424006019

31 Etafo NO, Bamisaye A, Bamidele MO, Renteria EV, Alli YA, Bakare OC, et al. Beyond the swipe: a review of photocatalytic antimicrobial biocompatible touchscreen technology. *Appl Mater Today.* 2025 Jun;44:102697.

32 Umoh GV, Gómez-Ovalle AE, Cruz MP, Leal-Pérez JE, Obot OU, Berrio HU, et al. Nanomechanical-ferroelastics behavior, and the low-temperature ferroelectric manifestation of $BiMnO_3$ thin films. *Phys Scr.* 2024 Feb 28 [cited 2025 Aug 7];99(3):035962. Available from: https://iopscience.iop.org/article/10.1088/1402-4896/ad2ad2

15

Metal Oxide Nanocatalysts in Sustainable Energy Production for Climate Change Mitigation

Joshua Ogboghena Akhigbe[1], Aisha Eniola Adegbite[1], Amina Yahaya[1], Juliana Okwena Pondei[1], Sesan Abiodun Aransiola[1,3], and Naga Raju Maddela[2]*

[1] *Department of Microbiology, Faculty of Science, University of Abuja, Abuja, Nigeria*
[2] *Departamento de Ciencias Biológicas, Facultad de Ciencias de la Salud, Universidad Técnica de Manabí, Portoviejo, Ecuador*
[3] *Centre for Artificial Intelligence & Multidisciplinary Innovation Studies, College of Accounting Sciences, University of South Africa, Pretoria, South Africa*

*Corresponding author: xavierakhigbe@gmail.com

15.1 Introduction

Globally and in any environment, climate is one of the most important parameters to take into consideration. Climate change refers to the changes that occur in the environmental factors of an area or location over a long period of time. The effects of climate change are felt over various sectors and eventually affect plant development and food distribution [1]. The most significant impact of climate change is an increase in global temperatures, leading to global warming [2]. The mean surface temperature of the Earth is expected to rise by 1.5–2 °C within 40–50 years [3]. As the problem of climate change exacerbates, it becomes imperative to find innovative solutions to it. Nanotechnology shows great promise for mitigating the challenges and problems introduced by climate change. Nanotechnology is concerned with materials, processes, and systems that operate at or less than a scale of 100 nm [4].

Nanotechnology is concerned with the creation of nanosized particles, the fact of which ultimately enhances their physical and chemical characteristics as well as their effectiveness biologically [5]. These nanoparticles, having at least one dimension between 1 and 100, possess unique physicochemical characteristics like small size, large surface area, and high reactivity and large surface area per unit volume, which proves useful in their interactions with other materials by providing an avenue for transporting and storing of clean energy fuels and greenhouse gas

Metal Oxide Nanocatalysts for Sustainable Energy Production, First Edition. Edited by Abayomi Bamisaye et al.
Published 2026 by Wiley-VCH GmbH.

adsorption [3]. Nanocatalysts are derived from nanoparticles of different dimensions and morphology, having a cross section of less than 100 nm [6]. Since the beginning of the 21st century, research on the application of nanotechnology to mitigate the effects of climate change and in pollution reduction has increased exponentially. Nanotechnology facilitates the production of nanoparticles, which promote faster chemical reactions, which in turn, increase efficiency and reduce input consumption, which lessens emissions. Nanotechnology also has applications in providing fossil fuel alternatives and reducing greenhouse gas emissions [7]. This technology provides an essential avenue for the enhancement of energy efficiency in various sectors. It employs nanostructures through nanoscale devices for problem-solving applications in various fields. Nanotechnology has useful applications in various fields, including agriculture, environment, energy, food, medicine, catalysis, material science, etc. [8]. It allows for the manipulation of matter at the nanoscale and energy storage devices augmented by nanotechnology are essential to the bypassing of intermittency problems that come with renewable energy sources, thus allowing for easier widespread adaptation [9].

Owing to reduction in surface area, nanoparticles are able to absorb pollutants more efficiently and catalyze reactions at a much faster rate [10]. These nanocatalysts are generally made up of metallic nanoparticles and semiconducting materials, and they possess potential for activity against organic contaminants [11]. Various metal-based nanomaterials have been studied for their application in contaminant remediation and pollution control. Nanomaterials made from metallic elements and oxides of these elements are excellent adsorbents and have fast kinetics [12].

Most of these nanocatalysts are photocatalysts, and their reaction with other compounds leads to the formation of new and useful nanocomposites, which promote the rate of contaminant degradation [13]. For example, Fe_2O_3 nanoparticles in combination with hydrogen peroxide act as catalysts and carry out the oxidation and breakdown of organic pollutants [14]. Photocatalysis is a process that employs nano-photocatalysts, and it is among the many unique reactions that occur in the presence of solar radiation [15, 16]. It is a promising process for controlling environmental issues and improving overall atmospheric health due to the presence of unspent solar energy on the Earth [17]. Photocatalysis can be applied in the photocatalytic degradation of toxic/harmful organic compounds and gases [18]. Recently, various metal oxide photocatalysts like titanium dioxide (TiO_2), zinc oxide (ZnO), copper oxide (CuO), etc., have been engineered as visible active photocatalysts. They have been improved upon and are very efficient in photocatalytic contaminant degradation in the presence of solar light [19].

Nanotechnology is also a promising technology for the conversion of greenhouse gases. Nanocatalysts promote the complete combustion of fuels by storing oxygen, which reduces greenhouse gas generation [3]. It involves the conversion of carbon dioxide (CO_2) and methane (CH_4) into valuable products through catalysis. Various sustainable nanocatalysts offer excellent catalytic activity, selectivity, and stability, all while reducing their impact on the environment [20]. Nanocatalysts have several advantages that make them crucial in greenhouse gas conversion. They have customizable surface properties, a large surface area, and improved catalyst activity [21].

By enhancing the composition of the nanocatalysts, their morphology and surface chemistry, it becomes possible to achieve effective conversion of carbon dioxide and methane into desired products, ensuring both high selectivity and yield. Nanotechnology, through nanocatalysis can be applied to energy processes optimization, sustainability, and climate change mitigation.

15.2 Environmental Impact of Climate Change

Climate change is a pressing, widespread issue that affects both natural and anthropogenic activities. It happens when the Earth's temperature rises and weather patterns become unpredictable, thus impacting ecosystems and communities around the world [22]. These recorded changes are attributed to anthropogenic activities like industrial processes, deforestation, and burning of fossil fuels, which all release greenhouse gases like CH_4 and CO_2 into the environment [23]. These emissions trap heat and lead to a variety of environmental issues, including hotter temperatures, irregular rainfall, and more frequent extreme weather events like floods and droughts [24, 25].

This problem becomes severe in dry areas like deserts and semiarid regions, where water evaporates faster than it is replaced by rainfall. For example, in Jordan, temperatures on the water surface could go up by 2.5–5.5 °C by the end of the 21st century. This could lead to shorter winters, hotter and drier summers, and a potential drop in rainfall by about 20% [23, 24]. In the north-western Himalayas, including Kashmir, scientists have observed that temperatures are rising, and there is less rainfall, along with extreme temperatures, whether high or low [24]. Glaciers and snowpacks, which melt and feed rivers during certain seasons, are melting quickly, making water shortages worse. Communities in the Himalayas rely on these glaciers for drinking water and farming.

This increasing strain has negative effects on several industries, including agriculture and public health, and could also lead to more competition for resources and potentially cause conflicts [26]. Climate change also profoundly impacts water quality. Rising temperatures reduce dissolved oxygen levels and encourage harmful algal blooms in lakes and reservoirs, leading to "dead zones" where oxygen is depleted [27]. Thermal pollution from increased water temperatures further deteriorates water conditions, making them uninhabitable for fish and aquatic organisms [28]. Coastal areas are particularly vulnerable, as rising sea levels allow saltwater to infiltrate freshwater sources, rivers, and estuaries, jeopardizing drinking water supplies and agricultural activities [25].

Changes in rainfall patterns, like when it rains more heavily or at different times, lead to unpredictable water availability [29]. Some areas get frequent, heavy rains that cause flooding, while others face prolonged droughts that mess up water cycles and supplies. This impacts infrastructure such as wastewater treatment plants. During floods, the excess water can overwhelm plants, reducing their ability to remove nitrogen effectively [28, 30]. During droughts, low water level means less dilution of waste, which can increase pollutants like BOD and COD, making treatment less effective, especially in winter [28]. Temperature also plays a role, whereby warmer weather in summer can help biological processes work better, improving treatment efficiency [28, 31].

Climate change affects not only water resources but also the ecosystems and human activities that rely on them. Forests are especially vulnerable as they face the combined pressures of air pollution and rising temperatures [32]. For example, on the Biga Peninsula, high sulfur pollution has been found to harm the chlorophyll-producing cells in conifers, causing them to shed leaves prematurely, grow less wood, and produce narrower growth rings [33]. Additionally, increased levels of atmospheric CO_2 contribute to drier conditions and cause climate zones to shift upward by 50 to 100 meters, which impacts the types of forests that can thrive there. This is already evident in changes such as insects like the pine processionary caterpillar moving to different areas as their typical habitats change [34].

Agriculture is a sector that is heavily impacted by the effects of climate change, facing many challenges that threaten livelihoods. When rainfall patterns change, rain-fed farmers are directly impacted, and water shortages can severely disrupt crop production and jeopardize food security [24, 25]. While increased CO_2 levels might boost crop growth in some cases, this potential benefit is often cancelled out by rising temperatures, water shortages, and soil degradation [23]. Marginalized communities, especially those who rely heavily on farming and natural resources, are at even greater risk, bearing the brunt of these climate impacts [22].

Tackling these environmental issues involves significant climate change adaptation efforts that are inherently political and often face resistance [22]. This opposition usually stems not from climate change denial but from concerns related to livelihoods, cultural traditions, perceived injustices, or exclusion from decision-making processes. For example, communities might oppose agricultural modernization, conflicting with traditional practices, or resist top-down adaptation projects, lacking transparency. Effective adaptation requires integrated strategies that address environmental, social, and political factors, empower marginalized groups, and foster inclusive decision-making [35, 24].

Green nanotechnology is an emerging, interdisciplinary field that provides innovative solutions to environmental problems and advances global sustainability [36, 37]. It employs principles of sustainable engineering and green chemistry in designing, producing, and using nanomaterials, focusing on eco-friendly design, nontoxic starting materials, energy-efficient manufacturing, and biodegradability at end-of-life [37, 38]. Essentially, climate change can be compared to a mighty, unpredictable river that overflows in some areas and diminishes in others, dramatically altering the landscape [29]. Traditional methods, like sandbags or small dams, offer only limited protection. In contrast, nanotechnology provides advanced tools, akin to microscopic engineers, that can reinforce riverbanks precisely, purify water at a molecular level, and improve energy harvesting from the river, provided we consider the sustainability of their creation and their long-term effects [39, 40]. Figure 15.1 gives a representation of some of the most significant impacts of climate change on the environment.

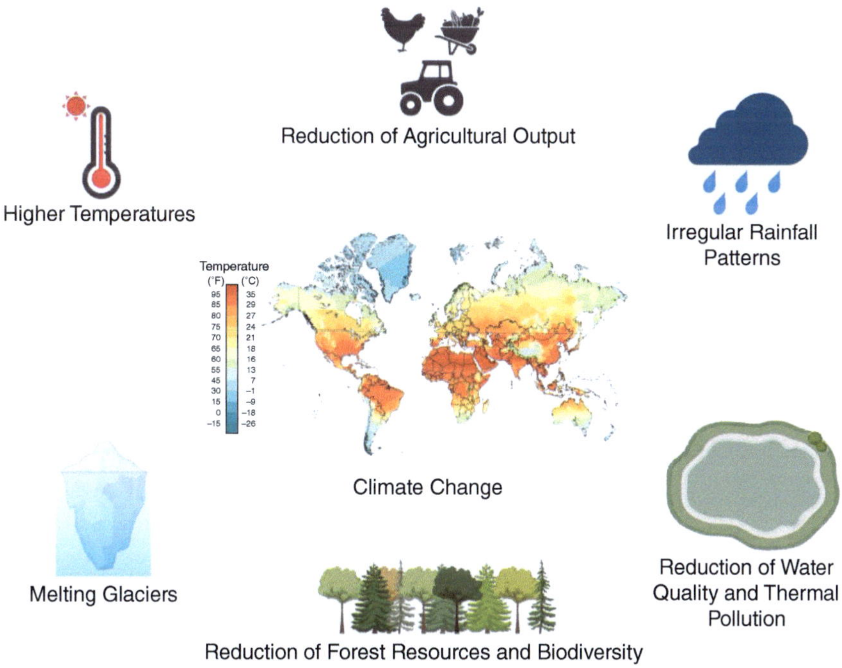

Figure 15.1 Environmental impact of climate change.

15.3 Metal Oxide Nanocatalysts (MONCs): An Overview

15.3.1 Introduction to Catalysis and Nanotechnology

Catalysis is a process that employs a substance (called a catalyst) to speed up a chemical reaction without being used up completely. Nano-catalysts are tiny catalysts that are less than 100 nm in size [41]. These particles can be made of carbon, metal, metal oxides, or organic substances. Given the fact that they are extremely small, these particles have a high surface area in relation to their volume. This means they provide more active sites where reactions can take place [41]. Catalysis is one of the first practical applications of nanoparticle technology. Various elements and compounds like titanium dioxide, silica, iron, and aluminum have been employed in the nanoscale for catalysis in the last two decades [42]. Nanoparticles are thought to be one of the more important contemporary catalyst materials, and they have various applications in industry, ranging from chemical manufacturing to energy transformation and storage applications [43].

15.3.2 Unique Properties of Metal Oxide Nanoparticles

Metal oxide nanoparticles, such as TiO_2, ZnO, and cerium oxide (CeO_2), have unique characteristics that make them highly effective when they act as catalysts [44]. One important feature is their large surface area compared to their size. Because they are so small, more of their atoms are exposed on the surface. This allows chemical reactions to happen more easily and more quickly [41]. Another key feature is their adjustable electronic structure. The energy levels inside these particles, known as the band gap, can change depending on the size and shape of the particles. This makes them suitable for reactions that depend on light or electron movement, such as photocatalysis [45]. These nanoparticles also have tiny defects, such as missing oxygen atoms. These defects can create highly reactive spots that help speed up chemical reactions. For example, CeO_2 and SnO_2 often have these useful defects [46]. The shape of the particles also plays an important role. Whether the nanoparticles are shaped like rods, spheres, or have a porous surface can affect how well they perform in different reactions [45].

Because of these properties, metal oxide nanoparticles are ideal for clean energy applications. They can absorb light more effectively, interact better with other substances, and help reactions happen faster and more efficiently. This makes them useful in processes like water splitting, CO_2 conversion, and fuel cell operation [41].

15.3.3 Common Types of MONCs

There are many types of metal oxide nanoparticles, but some are more commonly used than others in clean energy processes. These materials are chosen based on their stability, their reaction to light, their ability to be produced easily, and their performance in different types of reactions [47].

Based on the research done by Geldasa et al. [48], below are some of the most widely studied and applied MONCs.

15.3.3.1 TiO_2

Titanium dioxide is one of the most popular and widely used MONCs. In its nano form, it has the ability to absorb ultraviolet (UV) light and use it to speed up chemical reactions. This makes it really useful in photocatalysis, where light is used to help produce fuels from water or sunlight. TiO_2 is favored because it is stable, non-toxic, and affordable. It is often used in water purification, hydrogen production, and solar-driven processes [48].

15.3.3.2 ZnO

Zinc oxide shares many similar features with titanium dioxide, but compared to the latter, it behaves a bit differently under light. It also works well as a photocatalyst and is often used in solar cells, batteries, and hydrogen production systems. One of the reasons ZnO is widely used is that it is easy to shape into different forms like rods, tubes, or sheets, which can help improve its performance in different applications [48].

15.3.3.3 Iron Oxides (Fe_2O_3 and Fe_3O_4)

Iron oxide nanoparticles are cherished for their magnetic properties. They are commonly used in reactions that produce hydrogen or break down harmful substances in the environment. The magnetic nature of these particles allows for easy separation from liquids after a reaction is complete, which helps with recycling and reduces waste. Additionally, they are relatively cheap and environmentally friendly [48].

15.3.3.4 Copper Oxides (CuO and Cu_2O)

Copper oxides are particularly good at working under visible light, unlike TiO_2 and ZnO, which mostly respond to UV light. This makes them very useful for solar energy applications, including breaking down water in order to produce hydrogen or converting CO_2 into useful fuels. Copper-based catalysts are also more affordable than some of the more expensive materials [48, 47].

15.3.3.5 Calcium Oxide (CaO), Magnesium Oxide (MgO), and Zirconium Oxide (ZrO_2)

These metal oxides are used in the production of biodiesel from plant or animal oils. When made into nanoparticles, they help hasten the chemical reactions involved in converting fats into fuel. They are known for their stability and reusability, which makes them desirable for industrial-scale applications [48].

15.3.3.6 Mixed Metal Oxides and High-Entropy Oxides (HEOs)

Mixed metal oxides combine two or more metal elements to improve their catalytic performance. HEOs are a newer class of materials made from three or more different metals. These materials have many different active sites on their surface, which makes them very efficient in complex reactions. They are also known for being stable under harsh conditions and are often used in energy storage systems such as fuel cells and supercapacitors [48].

15.3.4 Synthesis and Characterization Techniques

15.3.4.1 Synthesis Methods

According to Negrescu et al. [49], these are some common ways to make metal oxide nanoparticles and their advantages:

- **Sol-Gel Method:** Metal salts are turned into a gel, then heated to form nanoparticles.
  Used for: TiO_2, ZnO, and CeO_2
  Advantage: It gives good control over the size and shape.
- **Hydrothermal Method:** This involves high-pressure heating in water to grow well-shaped crystals.
  Used for: ZnO, Fe_2O_3, and CuO
  Advantage: This produces uniform rods or spheres.
- **Microwave-Assisted Synthesis:** This method involves the use of microwaves for fast and even heating.
  Used for: ZnO and TiO_2
  Advantage: This is fast and energy-efficient.
- **Ultrasound-Assisted Synthesis:** This uses sound waves to break up particles and stop clumping.
  Used for: ZnO, CeO_2, and CuO
  Advantage: Better particle dispersion and shape control.
- **Pyrolysis (Thermal Decomposition):** This is the heating of precursors at high temperature to form oxides.

Used for: MgO, ZrO_2, and Fe_2O_3
Advantage: Simple and scalable.

- **Mechanochemical (Ball Milling):** This method involves the physical grinding of metal powders together to create mixed nanoparticles.
  Used for: Mixed metal oxides.
  Advantage: It is solvent-free and eco-friendly.

15.3.4.2 Characterization Techniques

According to Negrescu et al. [49], these are a few ways to test the properties of metal oxide nanoparticles:

- **XRD (X-Ray Diffraction):** This shows the crystal structure and confirms material composition.
- **SEM/TEM (Scanning/Transmission Electron Microscopy):** This provides images of particle shape, size, and surface.
- **BET (Brunauer–Emmett–Teller) Analysis:** This measures how much surface area the particles have.
- **XPS (X-Ray Photoelectron Spectroscopy):** This identifies surface elements and their oxidation states.
- **UV–Vis/PL (Ultraviolet–Visible and Photoluminescence):** This tests the way in which particles absorb light and separate electric charges.

15.3.5 Suitability of Metal Oxides for Energy Applications

Table 15.1 gives an outline of various ways metal oxide nanoparticles can be applied in sustainable energy creation.

Table 15.1 Applications of MONCs in sustainable energy.

Application	Common oxides used	Function	References
Hydrogen production	TiO_2, ZnO, Fe_2O_3, and Cu_2O	• Absorb light to split water into H_2 and O_2. • Surface defects boost reaction efficiency.	Awogbemi and Von Kallon [50]
CO_2 conversion	CuO, mixed oxides (e.g., CeO_2 and TiO_2)	• Convert CO_2 into fuels under light (photocatalysis).	Li et al. [51]
Fuel cells & batteries	NiO, Co_3O_4, Fe_2O_3, and ZnO	• Catalyze oxygen reactions. • ZnO nanorods improve storage performance.	Awogbemi and Von Kallon [50]
Biodiesel production	CaO, MgO, TiO_2, ZnO, and ZrO_2	• Speed up transesterification of oils to biodiesel. • Cost-effective and reusable.	Awogbemi and Von Kallon [50]
Environmental cleanup	ZnO, CuO, and Fe_2O_3	• Break down pollutants in water/air under light (photodegradation).	Li et al. [51]

15.4 MONCs Application in Sustainable Energy Production

15.4.1 Energy and Sustainable Energy Production

There is a crisis of energy in today's world unlike, what has ever been recorded. Increased global population, industrialization, urbanization, and commercialization have led to a global increase in demand for energy which traditional fuel sources cannot meet [52]. An over 33% increase in global energy consumption to 580 million terajoules (TJ) since 2002 has been recorded and that number is estimated to keep increasing hit 740 million TJ and eventually 900 million TJ by 2040 and 2050, respectively [53].

Fossil-based sources including coal, oil, and gas are used globally as the major sources of energy. They have domestic, commercial, and industrial applications. A setback, unfortunately, is the continuous release of carbon (IV) oxide into the atmosphere, which in many ways contributes to global warming, climate change, and environmental pollution, which has various adverse effects on human health [54]. The non-sustainable nature of the utilization of fossil fuels for energy has led to scientists, researchers, and other stakeholders seeking alternative sustainable, environmentally friendly, cost-efficient, and clean fuels to serve as sources of energy as fossil fuels provide about 80% of global energy [55]. A survey carried out in 2019 observed that 79% of the global economy depends on traditional energy sources from petroleum [56]. In the last decades, various alternatives have been proposed and developed, but in spite of these advancements, fossil fuels still remain the chief provider of global energy, although some progress in the integration of renewable energy sources has been recorded. It is expected that renewable energies will provide half of the global energy by 2040 and 70% of greenhouse gases will be reduced [57].

For the purpose of achieving and sustaining a clean ecosystem, strategies for increasing the amount of renewable energy in the global energy mix have to be implemented. One of such strategies is the investment in and development of technology to produce energy through biological means. MONCs have various applications in several of these processes.

15.4.2 MONCs Application in Energy Production

Generally, nanomaterials are applied to catalyzing the lignocellulosic biomass hydrolysis and in order to enhance their efficiency in biofuel production, they are immobilized [58]. Nanocatalysts are formed by the immobilization of enzymes onto metal oxide or magnetic nanoparticle complexes to form very efficient nanomaterial-enzyme systems. Various oxides of metallic elements have applications in energy production processes. This section will discuss their applications in the production of biodiesel.

15.4.2.1 Nanocatalysts for Biodiesel Production

Biodiesel stands out among the various renewable fuel sources owing to its various advantages. For one, it is possible to synthesize it from inexpensive sources like

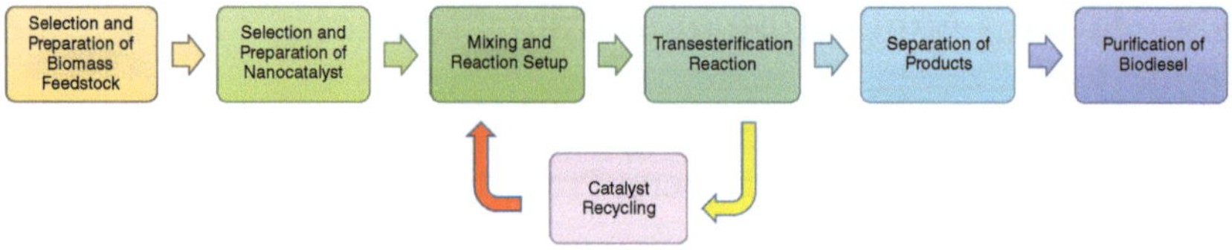

Figure 15.2 Steps in the biodiesel production process.

animal fats and algal lipids. Waste products from the food industry and poultry processing plants can also serve as a source for biodiesel. The product emits low CO levels, is biodegradable, has a high flash point, exceptional lubricant properties, does not have sulfur-containing compounds, and is generally less toxic than conventional diesel and other crude distillates [57, 59, 60] gave several methods for the conversion of biomass to biodiesel. They discussed techniques like gasification, liquefaction, pyrolysis, and the application of nanocatalysts as seen in Figure 15.2.

The application of nanotechnology to the production of biodiesel has become more common in recent years. Awogbemi et al. [54] stated that various processes like esterification and transesterification, purification, filtration, etc., are involved in the conversion of oil to pure biodiesel with MONCs via transesterification, the feedstock is converted to pure biodiesel at low temperature and ambient pressure. The reaction time is short and the alcohol: oil ratio is low. The process gives room for the recovery and reuse of heterogeneous catalysts, allows for easy product separation, and has a high conversion rate. The transesterification process has been proven, is scalable and sustainable, and is a relatively easily achieved technique for synthesizing biodiesel.

Owing to its ease of purification, low carbon footprint, easy recoverability and reusability metal oxide heterogeneous catalysts manufactured at the nanoscale have been deployed and have escalated the transesterification process. The fabrication of these catalysts to the nanoscale improved their performance by enhancing surface area, porosity, and increasing their capacity to tolerate acids and their thermal stability, while overcoming the challenges of low biomass conversion efficiency, fast deactivation, slow reaction in long-term usage, and mass transfer resistance faced by the macro catalysts [54]. Oxides of various metallic compounds have been successfully applied as catalysts to enhance the biodiesel production process. Metals like magnesium, calcium, titanium, zinc, and zirconium have been used.

15.4.2.1.1 Magnesium

Studies have shown that the transesterification of various oils with nanoscale MgO catalyst could yield high amounts of biodiesel. The transesterification of Ziziphus seed oil using magnesium oxide nanoparticles yielded 98% biodiesel [61]. Castor seed oil yielded 96.5% biodiesel with carbon-based magnesium oxide solid base catalysts [62]. In another study, MgO/Fe_2O_3-SiO_2 core-shell magnetic nanocatalyst was utilized to synthesize biodiesel from *Camelina Sativa* seed oil and methyl alcohol through transesterification. An impressive biodiesel yield of 99% was observed [63].

Table 15.2 Magnesium oxide nanocatalysts for biodiesel production.

Feedstock	Nanocatalyst	% Yield	References
Ziziphus seed oil	MgO	98	Saman et al. [61]
Camelina Sativa seed oil	MgO/Fe_2O_3-SiO_2	99	Rahimi et al. [63]
Castor seed oil	Carbon-based MgO	96.5	Du et al. [62]
Sunflower oil	$MgO/MgFe_2O_4$	92.9	Alaei et al. [64]
Waste cooking oil	GO@MgO	99.23	Aghel et al. [65]
Waste cooking oil	$MgO/MgSO_4$	98.8	Bora et al. [66]

Similar studies have also shown constructed MgO nanoparticles' ability to convert triglycerides into biodiesel. $MgO/MgFe_2O_4$ heterogeneous nanocatalyst was applied for the synthesis of biodiesel from sunflower oil. A product yield of 92.9% was recorded. The catalyst was very stable and displayed high reusability potential. The structure was used five times, and it remained stable [64]. Another study utilized graphene oxide-doped magnesium oxide (GO@MgO) to enhance biodiesel production from waste cooking oil. The maximum biodiesel purity recorded for MgO and GO@MgO was 93.84% and 99.23% respectively, at optimized conditions [65]. Similarly, a novel $MgO/MgSO_4$ nanocatalyst was applied in the transesterification and esterification of high free fatty acid-containing waste cooking oil into biodiesel in the presence of ultrasound. The synthesized nanocatalyst demonstrated high stability and reusability [66].

In these studies, triglycerides like waste cooking oil yielded high biodiesel levels that met international standards, and the nanoscale MgO catalyst was applied for about five cycles, and the efficiency of the reactions remained high. The catalyst was easily separated from the product, and regenerative processes were cost-efficient. The crude biodiesel synthesized from the regenerated catalysts also met international standards, proving the efficiency of the catalyst in the process [64–66]. All studies mentioned are summarized in Table 15.2.

15.4.2.1.2 Titanium

Titanium dioxide heterocatalysts have been applied to the optimization of biodiesel production in various ways in recent years. In a study of the catalytic role of titanium dioxide nanoparticles in the synthesis of biodiesel from *Carthamus tinctorius* L., under optimum conditions, a maximum yield of 95% biodiesel was recorded [67]. Another study examined the catalytic activity of TiO_2-Cu_2O nanoparticle composites for the synthesis of biodiesel from *Citrullus colocynthis* L. seed oil. TiO_2-Cu_2O nanoparticles had effective applications by catalyzing transesterification reactions between the oil and ethanol, and an above 60% conversion rate was recorded [68]. Similarly, for the production of biodiesel from palm oil, TiO_2 and Cu-impregnated TiO_2 were utilized, and under optimal conditions, a maximum product yield of 90.3% was reported [69].

Table 15.3 Titanium oxide nanocatalysts for biodiesel production.

Feedstock	Nanocatalyst	% Yield	References
Carthamus tinctorius L. oil	TiO_2	95	Jan et al. [67]
Citrullus colocynthis L. seed oil	TiO_2-Cu_2O	65	Patil et al. [68]
Palm oil	Cu-impregnated TiO_2	90.93	De and Boxi [69]
Used cooking oil	$TiO_2/PrSO_3H$	98.3	Gardy et al. [70]

A novel, $TiO_2/PrSO_3H$ solid acid nanocatalyst was used for simultaneous esterification and transesterification reactions from used cooking oil for biodiesel production and a yield of 98.3% was observed under optimum process conditions [70]. All studies mentioned are summarized in Table 15.3.

15.4.2.1.3 Zinc

Various studies have been conducted to investigate the efficiency of ZnO nanocatalysts in biodiesel production from various oils. ZnO was applied in the nanoscale to convert canola oil, castor oil, *Mesua ferrea* oil, vegetable oil, and sunflower oil, respectively, by [59, 71–73] and [74] into biodiesel. Each study recorded very high levels of product yield.

ZnO nanocomposites were also applied to biodiesel synthesis from waste cooking oil. Fe_3O_4/SiO_2@ZnO was synthesized and applied in the transesterification process for biodiesel synthesis from waste cooking oil. A product yield of 97.23% recorded [75]. Similarly, another study applied ZnO and ZnO@Al nanocatalysts for biodiesel production from waste frying oil [76]. It was observed that the Al-doped ZnO nanocatalyst had remarkable reusability in comparison with ZnO.

Similarly, a study of biodiesel production from waste cooking oil using ZnO nanocatalysts and H_2SO_4 showed that with only sulfuric acid as the catalyst, biodiesel yield was 71.3% and upon addition of the zinc oxide nanoparticles, the product yield increased up to 81.94% [77].

Activated carbon with zinc oxide/nickel oxide (AC@ZnO/NiO) was used to synthesize biodiesel from dairy waste scum oil. The study recorded a maximum yield of 97.81% [78]. An earlier study also produced biodiesel from dairy waste using ZnO. They reported a slightly lower yield of 94.8% under optimum conditions [79].

Biodiesel was produced at an impressive 99.18% yield from algal oil using novel calcium-doped zinc oxide nanocatalysts synthesized using a UV shaker. The catalyst was studied under various parameters and characterized for biodiesel production. Under optimum conditions, the maximum yield was recorded as 99.18% and the catalyst was able to be reused up to three times [80]. All studies mentioned are summarized in Table 15.4.

These studies all came to the conclusion that owing to various unique characteristics of ZnO nanoparticles, such as its ability to be reused for multiple cycles, high catalytic activity, relatively easy customization, biocompatibility, and ease of separation, make ZnO an effective compound for catalyzing the process.

Table 15.4 Zinc oxide nanocatalysts for biodiesel production.

Feedstock	Nanocatalyst	% Yield	References
Castor oil	ZnO	95.20	Baskar et al. [71]
Mesua ferrea oil	Co/ZnO	98.03	Borah et al. [72]
Canola oil	$ZnO/BiFeO_3$	95.43	Salimi and Hosseini [59]
Vegetable oil	ZnO	94.7	Rao et al. [73]
Sunflower oil	ZnO	71	Salim et al. [74]
Waste cooking oil	Fe_3O_4/SiO_2@ZnO	97.23	Maleki and Esmaeili [75]
Waste cooking oil	ZnO, ZnO@Al	91.79, 96.13	Maleki and Esmaeili [76]
Waste cooking oil	ZnO	81.94	Gada et al. [77]
Dairy waste scum oil	AC@ZnO/NiO	97.81	Maleki et al. [78]
Dairy waste scum bio-oil	ZnO	94.8	Sharma et al. [79]
Algae oil	Ca-doped-ZnO	99.18	Abdala et al. [80]

15.4.2.1.4 *Zirconium*

Various studies have evaluated the efficiency of ZrO_2 nanoparticles for biodiesel production. The transesterification of soybean to biodiesel with catalysis by heterogeneous ZrO_2 nanoparticles was investigated using zinc oxide-co-sodium zirconate ($ZnO{\cdot}Na_2ZrO_3$) nanocomposite. A product yield of 97% was recorded [81].

Zirconium dioxide in conjunction with three alkali metals (Li, Na, and K) for M_2ZrO_2 (M = Li, Na, and K) was applied as catalysts for biodiesel production and all series of the compound catalyst could produce biodiesel at a 98%–99% conversion rate in two hours [82].

In another study, calcia-stabilized zirconium oxide (CaO/ZrO_2) catalyst was prepared for the transesterification of oil to biodiesel. The maximum conversion of 96.99% was achieved, and even after being used for three cycles, conversion of 92.76% is still achieved [83].

These studies recorded very high biodiesel yields, and the catalyst used up was recovered and reapplied for multiple cycles, and there was no need for catalyst regeneration activities, and upon analysis, the synthesized biodiesel was found to meet industry and international standards and possess impressive fuel qualities.

Other studies investigating the catalysis of ZrO_2 for biodiesel production from different feedstocks observed high product yields and catalyst reusability of up to five cycles. Tannery waste was applied to produce biodiesel and a high biodiesel yield of over 90% was reported, and this was accredited to the higher catalytic activity of ZrO_2 nanoparticles heterogeneous catalyst [84]. All studies mentioned are summarized in Table 15.5.

15.4.2.1.5 *Sodium*

Sodium aluminate, supported on calcined alumina ($NaAlO_2/c\text{-}Al_2O_3$) at the nanoscale heterogeneous catalyst by response surface methodology (RSM) was

Table 15.5 Zirconium oxide nanocatalysts for biodiesel production.

Feedstock	Nanocatalyst	% Yield	References
Oil	M_2ZrO_2 (M = Li, Na, and K)	98–99	Dai et al. [82]
Soybean oil	$ZnO{\cdot}Na_2ZrO_3$	99	Rodríguez-Ramírez et al. [81]
Oil	CaO/ZrO_2	96.99	Li et al. [83]
Tannery waste sheep fat	$Fe\text{-}Mn\text{-}SO_4/ZrO_2$	>90	Booramurthy et al. [84]

Table 15.6 CaO nanocatalysts for biodiesel production.

Feedstock	Nanocatalyst	% Yield	References
Palm oil	$CaO\text{-}SiO_2$	93.01	Khazaai et al. [86]
Waste cooking oil	CaO	94.74	Cholapandian et al. [87]
Waste cooking oil	SiO_2 modified CaO	97.8	Ozor et al. [88]
Goat fat	CaO	93.72	Khalifeh and Esmaeili [89]
Chicken fat	$CaO/CuFe_2O_4$	94.54	Seffati et al. [90]

applied in the optimization of the biodiesel production and a product yield of 97.65% was achieved [85]. This study demonstrated the useful application of sodium oxide nanocatalyst in the biodiesel production process.

15.4.2.1.6 Calcium

CaO nanostructures have been extensively studied for the transformation via transesterification of waste edible oils to biodiesel. Using fabricated CaO in nanoscale structures for used oil transesterification improved the overall reaction efficiency by improving biodiesel yield and reducing reaction time and temperature. One study sourced waste eggshells and sea sand, which were applied in the catalyst ($CaO\text{-}SiO_2$) production for biodiesel synthesis from palm oil. It was observed that under optimal conditions, a maximum yield of 93.01% biodiesel was produced [86].

A different study synthesized CaO nanocatalyst from *Acalypha indica* for biodiesel production from waste cooking oil. Under optimal conditions, the highest biodiesel yield is 94.74% [87]. Similarly, biodiesel was produced from waste cooking oil using SiO_2 nanoparticles connected to the surface of modified CaO. The authors recorded a 97.8% product yield from waste cooking oil [88].

Replacing the oils with recovered goat fat yielded similar results. Biodiesel was produced from goat fat and under optimal conditions, the biodiesel yield recorded was 93.72% [89]. Similarly, biodiesel was produced efficiently from chicken fat (94.5% yield) using $CaO/CuFe_2O_4$ nanocatalyst with a catalyst content of just 3% [90]. All studies mentioned are summarized in Table 15.6.

15.5 MONCs Advantages Over Traditional Catalysts, Challenges, and Limitations

Oxides of metallic elements are a versatile group of compounds which can be utilized in many different fields of industry, including catalysis [91]. The partially filled d-shells of the metals and the availability of surface oxygen ligands contribute to the efficiency of these compounds as catalysts. In their application as catalysts or catalyst supports, these materials aid in the promotion of a variety of commercially relevant chemical reactions like isomerization, hydrogenation, dehydration, etc. [92]. Putting this class of compounds on the nanoscale does wonders for their catalytic efficiency. MONCs have a greater number of exposed active sites due to an increased surface-to-volume ratio, ultimately leading to an improvement in the rate of reaction. In addition to that, inherent properties of metal oxides can be amplified, and new ones can be discovered because of the changes that occur in the surface structure of the compounds as they are downsized to the nanoscale. This opens up possibilities for the engineering of the surface structure to produce specific properties, leading to improvement of overall efficiency [93].

15.5.1 Comparative Advantages of MONCs

An important feature of nanocatalysts is the presence of quantum effects. At the nanoscale, materials behave differently, which can lead to faster or more selective reactions [94]. The properties of magnetic metal oxide nanoparticles, in particular, can be adjusted depending on their shape and size. The various chemical, magnetic, and electronic properties of the nanoparticles are generally contingent on the size of the nanomaterial [91], and the manipulation of the size could lead to customization of the properties of the nanoparticles. Nanotechnology allows one to control the size and shape of catalysts to improve how they work, and this makes metal oxide nanoparticles particularly useful in clean energy, as they help reactions happen more efficiently while using less material [95]. The unique properties of metal oxide nanoparticles make them among the nanomaterials that are most widely applied in industry. These unique chemical and physical properties are ascribed to the high density of these materials and the limited size of corners and edges on their surface. Metal oxide nanoparticles possess;

- Superparamagnetic behavior.
- Nonlinear optical properties.
- Cold welding properties.
- More ductility at higher temperatures compared to coarse-grained ceramics.
- Unique catalytic, sensitivity, and selective activity.
- Lower melting point than their bulk counterpart of same composition [96].
- Speedy and precise chemical conversion and higher product yield.
- Simple catalyst separation and recovery [97].

A study [43] compared and differentiated nanocatalysts from homogeneous and heterogeneous catalysts. It is summarized in Table 15.7. Owing to their nano size,

Table 15.7 Comparative effectiveness of heterogeneous, homogeneous, and nanocatalysts.

Homogeneous	Heterogeneous	Nanocatalysts
Advantages		More active
More active	More stable	More stable
More selectivity	Easy recoverability	Easy recoverability
Moderate reaction time	Easily accessible	High efficiency
Disadvantages		Cost-effectiveness
Lower stability	Lower activity	Environmentally sustainable
Difficult to recover	Large reaction time	More selective

nanocatalysts' contact with the reactants is significantly increased (similar to homogeneous catalysis) and nanocatalysts' insolubility in the reactive solvents makes for easier isolation from the reaction mixture (similar to heterogeneous catalysis).

15.5.2 Challenges and Limitations of MONCs

Despite their various advantages, there has been some stagnation in the practical and industrialized application of metal oxide nanomaterials as catalysts. Perhaps most importantly, the lack of suitable legislation and regulation around the use of these materials presents a limitation for their real and practical application [98]. In that same vein, nanomaterials proliferate at exponential rates, and without a clear understanding of their environmental impacts, there is need for comprehensive research and regulatory measures around their widespread use and application [99].

Achieving uniform and reproducible control over particle size, shape, crystal phase, and chemical composition presents a challenge for MONC's application. These parameters critically determine catalytic activity and selectivity, and any heterogeneity can result in inconsistent or suboptimal catalytic performance [100]. Also, owing to their size, nanocatalysts may present more difficulty in recovery and separation, particularly from liquid-phase reaction mixtures [101]. Difficulties could also be experienced in catalyst regeneration if oxidizing thermal treatments are necessary for the removal of organic matter from the surface because the particles will inevitably grow and coalesce, which could lead to a reduction in the efficiency of the catalyst [102].

In addition to that, metal oxide-based catalysts are very complex materials, with characteristics like multiple valence states and crystal structures. They are also capable of promoting various transformations and these features complicate understanding and selectively applying them in catalysis [103, 104]. The nanoparticles of transition metals are quite unstable, as the highly active surface atoms are generally thermodynamically unstable, leading to severe nanoparticles sintering and this obstructs their practical applications in catalysis [105, 106]. There is also the issue of aggregation and agglomeration due to high surface energies. This causes a reduction of available surface area for reactions, loss of nano-specific properties, and ultimately a reduced catalytic performance over time [106].

Concerning applications in nanomedicine, several challenges block MONCs' clinical translation and application, although their unique physical characteristics offer better precision and efficiency in regulating redox processes when compared to traditional antioxidant therapies. Reproducibility and scalability in nanomaterials synthesis in order to precisely control the morphology, size, and surface properties of the engineered nanomaterials, which in turn guarantee minimum toxicity and biocompatibility in the long term is a major challenge. Another is the tackling of complex issues concerning biodistribution, clearance, and potential off-target effects in vivo [101]. Nanoparticles are also peculiarly adsorptive, exhibit fast diffusivities, and are unstable in critical conditions [107].

There are also concerns about the cytotoxicity and biocompatibility of these materials as interest has increased in metal and metal oxide nanoparticles applications in biomedical engineering. The physicochemical properties of metal and metal oxide nanoparticles are strongly linked to their resulting cytotoxic effects. In that manner, the cytotoxicity can be increased, decreased, or missing depending on the physicochemical properties (like shape, size, surface charge, dose, surface area, and solubility) of the nanomaterial [108]. It has also been found that nanoparticles of smaller size < 50 nm can enhance DNA damage and oxidation potential [109]. Figure 15.3 gives an outline of the major limitations facing MNOCs practical application.

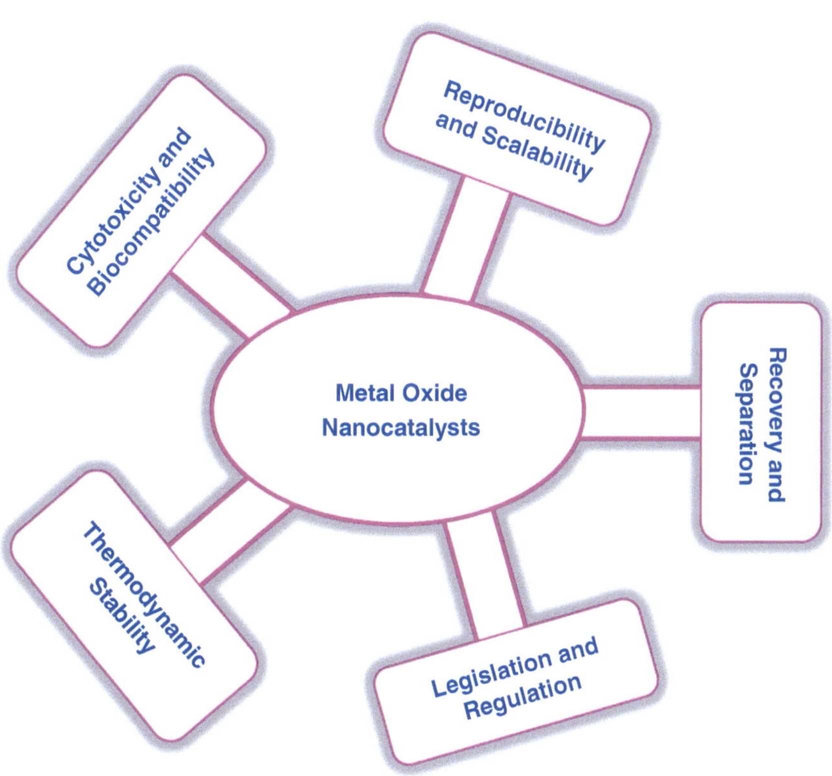

Figure 15.3 Challenges and limitations of MONCs.

15.6 Conclusion

Climate change is an insidious phenomenon, and much has to be done to reverse its harmful effects. Rising global temperatures, unstable and unpredictable weather, and rainfall patterns, among others, are some of the more harmful effects of climate change. While much is being done to combat and reverse the effects of climate change, much more has to be done to address the severity of this issue. MONCs have immense potential for real-world applications involving climate change mitigation and environmental sustainability, and although these materials are already being applied across various sectors, some challenges stifle their widespread applications, and there is need for more research to improve their effectiveness and optimize their processes. Improvement of this technology will work hand-in-hand with other technological advancements to improve environmental conditions and curb the effects of climate change.

References

1 Ahmad H, Zafar SA, Naeem MK, Shokat S, Inam S, Rehman MA, et al. Impact of pre-anthesis drought stress on physiology, yield-related traits, and drought-responsive genes in green super rice. *Front Genet.* 2022 Mar 24;13: 832542.

2 Kabir M, Habiba UE, Khan W, Shah A, Rahim S, De los Rios-Escalante PR, et al. Climate change due to increasing concentration of carbon dioxide and its impacts on environment in 21st century; a mini review. *J King Saud Univ Sci.* 2023 Jul 1;35(5): 102693.

3 Subramanian KS, Karthika V, Praghadeesh M, Lakshmanan A. Nanotechnology for mitigation of global warming impacts. In: Venkatramanan V, Shah S, Prasad R, editors. *Global Climate Change: Resilient and Smart Agriculture.* Singapore: Springer; 2020 Sep 19. p. 315–36.

4 Porkodi G, Anand G. Nanotechnology and its application in water treatment: a review. *Int J Environ Agric Res.* 2020;6(2):59–66.

5 Chausali N, Saxena J, Prasad R. Recent trends in nanotechnology applications of bio-based packaging. *J Agr Food Res.* 2022 Mar 1;7: 100257.

6 Polshettiwar V, Varma RS. Green chemistry by nano-catalysis. *Green Chem.* 2010;12(5):743–54.

7 Rai M, dos Santos JC, Soler MF, Franco Marcelino PR, Brumano LP, Ingle AP, et al. Strategic role of nanotechnology for production of bioethanol and biodiesel. *Nanotechnol Rev.* 2016 Apr 1;5(2): 231–50.

8 Chausali N, Saxena J, Prasad R. Nanobiochar and biochar based nanocomposites: advances and applications. *J Agr Food Res.* 2021 Sep 1;5: 100191.

9 Muhammad M, Arooj A, Ullah S, Wahab A, Nazim M, Basit A. Future of nanotechnology and climate change. In: Abdel Latef AAH, Zayed EM, Omar AA, editors. *Bioremediation and Nanotechnology for Climate Change Mitigation.* Singapore: Springer; 2025 May 1. p. 621–52.

10 Mehndiratta P, Jain A, Srivastava S, Gupta N. Environmental pollution and nanotechnology. *Environ Pollut*. 2013 Apr 1;2(2): 49.

11 Kuna E, Mrdenovic D, Jönsson-Niedziółka M, Pieta P, Pieta IS. Bimetallic nanocatalysts supported on graphitic carbon nitride for sustainable energy development: the shape-structure–activity relation. *Nanoscale Adv*. 2021;3(5):1342–51.

12 Santhosh C, Velmurugan V, Jacob G, Jeong SK, Grace AN, Bhatnagar A. Role of nanomaterials in water treatment applications: a review. *Chem Eng J*. 2016 Dec 15;306: 1116–37.

13 Ningthoujam R, Singh YD, Babu PJ, Tirkey A, Pradhan S, Sarma M. Nanocatalyst in remediating environmental pollutants. *Chem Phys Impact*. 2022 Jun 1;4: 100064.

14 Lu F, Astruc D. Nanocatalysts and other nanomaterials for water remediation from organic pollutants. *Coord Chem Rev*. 2020 Apr 1;408: 213180.

15 Chakraborty A, Ruzimuradov O, Gupta RK, Cho J, Prakash J. TiO_2 nanoflower photocatalysts: synthesis, modifications and applications in wastewater treatment for removal of emerging organic pollutants. *Environ Res*. 2022 Sep 1;212: 113550.

16 Chen Z, Sun S, Prakash J. Design and engineering of graphene nanostructures as independent solar-driven photocatalysts for emerging applications in the field of energy and environment. *Mol Syst Des Eng*. 2022;7(3):213–38.

17 Gupta T, Cho J, Prakash J. Hydrothermal synthesis of TiO_2 nanorods: formation chemistry, growth mechanism, and tailoring of surface properties for photocatalytic activities. *Mater Today Chem*. 2021 Jun 1;20: 100428.

18 Rajput V, Gupta RK, Prakash J. Engineering metal oxide semiconductor nanostructures for enhanced charge transfer: fundamentals and emerging SERS applications. *J Mater Chem C*. 2022;10(1):73–95.

19 Kumar P, Mathpal MC, Prakash J, Viljoen BC, Roos WD, Swart HC. Band gap tailoring of cauliflower-shaped CuO nanostructures by Zn doping for antibacterial applications. *J Alloys Compd*. 2020 Aug 15;832: 154968.

20 Ahmed M, Ahmad KA, Vo DV, Yusuf M, Haq A, Abdullah A, et al. Recent trends in sustainable biodiesel production using heterogeneous nanocatalysts: function of supports, promoters, synthesis techniques, reaction mechanism, and kinetics and thermodynamic studies. *Energy Convers Manag*. 2023 Mar 15;280: 116821.

21 Ameen M, Zafar M, Ramadan MF, Ahmad M, Makhkamov T, Bokhari A, et al. Conversion of novel non-edible *Bischofia javanica* seed oil into methyl ester via recyclable zirconia-based phyto-nanocatalyst: a circular bioeconomy approach for eco-sustenance. *Environ Technol Innov*. 2023 May 1;30: 103101.

22 Mills-Novoa M, Mikulewicz M. The promise of resistance: a new lens for climate change adaptation research and practice. *Wiley Interdiscip Rev Clim Change*. 2025 Jan; 16(1):e922.

23 Kumler A, Kravitz B, Draxl C, Vimmerstedt L, Benton B, Lundquist JK, et al. Potential effects of climate change and solar radiation modification on renewable energy resources. *Renew Sustain Energy Rev*. 2025 Jan 1;207: 114934.

24 Malik IH, Ford JD. Monitoring climate change vulnerability in the Himalayas. *Ambio*. 2025 Jan;54(1): 1–9.

25 Ray RL, Tikuye BG. Impact of climate change on surface water resources. In: *Limnology – Surface and Ground Water Ecosystems*. IntechOpen; 2025.
26 Verma S. Climate change in Himalayas: impacts on lives and livelihoods. *Int J Sci Res Arch*. 2023;9(1):419–25.
27 Salila E, Sharma MP, Singh R. Climate change impacts on water quality of water bodies-a review. *Int J Environ Clim Chang*. 2020;10(3):26–37.
28 Abdulla F, Farahat S. Impact of climate change on the performance of wastewater treatment plant: case study central irbid WWTP (Jordan). *Procedia Manuf.* 2020 Jan 1;44: 205–12.
29 Dechezleprêtre A, Fabre A, Kruse T, Planterose B, Sanchez Chico A, Stantcheva S. Fighting climate change: international attitudes toward climate policies. *Am Econ Rev*. 2025 Apr 1;115(4): 1258–300.
30 Nazari R, Vasiliadis H, Karimi M, Fahad MG, Simon S, Zhang T, et al. Hydrodynamic study of the impact of extreme flooding events on wastewater treatment plants considering total water level. *Nat Hazards Rev*. 2022 Feb 1;23(1): 04021056.
31 Qiu J, Shen Z, Xie H. Drought impacts on hydrology and water quality under climate change. *Sci Tot Environ*. 2023 Feb 1;858: 159854.
32 Zhang M, Liu S, Jones J, Sun G, Wei X, Ellison D, et al. Managing the forest-water nexus for climate change adaptation. *For Ecol Manag*. 2022 Dec 1;525: 120545.
33 Ejaz H, Bibi E, Ali W, Ahmad I, Lashari A, Faiz H, et al. Sulphur and particulate matter affecting on soil and underground plants. *J Agric Appl Biol*. 2022;3:40–9.
34 Mansoor S, Farooq I, Kachroo MM, Mahmoud AE, Fawzy M, Popescu SM, et al. Elevation in wildfire frequencies with respect to the climate change. *J Environ Manag*. 2022 Jan 1;301: 113769.
35 Rasmussen JB. Advancing environmental justice through the integration of traditional ecological knowledge into environmental policy. *Challenges*. 2023 Jan 11;14(1): 6.
36 Abdel-Fatah MA, Ewies EF. Green nanotechnology for sustainable ecosystems: innovations in pollution remediation and resource recovery. *Egypt J Chem*. 2025 Oct 1;68(10): 621–33.
37 Islam FS, Islam MN. AI-driven integration of nanotechnology and green nanotechnology for sustainable energy and environmental remediation. *J Eng Res Reports*. 2025 Jul 11;27(7): 260–311.
38 Channappagoudra M, Samantaray S, Mishra AK, Juneja G, Devarajan Y, Chand K. Transforming waste to energy: nanocatalyst innovations driving green hydrogen production. *Rev Inorg Chem*. 2025 Jun 13.
39 Hasan N, Muthu M, Hakami O, Gopal J. Assessing the sustainability of energy-related nanomaterial synthesis: emphasizing the need for energy-efficient nanomaterial preparation techniques. *Energies*. 2025 Jan 23;18(3): 523.
40 Mohammed ZS. Advancing bioenergy and biofuels: role of nanotechnology in sustainable energy systems. *J Environ Nanotechnol*. 2025;14(2):82–95.
41 Khan Y, Sadia H, Ali Shah SZ, Khan MN, Shah AA, Ullah N, et al. Classification, synthetic, and characterization approaches to nanoparticles, and their applications in various fields of nanotechnology: a review. *Catalysts*. 2022 Nov 8;12(11): 1386.

42 Yan K, Chen A. Efficient hydrogenation of biomass-derived furfural and levulinic acid on the facilely synthesized noble-metal-free Cu–Cr catalyst. *Energy* 2013 Sep 1;58: 357–63.

43 Somwanshi SB, Somvanshi SB, Kharat PB. Nanocatalyst: a brief review on synthesis to applications. *J Phys Conf Ser.* 2020 Oct 1;1644(1): 012046.

44 Wolski L, Grzelak K, Muńko M, Frankowski M, Grzyb T, Nowaczyk G. Insight into photocatalytic degradation of ciprofloxacin over CeO_2/ZnO nanocomposites: unravelling the synergy between the metal oxides and analysis of reaction pathways. *Appl Surf Sci.* 2021 Oct 15;563: 150338.

45 Zavašnik J, Šestan A, Shvalya V. Microscopic techniques for the characterisation of metal-based nanoparticles. *Compr Anal Chem.* 2021 Jan 1;93: 241–84.

46 Vijayakumar J, Savchenko TM, Bracher DM, Lumbeeck G, Béché A, Verbeeck J, et al. Absence of a pressure gap and atomistic mechanism of the oxidation of pure Co nanoparticles. *Nat Commun.* 2023 Jan 12;14(1): 174.

47 Sharma RK, Bandichhor R, Mishra V, Sharma S, Yadav S, Mehta S, et al. Advanced metal oxide-based nanocatalysts for the oxidative synthesis of fine chemicals. *Mater Adv.* 2023;4(8):1795–830.

48 Geldasa FT, Kebede MA, Shura MW, Hone FG. Experimental and computational study of metal oxide nanoparticles for the photocatalytic degradation of organic pollutants: a review. *RSC Adv.* 2023;13(27):18404–42.

49 Negrescu AM, Killian MS, Raghu SN, Schmuki P, Mazare A, Cimpean A. Metal oxide nanoparticles: review of synthesis, characterization and biological effects. *J Funct Biomater.* 2022 Dec 5;13(4): 274.

50 Awogbemi O, Von Kallon DV. Recent advances in the application of nanomaterials for improved biodiesel, biogas, biohydrogen, and bioethanol production. *Fuel.* 2024 Feb 15;358: 130261.

51 Li X, Xiong J, Tang Z, He W, Wang Y, Wang X, et al. Recent progress in metal oxide-based photocatalysts for CO_2 reduction to solar fuels: a review. *Molecules.* 2023 Feb 9;28(4): 1653.

52 Periasamy P, Mohanta YK. Global energy crisis: need for energy conversion and storage. In: Chakrabartty I, Hakeem KR, editors. *Green Nanomaterials in Energy Conversion and Storage Applications.* Apple Academic Press; 2024 Feb 6. p. 45–73.

53 The World Counts. 2024. Available from: https://www.theworldcounts.com/challenges/energy/global-energy-consumption

54 Awogbemi O, Ojo AA, Adeleye SA. Advancements in the application of metal oxide nanocatalysts for sustainable biodiesel production. *Discov Appl Sci.* 2024 May 7;6(5): 250.

55 Li A, Sun Y, Yao T, Han H. Earth-abundant transition-metal-based electrocatalysts for water electrolysis to produce renewable hydrogen. *Chem Eur J.* 2018 Dec 10;24(69): 18334–55.

56 Zhu J, Hu L, Zhao P, Lee LY, Wong KY. Recent advances in electrocatalytic hydrogen evolution using nanoparticles. *Chem Rev.* 2019 Oct 28;120(2): 851–918.

57 Banković–Ilić IB, Miladinović MR, Stamenković OS, Veljković VB. Application of nano CaO–based catalysts in biodiesel synthesis. *Renew Sustain Energy Rev.* 2017 May 1;72: 746–60.

58 Leo VV, Singh BP. Prospectus of nanotechnology in bioethanol productions. In: Srivastava N, Srivastava M, Pandey H, Mishra PK, Ramteke PW, editors. *Green Nanotechnology for Biofuel Production*. Cham: Springer International Publishing; 2018 Mar 20. p. 129–39.

59 Salimi Z, Hosseini SA. Study and optimization of conditions of biodiesel production from edible oils using $ZnO/BiFeO_3$ nano magnetic catalyst. *Fuel*. 2019 Mar 1;239: 1204–12.

60 Akia M, Yazdani F, Motaee E, Han D, Arandiyan H. A review on conversion of biomass to biofuel by nanocatalysts. *Biofuel Res J*. 2014 Mar 1;1(1): 16–25.

61 Saman S, Balouch A, Talpur FN, Memon AA, Mousavi BM, Verpoort F. Green synthesis of MgO nanocatalyst by using Ziziphus mauritiana leaves and seeds for biodiesel production. *Appl Organomet Chem*. 2021 May; 35(5):e6199.

62 Du L, Li Z, Ding S, Chen C, Qu S, Yi W, et al. Synthesis and characterization of carbon-based MgO catalysts for biodiesel production from castor oil. *Fuel*. 2019 Dec 15;258: 116122.

63 Rahimi T, Kahrizi D, Feyzi M, Ahmadvandi HR, Mostafaei M. Catalytic performance of MgO/Fe_2O_3-SiO_2 core-shell magnetic nanocatalyst for biodiesel production of Camelina sativa seed oil: optimization by RSM-CCD method. *Ind Crops Prod*. 2021 Jan 1;159: 113065.

64 Alaei S, Haghighi M, Rahmanivahid B, Shokrani R, Naghavi H. Conventional vs. hybrid methods for dispersion of MgO over magnetic Mg–Fe mixed oxides nanocatalyst in biofuel production from vegetable oil. *Renew Energy*. 2020 Jul 1;154: 1188–203.

65 Aghel B, Gouran A, Parandi E, Jumeh BH, Nodeh HR. Production of biodiesel from high acidity waste cooking oil using nano GO@ MgO catalyst in a microreactor. *Renew Energy*. 2022 Nov 1;200: 294–302.

66 Bora AP, Konda LD, Pasupuleti S, Durbha KS. Synthesis of $MgO/MgSO_4$ nanocatalyst by thiourea–nitrate solution combustion for biodiesel production from waste cooking oil. *Renew Energy*. 2022 May 1;190: 474–86.

67 Jan HA, Saqib NU, Khusro A, Sahibzada MU, Rauf M, Alghamdi S, et al. Synthesis of biodiesel from *Carthamus tinctorius* L. oil using TiO_2 nanoparticles as a catalyst. *J King Saud Univ Sci*. 2022 Nov 1;34(8): 102317.

68 Patil A, Baral SS, Dhanke P, Kore V. Biodiesel production using prepared novel surface functionalised TiO_2 nano-catalyst in hydrodynamic cavitation reactor. *Mater Today Proc*. 2020 Jan 1;27: 198–203.

69 De A, Boxi SS. Application of Cu impregnated TiO_2 as a heterogeneous nanocatalyst for the production of biodiesel from palm oil. *Fuel*. 2020 Apr 1;265: 117019.

70 Gardy J, Hassanpour A, Lai X, Ahmed MH, Rehan M. Biodiesel production from used cooking oil using a novel surface functionalised TiO_2 nano-catalyst. *Appl Catal B Environ*. 2017 Jun 15;207: 297–310.

71 Baskar G, Selvakumari IA, Aiswarya RJ. Biodiesel production from castor oil using heterogeneous Ni doped ZnO nanocatalyst. *Bioresour Technol*. 2018 Feb 1;250: 793–8.

72 Borah MJ, Devi A, Borah R, Deka D. Synthesis and application of Co doped ZnO as heterogeneous nanocatalyst for biodiesel production from non-edible oil. *Renew Energy*. 2019 Apr 1;133: 512–9.

73 Rao AK, Dudhe P, Chelvam V. Role of oxygen defects in basicity of Se doped ZnO nanocatalyst for enhanced triglyceride transesterification in biodiesel production. *Catal Commun.* 2021 Jan 15;149: 106258.

74 Salim SM, Izriq R, Almaky MM, Al-Abbassi AA. Synthesis and characterization of ZnO nanoparticles for the production of biodiesel by transesterification: kinetic and thermodynamic studies. *Fuel.* 2022 Aug 1;321: 124135.

75 Maleki B, Esmaeili H. Application of Fe_3O_4/SiO_2@ ZnO magnetic composites as a recyclable heterogeneous nanocatalyst for biodiesel production from waste cooking oil: response surface methodology. *Ceram Int.* 2023 Apr 1;49(7): 11452–63.

76 Maleki B, Esmaeili H. Ultrasound-assisted conversion of waste frying oil into biodiesel using Al-doped ZnO nanocatalyst: box-Behnken design-based optimization. *Renew Energy.* 2023 Jun 1;209: 10–26.

77 Gada P, Dusi SK, Abrar I. Enhanced biodiesel production from waste cooking oil using ZnO nanocatalyst. In: *International Conference on Nanotechnology: Opportunities and Challenges.* Singapore: Springer; 2022 Nov 28. p. 47–54.

78 Maleki B, Venkatesh YK, Talesh SS, Esmaeili H, Mohan S, Balakrishna GR. A novel biomass derived activated carbon mediated AC@ ZnO/NiO bifunctional nanocatalyst to produce high-quality biodiesel from dairy industry waste oil: CI engine performance and emission. *Chem Eng J.* 2023 Jul 1;467: 143399.

79 Sharma U, Muralidharan NG, Vijayalakshmi S, Ranjitha J. Process optimisation and kinetics study of biodiesel production from dairy waste scum using ZnO heterogeneous nanocatalyst. *IOP Conf Ser Mater Sci Eng.* 2020 Sep 1;923(1): 012072.

80 Abdala E, Nur O, Mustafa MA. Efficient biodiesel production from algae oil using Ca-doped ZnO nanocatalyst. *Ind Eng Chem Res.* 2020 Oct 16;59(43): 19235–43.

81 Rodríguez-Ramírez R, Sosa-Rodríguez FS, Vazquez-Arenas J. Zinc oxide-co-sodium zirconate: a fast heterogeneous catalyst for biodiesel production from soybean oil. *J Environ Chem Eng.* 2022 Aug 1;10(4): 108191.

82 Dai YM, Li YY, Lin JH, Kao IH, Lin YJ, Chen CC. Applications of M_2ZrO_2 (M= Li, Na, K) composite as a catalyst for biodiesel production. *Fuel.* 2021 Feb 15;286: 119392.

83 Li H, Wang Y, Ma X, Guo M, Li Y, Li G, et al. Synthesis of CaO/ZrO_2 based catalyst by using UiO-66 (Zr) and calcium acetate for biodiesel production. *Renew Energy.* 2022 Feb 1;185: 970–7.

84 Booramurthy VK, Kasimani R, Pandian S, Ragunathan B. Nano-sulfated zirconia catalyzed biodiesel production from tannery waste sheep fat. *Environ Sci Pollut Res.* 2020 Jun;27(17): 20598–605.

85 Zhang Y, Niu S, Lu C, Gong Z, Hu X. Catalytic performance of $NaAlO_2/\gamma\text{-}Al_2O_3$ as heterogeneous nanocatalyst for biodiesel production: optimization using response surface methodology. *Energy Convers Manag.* 2020 Jan 1;203: 112263.

86 Khazaai SN, Yiting S, Rahim MH, Ibrahim ML, Maniam GP. Application of waste catalyst, CaO-SiO_2 in the transesterification of palm oil. *Malays J Anal Sci.* 2021 Jun 27;25(3): 490–3.

87 Cholapandian K, Gurunathan B, Rajendran N. Investigation of CaO nanocatalyst synthesized from *Acalypha indica* leaves and its application in biodiesel production using waste cooking oil. *Fuel.* 2022 Mar 15;312: 122958.

88 Ozor PA, Aigbodion VS, Sukdeo NI. Modified calcium oxide nanoparticles derived from oyster shells for biodiesel production from waste cooking oil. *Fuel Comm.* 2023 Mar 1;14: 100085.

89 Khalifeh R, Esmaeili H. Biodiesel production from goat fat using calcium oxide nanocatalyst and its combination with diesel fuel to improve fuel properties. *Int J Sustain Eng.* 2021 Sep 3;14(5): 1122–31.

90 Seffati K, Honarvar B, Esmaeili H, Esfandiari N. Enhanced biodiesel production from chicken fat using $CaO/CuFe_2O_4$ nanocatalyst and its combination with diesel to improve fuel properties. *Fuel.* 2019 Jan 1;235: 1238–44.

91 Chavali MS, Nikolova MP. Metal oxide nanoparticles and their applications in nanotechnology. *SN Appl Sci.* 2019 Jun;1(6): 607.

92 Védrine JC. Importance, features and uses of metal oxide catalysts in heterogeneous catalysis. *Chin J Catal.* 2019 Nov 1;40(11): 1627–36.

93 Jia J, Qian C, Dong Y, Li YF, Wang H, Ghoussoub M, et al. Heterogeneous catalytic hydrogenation of CO_2 by metal oxides: defect engineering–perfecting imperfection. *Chem Soc Rev.* 2017;46(15):4631–44.

94 Kambhampati P. Nanoparticles, nanocrystals, and quantum dots: what are the implications of size in colloidal nanoscale materials? *J Phys Chem Lett.* 2021 May 13;12(20): 4769–79.

95 Soltys L, Olkhovyy O, Tatarchuk T, Naushad M. Green synthesis of metal and metal oxide nanoparticles: principles of green chemistry and raw materials. *Magnetochemistry.* 2021 Oct 31;7(11): 145.

96 Castro Alarcon N, Herrera Arizmendi JL, Luis Alberto MC, Guzman IP, Perez Centeno A, Santana Aranda MA. Antibacterial activity of nanoparticles of titanium dioxide, intrinsic and doped with indium and iron. *J Med Chem Tox.* 2016;4(4):55–62.

97 Li Z, Li M, Bian Z, Kathiraser Y, Kawi S. Design of highly stable and selective core/yolk–shell nanocatalysts—a review. *Appl Catal B Environ.* 2016 Jul 5;188: 324–41.

98 Sharma R, Watanabe T, Kumar A. Estimation of risk to soil and human health during irrigation using ZnO nanoparticles-containing water. *J Environ Chem Eng.* 2023 Dec 1;11(6): 111230.

99 Gayathiri E, Prakash P, Pandiaraj S, Ramasubburayan R, Gaur A, Sekar M, et al. Investigating the ecological implications of nanomaterials: unveiling plants' notable responses to nano-pollution. *Plant Physiol Biochem.* 2024 Jan 1;206: 108261.

100 Karadaghi LR, Williamson EM, To AT, Forsberg AP, Crans KD, Perkins CL, et al. Multivariate Bayesian optimization of CoO nanoparticles for CO_2 hydrogenation catalysis. *J Am Chem Soc.* 2024 May 10;146(20): 14246–59.

101 Araujo SF, da Silva Fonseca RA, Araujo AS. Catalytic nanomedicine: antioxidant action and clinical benefits using cerium oxide nanoparticles. *Precis Nanomed.* 2025 Apr 7;8(2): 1473–81.

102 Skrodczky K, Antunes MM, Zhu Q, Valente AA, Pinna N, Russo PA. Single-step formation of metal oxide nanostructures wrapped in mesoporous silica and silica–niobia catalysts for the condensation of furfural with acetone. *Nanomaterials.* 2023 Nov 29;13(23): 3046.

103 Li K, Li X, Li L, Chang X, Wu S, Yang C, et al. Nature of catalytic behavior of cobalt oxides for CO_2 hydrogenation. *JACS Au.* 2023 Feb 1;3(2): 508–15.

104 Have IC, Kromwijk JJ, Monai M, Ferri D, Sterk EB, Meirer F, et al. Uncovering the reaction mechanism behind CoO as active phase for CO_2 hydrogenation. *Nat Commun.* 2022 Jan 14;13(1): 324.

105 Zhang J, Wang L, Zhang B, Zhao H, Kolb U, Zhu Y, et al. Sinter-resistant metal nanoparticle catalysts achieved by immobilization within zeolite crystals via seed-directed growth. *Nat Catal.* 2018 Jul;1(7): 540–6.

106 Xiong P, Xu Z, Wu TS, Yang T, Lei Q, Li J, et al. Synthesis of core@ shell catalysts guided by Tammann temperature. *Nat Commun.* 2024 Jan 10;15(1): 420.

107 Bindhu MR, Umadevi M, Micheal MK, Arasu MV, Al-Dhabi NA. Structural, morphological and optical properties of MgO nanoparticles for antibacterial applications. *Mater Lett.* 2016 Mar 1;166: 19–22.

108 Nikzamir M, Akbarzadeh A, Panahi Y. An overview on nanoparticles used in biomedicine and their cytotoxicity. *J Drug Deliv Sci Technol.* 2021 Feb 1;61: 102316.

109 Gowda BJ, Ahmed MG, Chinnam S, Paul K, Ashrafuzzaman M, Chavali M, et al. Current trends in bio-waste mediated metal/metal oxide nanoparticles for drug delivery. *J Drug Deliv Sci Technol.* 2022 May 1;71: 103305.

16

Metal/Bimetallic Oxide Nanocatalysts for CO_2 Conversion

Mustapha Abdulsalam[1], Innocent Ojeba Musa[1], Miracle Uwa Livinus[2], Salam Olaitan Lateefat[3], Ganiyat Omotayo Ibrahim[4], Ibrahim Abdulrazaq[5], Abdulhakeem Idris Abdulhakeem[6], Ajadi Ibrahim[7], and Muhammed Abdulazeez[1]*

[1] *Department of Microbiology, Skyline University Nigeria, Kano, Nigeria*
[2] *Department of Biochemistry, Skyline University Nigeria, Kano, Nigeria*
[3] *Department of Chemistry and Molecular Biology, University of Gothenburg, Gothenburg, Sweden*
[4] *Department of Chemistry, Nottingham Trent University, Nottingham, UK*
[5] *Department of Microbiology, University of Ilorin, Ilorin, Nigeria*
[6] *Department of Biodiversity and Bioresources, Al-Farabi Kazakh National University, Almaty, Kazakhstan*
[7] *Department of Plant and Environmental Biology, Kwara State University, Malete, Nigeria*

*Corresponding author: mustypeace2006@gmail.com

16.1 Introduction

The relentless accumulation of atmospheric carbon dioxide (CO_2) has become one of the most pressing environmental and socioeconomic challenges of our era. Fueled predominantly by anthropogenic activities—especially fossil fuel combustion, deforestation, and large-scale industrial processes—the current CO_2 concentration in the atmosphere has surpassed 420 ppm, marking an unprecedented rise in recent human history [1]. This escalation not only exacerbates global warming but also triggers a cascade of climate-related crises, including ocean acidification, ecosystem disruptions, and public health challenges [2]. Given the severity of the problem, scientists, engineers, and policymakers are collectively exploring solutions that extend beyond emission reductions. Among the emerging strategies, CO_2 capture and utilization (CCU) stands out as a promising and potentially circular approach. Rather than treating CO_2 merely as a waste product to be sequestered underground, CCU aims to convert it into value-added fuels, chemicals, and materials. This paradigm shift—from carbon liability to carbon resource—opens up possibilities for sustainable industrial symbiosis and climate-positive technologies [3].

Metal Oxide Nanocatalysts for Sustainable Energy Production, First Edition. Edited by Abayomi Bamisaye et al.
Published 2026 by Wiley-VCH GmbH.

The conversion of CO_2 electrochemically or thermochemically is, however, inherently difficult. The molecule is thermodynamically quite stable and kinetically inert, and thus hard to activate in mild conditions. It requires a sophisticated catalyst that is able to reduce activation energy, direct product selectivity, and operate productively in either aqueous or gaseous conditions. A robust and highly versatile solution to these problems is provided by nanocatalysis, and especially the use of metal and bimetallic oxide nanostructures. There are novel quantum, surface, and electronic properties that occur at the nanoscale that are not found in the bulk materials. Such characteristics are carefully amenable to control by controlling the size, shape, composition, and structure of the particles—allowing the construction of catalysts of high reactivity with excellent selectivity [4]. As an example, the use of platinum (Pt)-based nanocatalysts has proven invaluable in a range of electrocatalytic reactions, such as in the oxidation of formic acid, the reduction of oxygen, and hydrogen evolution, because of their high surface area and ideal electronic structure [5]. Over the past few years, scientists have started to apply the use of Pt-based multimetallic and oxide-supported nanocatalysts also to reduce CO_2. Scientists have achieved the cost-effectiveness and effectiveness of the catalytic activity of Pt-based systems by including transition metals Ni, Co, Fe, and Cu in bimetallic alloys or core-shell complexes [6]. Such bimetallic pairs provide synergistic effects, that is, the electronic and geometric interactions between the reactants modify the adsorption energies of reaction intermediates and enhance the catalytic turnover frequencies [7].

Furthermore, Pt nanoparticles show greater stability after they are incorporated into oxide forms such as perovskite types (ABO_3) and spinel ferrites (e.g., $CoFe_2O_4$), further augmenting their structural stability and catalytic performance. These oxides also induce a dramatic electronic effect by modulating their local electronic environment through the metal-support interactions (MSIs), thus facilitating an enhanced electron transfer and catalytic stability over CO_2 reduction reactions (CO_2RRs) [8]. Complex nanostructures have been successfully made via the modified polyol process, a wet chemical procedure that is distinguished by simplicity, versatility, and scalability. Through this procedure, the controlled growth of the particles was made possible, and uniform alloyed or core-shell nanocatalysts with controllable surface chemistry could be created [9]. To illustrate, Pt-based core-shell structures with a dispersed thin transition metal Pt core (e.g., Ni@Pt or Fe@Pt) reduce Pt loading by a magnitude and maintain or even increase the performance of the catalyst [10].

Beyond the laboratory, the real-world integration of these catalysts into low-temperature fuel cells, photoelectrochemical (PEC) cells, and CO_2 electrolyzers is gaining momentum. Particularly compelling is their potential to facilitate the electrochemical reduction of CO_2 (CO_2RR) into valuable products such as CO, formate, methane (CH_4), ethylene (C_2H_4), and alcohols—compounds that can reenter existing industrial value chains [10]. This not only offsets fossil fuel use but also helps build a net-zero carbon economy, where CO_2 is continuously recycled and valorized. However, the global urgency of mitigating CO_2 emissions aligns strongly with the need for innovative nanocatalysts that can drive its transformation into valuable commodities. Through the intersection of nanotechnology, materials science, and green chemistry, metal, and bimetallic oxide nanocatalysts represent a pivotal frontier in the fight against climate change. This chapter aims to unravel the design, synthesis, and application of such nanostructures for CO_2 conversion, offering insights into both fundamental mechanisms and practical implementations.

16.2 Fundamentals of CO_2 Electrochemical Reduction (CO_2RR)

16.2.1 Overview of CO_2RR Mechanisms

The reduction reaction occurs between an electrode (typically the cathode) and an electrolyte at a specified potential. Most reactions are carried out in either aqueous or gas-supported electrochemical cells, and the supply of protons is provided by the water, which is also the solvent. In short, the CO_2 molecules are adsorbed onto the surface of the catalyst and activated, resulting in the molecules being trafficked with waving electrons and protons, in turn producing reduced carbon products. The CO_2RR mechanism, however, is far from straightforward. CO_2 is a linear, highly oxidized molecule with a strong C=O bond (~750 kJ/mol), making its activation energy-intensive [11]. The general reaction pathway involves multiple steps, including adsorption of CO_2, formation of reactive intermediates (such as *COOH or *HCOO), and their further reduction or desorption. Depending on the pathway, the reaction can yield different products, often influenced by the choice of catalyst, applied potential, pH, and electrolyte composition.

For example:

- On gold and silver, CO_2 is selectively reduced to CO.
- Tin and bismuth favor formate formation.
- Copper, uniquely, facilitates the production of hydrocarbons and oxygenates such as methane, ethylene, and ethanol, though often at the expense of selectivity [12].

16.2.2 Thermodynamic and Kinetic Challenges

From a thermodynamic perspective, CO_2 reduction reactions must compete with the hydrogen evolution reaction (HER)—a parallel reaction where protons in the electrolyte are reduced to hydrogen gas (H_2). HER typically has a lower overpotential, making it a dominant and often undesirable side reaction, especially in aqueous electrolytes [13]. This competition reduces CO_2RR efficiency and makes selective product generation a significant challenge. Kinetically, CO_2 is an inert molecule, and its initial adsorption and activation on the catalyst surface require high energy input. The formation of the *COOH or *HCOO intermediate is typically the rate-determining step (RDS) for most products. Additionally, multi-electron transfer processes (e.g., for forming CH_4 or C_2H_4) involve several intermediate steps and proton-coupled electron transfers, which further complicate the reaction dynamics and slow the overall kinetics [14].

Moreover, the reaction environment—such as local pH, ion concentrations, and diffusion limitations—can significantly alter the reaction energetics and favor competing reactions. The fine-tuning of these factors remains a persistent bottleneck in CO_2RR technology.

$$j = j_0 \left[\exp\left(\frac{\alpha n F \eta}{RT} \right) - \exp\left(-\frac{(1-\alpha) n F \eta}{RT} \right) \right] \tag{1}$$

where:

- j: current density
- j_0: exchange current density
- α: charge transfer coefficient
- n: number of electrons transferred
- F: Faraday constant
- η: overpotential
- R: universal gas constant
- T: temperature (in K)

16.2.3 Role of Catalyst Composition, Structure, and Morphology

One of the most powerful levers for improving CO_2RR performance is the design of the catalyst itself. Three key aspects influence catalytic activity and selectivity:

a. **Composition**
The elemental makeup of a catalyst determines how it binds CO_2 and its intermediates. Single-metal catalysts often suffer from limited selectivity or stability. For example, while platinum is excellent for HER, it is poor for CO_2RR due to the strong binding of reaction intermediates that leads to poisoning [15]. Introducing bimetallic or multimetallic combinations, such as Pt–Ni, Pt–Fe, or Cu–Sn, can create synergistic effects, where one metal modulates the electronic environment of the other, altering adsorption energies and reaction barriers [16]. This is the rationale behind many recent advances in Pt-based multimetallic nanocatalysts.

b. **Structure**
The crystal facets exposed by a nanocatalyst significantly influence its reactivity. For instance, Cu(100) facets are known to promote C—C coupling and favor the formation of C_2 products like ethylene, while Cu(111) surfaces are more selective for methane [17]. Therefore, controlling the exposed crystallographic planes—via shape-controlled synthesis—is critical for tuning product distribution. In alloy and core-shell nanostructures, electronic interactions at the interface can create strain effects that shift the d-band center of active metals, leading to favorable binding and improved kinetics [18].

c. **Morphology and Size**
Nanoparticles with high surface area-to-volume ratios offer more active sites per unit mass. Morphological control (e.g., nanowires, hollow structures, and

dendrites) not only enhances surface accessibility but also affects mass transport properties, which are vital for reducing concentration gradients and facilitating intermediate stabilization [19]. The MSIs further enhance the catalytic performance, in which oxide support (evolved to perovskites and ferrites) is used along with metallic nanoparticles. Such interactions can stabilize some oxidation states, boost the charge transfers, and prevent any catalyst degradation [20].

16.3 Design and Synthesis Principles for Nanocatalysts

The development of nanocatalysts to reduce CO_2 by electrochemical reactions is no longer only a matter of selecting a material. It is smart atomic engineering of the space around, the combination of surface structure, composition, and electronic functionality that determines the activation, conversion, or release of CO_2 molecules. Since multi-electron processes are involved during the CO_2 reduction reaction, and due to the presence of competitive pathways to such a reaction, it is imperative to develop catalysts that are precise and purposeful. Recent advances show that catalytic performance hinges on both design and synthesis. While structural concepts such as bimetallic alloys, core-shell architectures, and oxide-supported systems offer the blueprint, it is synthesis methods like the modified polyol and sol-gel routes that bring these ideas to life with tunable morphology, particle size, and surface chemistry. Figure 16.1 shows a core-shell Pt-based bimetallic nanocatalyst designed for CO_2 electroreduction. Charge transfer from the core (e.g., Ni, Fe, or Co) to the Pt shell enhances CO_2 adsorption and stabilizes intermediates like *COOH, promoting CO formation. This structure improves both selectivity and catalyst stability.

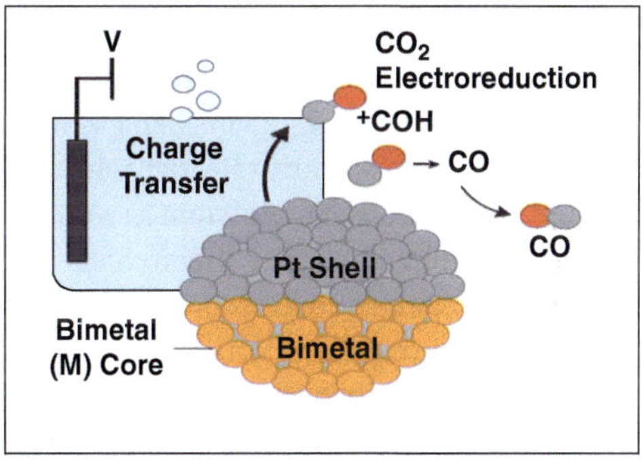

Figure 16.1 Core-shell Pt-based bimetallic nanocatalyst showing charge transfer and intermediate formation during CO_2 electroreduction.

16.3.1 Single Metals vs. Bimetallic and Multimetallic Systems

Early studies in CO_2RR emphasized monometallic catalysts—Au, Ag, Cu, Sn, and In—each with unique selectivity profiles. For instance, gold and silver predominantly produce CO, while tin and indium favor formate. Copper remains the only pure metal known to produce multicarbon products like ethylene or ethanol [21]. Yet, despite their simplicity, monometallic catalysts are often constrained by high overpotentials and poor long-term durability.

In contrast, bimetallic and multimetallic catalysts provide a more flexible platform. Through alloying or phase separation, they offer tunability in intermediate binding energies, improved selectivity, and, in many cases, lower Pt usage. For example, Pt–Ni, Pt–Co, and Pt–Fe alloys leverage the synergy between noble and base metals to modulate electron density and optimize reaction pathways [22]. Notably, Cu–In and Cu–Sn systems have also shown improved selectivity for formate due to enhanced proton delivery and intermediate stabilization [23]. Multimetallic systems such as Pt–Fe–Ni or Cu–Ag–In further expand the design landscape. These materials mimic enzyme-like complexity, creating microenvironments with tailored reactivity. However, their synthesis requires advanced strategies that allow simultaneous control over multiple elements, crystal phases, and surface distributions—something conventional synthesis methods often struggle to achieve.

16.3.2 Synthesis: Polyol and Sol-Gel Routes for Tailored Nanostructures

The modified polyol technique has also become the benchmark technique for the synthesis of high-quality metallic and bimetallic nanoparticles. In this case, polyhydric alcohols (e.g., ethylene glycol) are used as a solvent, as well as a reducing agent. Metal precursors under regulated heating are reduced into zerovalent atoms and are further reduced into nucleated nanoparticles. The size, morphology, and even facets of the nanoparticles can be tuned by the researchers by manipulation of various parameters like temperature, PH, and precursor concentration, as well as using shape-directing agents like PVP or capping molecules [24]. Such a strategy of synthesis is particularly effective when it is used in complex systems. As just one example, a Pt and Ni co-reduction in a polyol system can result in either homogeneous composition alloying or hierarchical core-shell nanoparticle structure, depending on both kinetic control and the reactivity of the precursor starting materials [25]. Similarly, oxide supports (e.g., spinels and perovskites) with high surface area and defined porosity can be generated using the sol-gel route. These catalysts can then, in turn, be utilized to fix (immobilize) metal nanoparticles, thus giving rise to hybrid catalysts that have strong MSIs.

16.3.3 Influence of Core-Shell Architectures

The architectures studied are numerous, yet core-shell nanostructures can be mentioned to have a balance between cost and performance. The thin shell is a noble metal (usually Pt), and the cheaper, more abundant metal core can be something like Ni or Fe. This scheme minimizes Pt consumption and allows taking advantage of the electronic properties of the core to tune the shell surface properties. Transferring electrons across the core shell interface can transform the d-band center of the shell metal, shifting binding energies and accelerating reaction kinetics. On top of that, lattice mismatch-induced strain effects might also add to the property of catalytic activity [26]. As an illustration, Ni@Pt core-shell nanoparticles prepared by the polyol method demonstrate improved activity and durability, in particular, due to correspondingly precise control over the thickness of shells, achieving the optimal balance between structural stability and electronic connections [23]. Such nanostructures have also been confirmed by the use of electrochemical techniques, such as cyclic voltammetry (CV), CO stripping, and electrochemical surface area (ECSA) measurements, which imply they are impervious to deactivation and favorable CO_2RR routes [25].

16.3.4 Oxide Support Strategies (Perovskites, Ferrites, and Beyond)

Catalyst performance does not depend on active metal alone—the support material plays a vital, often underappreciated, role. Oxides such as $LaNiO_3$ (perovskites), $CoFe_2O_4$ (spinels), and CeO_2 (ceria) are more than scaffolds; they stabilize metal particles, improve charge transport, and create MSIs that boost activity and selectivity. These supports can introduce oxygen vacancies, enhance intermediate adsorption, and suppress unwanted reactions like HER. For instance, Pt supported on $LaNiO_3$ has shown improved CO_2RR performance due to stronger electronic interactions and better intermediate stabilization [27]. Similarly, $CoFe_2O_4$ can enhance dispersion and enable magnetic recovery of spent catalysts—key for industrial reusability [28].

16.3.5 Case Examples: Bringing It All Together

Several recent studies highlight how synthesis and design go hand-in-hand:

- Using the polyol method, Al Harthi et al. [29] created core-shell particles with tunable thickness, demonstrating excellent CO_2RR selectivity and structural integrity.
- Orlić et al. [30] reviewed how co-reduction synthesis methods yield alloyed systems with long-term stability in acidic media, selectively producing CO or formate.
- The integration of Pt nanoparticles on oxide supports via polyol routes yielded hybrid systems that showed enhanced durability, oxygen vacancy generation, and improved electron conductivity [31].

16.4 Electrocatalytic Performance in CO_2 Conversion

16.4.1 Faradaic Efficiency and Benchmarking: Pt vs. Pt-Alloys

Faradaic efficiency measures how effectively electrical energy is used to produce a specific product. In CO_2RR, achieving high FE for targeted products such as CO, formate ($HCOO^-$), CH_4, or C_2H_4 remains challenging. This is primarily due to the complexity of reaction pathways and the ever-competing HER—especially on catalysts like platinum that have a strong affinity for proton reduction. Pure Pt, although exceptional for HER, is relatively poor at reducing CO_2. In aqueous electrolytes, the majority of charge on Pt is consumed in hydrogen generation, leading to CO_2RR faradaic efficiencies typically below 10% [32]. For this reason, researchers have shifted attention toward Pt-based alloys, where the introduction of transition metals like Ni, Fe, or Co alters Pt's surface electronic properties and reduces its hydrogen-binding strength. Take, for example, Pt–Ni catalysts. These materials exhibit Faradaic efficiencies exceeding 40% toward CO, a dramatic improvement over pure Pt, while simultaneously suppressing HER [33]. The alloying process adjusts the d-band center of Pt, weakening hydrogen adsorption and making room for *COOH intermediate formation—a key step in CO generation. Other bimetallics like Pt–Fe and Pt–Co demonstrate similar enhancements in selectivity and electrochemical activity.

What truly underscores the value of alloying is revealed through performance benchmarking. Studies comparing monometallic Pt and Pt-alloys consistently show that the latter outperform in nearly every key metric—higher current density, better selectivity, more favorable onset potentials, and improved tolerance to deactivation. For instance, Pt–Fe nanoparticles synthesized via the modified polyol method have been shown to exhibit nearly three times the current density of commercial Pt/C, with significantly better stability under continuous operation [34]. Further, core-shell architectures, such as Ni@Pt, take this a step further by localizing Pt atoms on the surface while modifying their electronic behavior through subsurface alloying. These designs optimize both performance and material efficiency, reducing precious metal usage without sacrificing catalytic effectiveness [35].

These benchmarking results have been confirmed across a variety of experimental conditions—H-type cells, flow cells, gas diffusion electrodes—using both neutral and alkaline electrolytes. In nearly all cases, Pt-alloys not only improve reaction kinetics (evidenced by lower Tafel slopes) but also offer greater resistance to CO poisoning, a common degradation pathway for CO_2RR catalysts.

16.4.2 Role of Oxide Supports in Enhancing CO_2RR

Beyond alloying, another way to improve electrocatalytic performance is by engineering the support material. In heterogeneous catalysis, supports are no longer passive frameworks; they are active players. Oxides like $CoFe_2O_4$, $LaNiO_3$, and CeO_2 enhance performance through MSIs that impact charge transfer, electron density, and active site accessibility. For example, Pt nanoparticles supported on spinel ferrites like $CoFe_2O_4$ exhibit better CO_2 adsorption and improved product selectivity due to electron exchange at the metal oxide interface [36]. Such supports often have high surface area, excellent redox properties, and the ability to generate oxygen vacancies—all of which contribute to higher catalytic activity and better product control. Perhaps more importantly, oxide supports can suppress HER, particularly in aqueous systems. They achieve this either by lowering local proton availability or by stabilizing CO_2RR intermediates preferentially over hydrogen. For instance, when Pt is combined with $LaNiO_3$ perovskite supports, faradaic efficiencies above 60% for CO or formate production have been recorded, with consistent performance over extended cycling [37]. These supports also offer material advantages: improved thermal and chemical stability, reusability, and in some cases, magnetic separability for catalyst recycling. Some studies have even found that oxide components influence product selectivity, tilting the balance between CO and $HCOO^-$ depending on their surface acidity and coordination geometry [38].

16.5 Catalyst Stability and Deactivation Pathways

16.5.1 CO Poisoning and Structural Degradation

One of the most notorious culprits behind catalyst deactivation in CO_2RR is carbon monoxide (CO) poisoning. CO is not only a valuable product in its own right—it is also a reaction intermediate in most CO_2 reduction pathways. Unfortunately, it binds very strongly to the surface of noble metals such as Pt, occupying active sites and preventing further CO_2 adsorption and conversion [38]. This phenomenon is especially pronounced under mildly reducing conditions, where CO desorption is kinetically slow. Bimetallic catalysts, especially Pt–Ru, offer a clever workaround. Here, ruthenium promotes the formation of OH species at lower potentials, which in turn helps oxidize adsorbed CO into CO_2, effectively regenerating the Pt surface. This is known as the bifunctional mechanism, and it has been instrumental in improving catalyst longevity, especially in low-temperature fuel cells and CO_2RR systems [39]. Besides poisoning, structural degradation is another persistent issue. Nanoparticles, especially those below 5 nm, are thermodynamically unstable. Over time, they tend to undergo Ostwald ripening, where smaller particles dissolve and redeposit onto larger ones, leading to a loss in surface area [40]. There is also the potential of the alloyed transition metals, such as Ni or Fe, to be leached out at highly reducing or acidic/alkaline conditions, leading to the collapse of the structure and inactivity [41]. Another process, which is less dramatic but also destructive, is surface restructuring, or the modification of morphology or crystallographic facets of a nanoparticle. The selectivity is lost and performance is compromised long-term when catalytically active facets are lost or reshaped going into operation [42].

16.5.2 Strategies for Enhancing Durability

To address these stability challenges, researchers have devised several approaches grounded in materials science, surface chemistry, and electrochemical engineering.

a. **Alloying and Core-Shell Design**
   Addition of Pt to transition metal additions (e.g., Ni, Co, Fe) not only modifies the electronic structure but also improves mechanical and chemical durability. To give an example, Ni@Pt core-shell nanoparticles prepared under the polyol process presented greater CO tolerance and structural integrity, which is attributed to lattice strain and to the charge redistribution at the core-shell interface [43]. Such designs help in the increase in the use of Pts, which are economical to use and also environmentally friendly.

b. **Use of Oxide Supports**
Stability is significantly improved by anchoring nanoparticles on chemically robust oxide supports, such as $CoFe_2O_4$, CeO_2, or $LaNiO_3$. These supports help prevent nanoparticle migration and sintering while providing additional functionality—such as oxygen vacancies and redox buffering—that can facilitate intermediate stabilization and CO_2 activation [44]. They also act as physical barriers that prevent metal leaching and aggregation.

c. **Surface Modification and Functional Layers**
Another approach involves surface functionalization, where ligands, surfactants, or atomic layers are used to selectively passivate unstable sites or block poisoning species. For example, plasma-treated Pt surfaces or self-assembled monolayers (SAMs) can enhance CO_2 selectivity and reduce CO binding without sacrificing activity [45].

d. **Electrochemical Activation and Regeneration**
Pretreatment through CV or potential cycling can activate the catalyst surface by removing weakly bound impurities and reorganizing surface atoms into more stable configurations. Some bimetallic systems even show self-healing behavior, where surface restructuring under electrochemical stress leads to the reformation of active sites [46].

e. **Defect and Vacancy Engineering**
Intentional introduction of oxygen vacancies or metal defects in the catalyst or its support can enhance durability by facilitating charge transfer and reducing susceptibility to structural collapse. In Pt-oxide composites, such defects often serve as active anchoring points that improve both electronic coupling and catalyst dispersion [47].

16.6 Synergistic Effects in Bimetallic/Oxide Systems

16.6.1 Charge Redistribution and Orbital Interactions

When two or more metals are combined at the nanoscale, their electronic environments change dramatically. One of the most impactful effects is charge redistribution, where electrons are transferred between different metals due to differences in electronegativity, oxidation potential, and coordination environment. For example, in Pt–Ni or Pt–Fe bimetallic catalysts, electrons typically flow from the less noble metal (like Ni or Fe) toward platinum. This shift modifies the d-band center of Pt, weakening its interaction with reaction intermediates like CO and promoting faster desorption—critical for avoiding CO poisoning, a common issue in CO_2RR [48]. This effect directly enhances selectivity for CO_2-derived products over unwanted hydrogen evolution.

Orbital hybridization further complements this behavior. In Pt–Co or Pt–Ni alloys, the d-orbitals of both metals overlap, creating new energy states that facilitate multi-electron transfer processes needed for CO_2RR. As shown by Do and Lee [49],

Pt_3Ni and Pt_3Co surfaces display stronger binding to key intermediates like *COOH or *CHO without leading to excessive stabilization—a delicate balance for efficient catalysis. From a practical standpoint, these synergistic effects allow researchers to fine-tune catalysts without needing exotic elements. It's a shift toward intelligent design: using common transition metals to engineer high-performance, affordable systems that still benefit from the catalytic excellence of platinum.

16.6.2 Experimental Validation and DFT Insights

On the one hand, theoretical projections are effective; on the other hand, experimental data are the driving force in catalysis studies. A plethora of literature supports the argument that Pt-based multimetallic nanoparticles are more useful than Pt monometallic catalysts on the basis of activity and stability. In a recent study by Hang et al. [43], Pt–Ni and Pt–Fe nanoparticles synthesized via the modified polyol method showed superior resistance to CO poisoning and enhanced ECSA. This was linked to atomic-scale charge redistribution observed via X-ray photoelectron spectroscopy (XPS) and transmission electron microscopy (TEM). Notably, Pt 4f peaks in XPS are shifted in binding energy, suggesting an electron-rich Pt environment due to alloying with Ni or Fe.

In another example, Pt@Pd core-shell structures displayed remarkable CO_2RR performance. DFT calculations revealed that the lattice strain and orbital mixing between core and shell atoms optimized the energy levels for intermediate adsorption and product desorption [50]. These results were supported by higher-resolution TEM and CV techniques, in which there was increased activity and stability as compared with traditional Pt catalysts. Furthermore, MSIs of oxide-supported catalysts provide an additional dimension of synergy (Table 16.1). In $Pt/CoFe_2O_4$ or $Pt/LaNiO_3$ systems, the oxide not only stabilizes the metal nanoparticles but also donates electrons or creates oxygen vacancies that serve as secondary active sites. These supports can reduce activation energy for CO_2 reduction, enhance CO_2 adsorption, and improve resistance to deactivation [51]. Density functional theory (DFT) simulations have been instrumental in predicting how bimetallic systems shift the binding energies of key intermediates. These simulations often calculate the corrected Gibbs free energy of adsorption and reaction steps using the expression:

$$\Delta G = \Delta E + \Delta \text{ZPE} - T\Delta S + \Delta G_{\text{solvation}} + \Delta G_{\text{field}} \tag{2}$$

Where:

- ΔG: Gibbs free energy change
- ΔE: reaction energy from DFT
- ΔZPE: change in zero-point energy
- T: temperature
- ΔS: change in entropy
- $\Delta G_{\text{solvation}}$: solvation correction
- ΔG_{field}: correction for external electric field

Table 16.1 Bimetallic and oxide-supported catalysts, their synergistic mechanisms, and corresponding performance indicators.

Catalyst system	Synergy type	Key effect	Faradaic efficiency	References
Pt–Ni alloy	Charge redistribution	Improved *COOH binding and reduced HER	~40–55% for CO	Gao et al. [52]
Pt–Fe alloy	Orbital hybridization	Enhanced *COOH stability	~47% for formate	Sui et al. [53]
Pt@Pd core-shell	Strain + orbital coupling	Tuned d-band and lower overpotential	50–60% for CO	Li et al. [54]
Pt/$CoFe_2O_4$ hybrid	MSI	Oxygen vacancy generation and MSI effect	65% for CO/formate	Luo et al. [55]
Pt/$LaNiO_3$ composite	Electronic coupling	Higher ECSA and suppressed CO adsorption	>70% CO at −0.8 V	Nasim et al. [56]

16.7 Applications and Integration in Energy Systems

16.7.1 Hybrid Fuel Cell–CO_2RR Systems

One of the more innovative directions in renewable energy technology is coupling CO_2 electroreduction (CO_2RR) with fuel cell systems. In such hybrid devices, CO_2 is first reduced to fuels like CO, methane, or methanol, which are then used to power proton exchange membrane fuel cells (PEMFCs) or direct methanol fuel cells (DMFCs). The cycle is clean, closed, and may be highly efficient. An example is that a Pt-based bimetallic catalyst like Ni_3Pt or Pt-Fe is synthesized using the modified polyol method that can allow high CO_2RR activity and also fit perfectly into the fuel cell structures [57]. The materials provide tunable surface electronic states, enhancing CO_2 activation with improved minimization of the HER, which is otherwise competing under comparable potentials [58].

This synergy is not just theoretical. Studies have shown that Pt-based alloys on carbon supports or embedded in oxide matrices (e.g., Pt/$CoFe_2O_4$) can function as dual-purpose electrocatalysts—working on the cathode side of a CO_2RR electrolyzer and as the anode in a PEM fuel cell [59]. This kind of electrochemical integration drastically improves energy efficiency and material utilization.

A representative cycle can be outlined as:

$$CO_2 + H_2O \rightarrow \text{(electrocatalysis)} \rightarrow CH_3OH \text{ or}$$
$$CO \rightarrow \text{(fuel cell)} \rightarrow H_2O + \text{electricity}$$

This model offers carbon-neutral energy storage and release, a concept often referred to as "chemical batteries" or CO_2-to-fuel loops.

Figure 16.2 shows how CO_2 electroreduction integrates with renewable systems like artificial photosynthesis and hybrid fuel cells. Sunlight drives water splitting, and the generated electrons reduce CO_2 at a Pt-based catalyst. The resulting CO powers a fuel cell, creating a closed-loop, sustainable energy cycle.

16.7.2 Smart Electrocatalysts for Artificial Photosynthesis

Beyond hybrid devices, another area where Pt-based nanocatalysts are thriving is artificial photosynthesis—systems designed to mimic the natural conversion of sunlight, water, and CO_2 into energy-rich compounds (Table 16.2). In these systems, catalysts must perform under fluctuating light intensity, temperature, and electrolyte conditions. Researchers have responded by designing smart electrocatalysts, which dynamically adjust their electronic structure or catalytic behavior depending

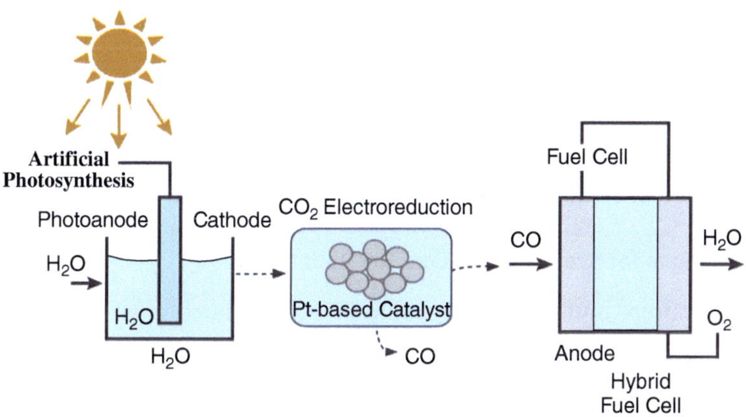

Figure 16.2 Integration of CO_2 electroreduction with renewable energy systems.

Table 16.2 Integrated applications of Pt-based nanocatalysts in CO_2-to-energy systems.

Application strategy	Catalyst system	Product focus	Key advantages	References
Hybrid fuel cell–CO_2RR loop	Pt–Ni alloy, Pt–Fe/C	CO, CH_4, and CH_3OH	Energy recycling, high selectivity	Ross et al. [63]
Artificial photosynthesis cell	Pt/$LaNiO_3$ perovskite	CO, HCOOH	Light-enhanced performance	Dogutan and Nocera [64]
Tandem PEC-electrocatalysis	Pt@Pd on TiO_2	CH_3OH, CO	Dual-function photoelectrocatalysis	Tang et al. [65]
Fuel-cell compatible catalyst	Pt/$CoFe_2O_4$ nanohybrids	H_2 + CO utilization	Long-term durability and low Pt loading	Islam et al. [66]

on external stimuli. For instance, core-shell Pt@Pd structures or Pt–Ni bimetallic particles on perovskite oxide supports have shown high solar-to-fuel conversion efficiency and stability under simulated solar irradiation [60].

In one study, Pt nanoparticles anchored on $LaNiO_3$ perovskites exhibited stable CO production at Faradaic efficiencies exceeding 60%, even under light-assisted electrochemical operation [61]. The oxide support not only stabilized the nanoparticles but also enhanced charge separation and CO_2 adsorption capacity, crucial for sustained operation.

These artificial photosynthetic platforms often operate in PEC tandem cells, where one half-cell performs light-driven water oxidation and the other half reduces CO_2 to fuels such as formate or methanol.

A simplified reaction pathway is:

$$CO_2 + 2H^+ + 2e^- \xrightarrow{\text{Pt-based catalyst}} CO + H_2O \text{ or,}$$
$$CO_2 + 6H^+ + 6e^- \rightarrow CH_3OH + H_2O$$

These reactions are enabled at potentials as low as −0.3 V vs. RHE when using optimized Pt-alloy systems with tailored d-band centers [62].

16.8 Future Perspectives and Research Roadmap

16.8.1 Challenges for Scale-Up and Commercialization

Scaling a catalyst from milligram batches in a lab to kilograms in an industrial reactor is far from trivial. Many of the synthesis methods used for Pt-based nanocatalysts—such as the modified polyol route—require tight control over temperature, pH, precursor purity, and reducing atmosphere [43]. Replicating such precision in large-scale reactors remains a significant barrier.

Moreover, cost is a substantial limitation. Platinum, while an exceptional catalyst, remains expensive and scarce. Strategies such as core-shell architectures (e.g., Ni@Pt) can reduce Pt loading, but the scalability of such precise structures is still technically challenging [26]. Furthermore, equipment degradation in industrial CO_2RR systems due to CO poisoning, corrosion, and fluctuating electrochemical conditions must be considered during catalyst design and engineering [33].

Another hurdle is standardization. There is a pressing need for globally accepted metrics and protocols for benchmarking CO_2RR catalysts. For example, variations in electrolyte composition, cell geometry, and gas diffusion layer materials often make performance comparison difficult across different research groups [67]. Table 16.3 summarizes some of the main scale-up obstacles.

16.8.2 AI/ML-Driven Catalyst Discovery and Process Optimization

As experimental bottlenecks persist, artificial intelligence (AI) and machine learning (ML) are emerging as powerful tools in the search for next-generation catalysts. These tools allow researchers to rapidly screen thousands of bimetallic

Table 16.3 Some of the main scale-up obstacles.

Challenge	Description	Potential mitigation	References
Cost and Pt scarcity	The high cost of platinum impedes large-scale adoption	Alloying, core-shell strategies, and non-Pt substitutes	Ren et al. [68]
Synthetic reproducibility	Lab conditions are challenging to replicate at scale	Continuous-flow synthesis and green chemistry routes	Lovato et al. [69]
Product separation	Gaseous/liquid products require complex purification	Integration with downstream membrane technologies	Vermaak et al. [70]
Long-term stability	CO poisoning and sintering during prolonged use	Surface modification, oxide supports, and in-situ regeneration	Liu et al. [71]

combinations, simulate surface interactions, and even predict catalytic performance based on structural features—all at a fraction of the cost and time of traditional methods. For instance, ML algorithms trained on DFT-calculated adsorption energies have been used to predict *CO binding energies across hundreds of metal alloy surfaces, guiding researchers toward optimal catalyst compositions [42]. Neural networks and genetic algorithms can even propose novel core-shell or high-entropy alloy structures that exhibit favorable energy profiles for CO_2RR [18]. The experimental study that was followed up with the use of AI revealed that Pt-Co-Zn ternary alloys had better CO selectivity due to synergetic d-band interactions and strain effect on the surface [72]. In the absence of AI, such an attestation could have been overlooked, given the mystifying permutations of possibilities. The possibility does not stop at the design of catalysts. The AI-driven control systems can also be helpful in process optimization (e.g., electrochemical cell configuration, control over flow rate, and low-level stability). These intelligent systems can tune themselves in real-time, that is, the current density and the conditions of the reaction can be maximized using real-time feedback.

16.8.3 Sustainability and Life-Cycle Considerations

Designing the perfect catalyst means nothing if it comes at the expense of environmental sustainability. Life cycle assessment (LCA) is essential for ensuring that the carbon footprint of producing and deploying these catalysts is lower than the CO_2 they aim to mitigate. Several concerns arise in this context. First, the use of rare or toxic metals—even in small quantities—raises ethical and environmental issues. Second, the energy-intensive synthesis of some nanocatalysts, including high-temperature annealing or inert atmosphere processing, can offset the carbon gains achieved through CO_2RR unless renewable energy is used. Third, the disposability and recycling of spent catalysts must be addressed. While Pt-based systems are often recyclable, transition metals like Ni or Fe may leach into the environment unless immobilized on stable supports. Researchers are now exploring green synthesis methods, such as aqueous-phase reduction, and using biodegradable surfactants in place of traditional capping agents like PVP [73].

16.9 Conclusion

The challenge of climate change often feels immense, so large that it's easy to be overwhelmed by the scale of the problem. But what this chapter has shown is that science continues to offer real, tangible paths forward. Among these, one of the most promising is the use of nanocatalysts to transform CO_2 from a problematic waste into a valuable resource. It's a shift in perspective—from seeing CO_2 as an end-product of destruction to viewing it as the raw material of a sustainable future. Throughout the chapter, we've explored how bimetallic and oxide-supported nanocatalysts—particularly those based on platinum and its alloys—are reshaping the field of CO_2 electroreduction. These materials go far beyond their monometallic predecessors. Their enhanced selectivity, efficiency, and adaptability arise from carefully tuned structure-activity relationships. From synthesis to surface chemistry, these catalysts are engineered to address long-standing issues like CO poisoning and catalyst degradation under real-world conditions.

Several themes emerged as key takeaways. First, the synergistic behavior of bimetallic systems such as Pt–Ni and Pt–Fe consistently leads to improved catalytic performance: higher activity, better selectivity, and lower operating overpotentials. Second, core-shell designs and oxide-based supports, such as $CoFe_2O_4$ and $LaNiO_3$, not only offer structural integrity but also multifunctional properties that facilitate prolonged CO_2 conversion. Third, the new characterization techniques, which are XPS, TEM, and DFT, allow us to have a better idea of what counts at the atomic scale, and thus our design can be rational, rather than trial-and-error. Fourth, finding ways to use these catalysts in vastly expanded systems, such as in fuel cells or artificial photosynthesis systems, or hybrid systems, will lead to a path toward a circular, carbon-neutral economy. And lastly, we have the emergence of AI and ML; the discovery process itself is changing, and we can move through the catalyst space quicker and smarter than ever before.

Nevertheless, it is not just performance that suffices. We scientists and engineers, along with all those who care about the planet's future, must also consider the practical realities. Is it possible to make these catalysts industrially? Is it possible to produce them ethically and sustainably? Are they going to work in real systems with renewable electricity and in industrial hostile conditions? The ideal future nanocatalyst must fulfill several key requirements: scalability, low cost, long-lasting performance, and sustainability. The future of next-generation CO_2 conversion nanocatalysts is also getting clearer. They will be smarter, programmed by algorithms instead of guessing. They will be developed sustainably and are end-of-life recyclable. They will also be multi-purpose, meaning they can be selectively used to generate fuels and chemicals based on need, geography, and application. Eventually, such nanocatalysts will not only be extensions of materials science. They are instruments of change (not only to structures and energy regimes but to our interaction with carbon itself). We have a long way to go, but every new heartbreak brings us one step closer to the future and makes it a bit more attainable.

References

1 Nunes LJ. The rising threat of atmospheric CO_2: a review on the causes, impacts, and mitigation strategies. *Environments*. 2023;10(4):66.

2 Abdulsalam M, Innocent MO, Livinus MU, Abdulhakeem AI, Danjuma FA, Khadijah A, et al. Community health and vulnerability. In: Babaniyi BR, Babaniyi EE, editors. *The Interplay of Pesticides and Climate Change: Environmental Dynamics and Challenges*. Cham: Springer Nature; 2025. p. 443–67.

3 Aresta M, Dibenedetto A, Angelini A. Catalysis for the valorization of exhaust carbon: from CO_2 to chemicals, materials, and fuels. Technological use of CO_2. *Chem Rev*. 2014;114(3):1709–42.

4 Campbell CT, Sauer J. Introduction: surface chemistry of oxides. *Chem Rev*. 2013;113(6):3859–62.

5 Huang L, Zaman S, Tian X, Wang Z, Fang W, Xia BY. Advanced platinum-based oxygen reduction electrocatalysts for fuel cells. *Acc Chem Res*. 2021;54(2):311–22.

6 Tian H, Wu D, Li J, Luo J, Jia C, Liu Z, et al. Rational design of ternary platinum-based electrocatalysts for effective methanol oxidation reaction. *J Energy Chem*. 2022;70:230–5.

7 Montoya JH, Shi C, Chan K, Nørskov JK. Theoretical insights into a CO dimerization mechanism in CO_2 electroreduction. *J Phys Chem Lett*. 2015;6(11):2032–7.

8 Zhu MN, Zhang BW, Gao MR, Sui PF, Xu C, Gong L, et al. Electrochemically reconstructed perovskite with cooperative catalytic sites for CO_2-to-formate conversion. *Appl Catal B Environ*. 2022;306:121101.

9 Rusdan NA, Timmiati SN, Isahak WN, Yaakob Z, Lim KL, Khaidar D. Recent application of core-shell nanostructured catalysts for CO_2 thermocatalytic conversion processes. *Nanomater*. 2022;12(21):3877.

10 Jouny M, Luc W, Jiao F. General techno-economic analysis of CO_2 electrolysis systems. *Ind Eng Chem Res*. 2018;57(6):2165–77.

11 Hori YI. Electrochemical CO_2 reduction on metal electrodes. In: Vayenas CG, White RE, Gamboa-Adelco ME, editors. *Modern Aspects of Electrochemistry, vol. 42*. New York: Springer; 2008. p. 89–189.

12 Hori Y, Wakebe H, Tsukamoto T, Koga O. Electrocatalytic process of CO selectivity in electrochemical reduction of CO_2 at metal electrodes in aqueous media. *Electrochim Acta*. 1994;39 (11–12):1833–9.

13 Zou X, Gu J. Strategies for efficient CO_2 electroreduction in acidic conditions. *Chin J Catal*. 2023;52:14–31.

14 Schreiber MW. Industrial CO_2 electroreduction to ethylene: main technical challenges. *Curr Opin Electrochem*. 2024;44:101438.

15 Peterson AA, Nørskov JK. Activity descriptors for CO_2 electroreduction to methane on transition-metal catalysts. *J Phys Chem Lett.* 2012;3(2):251–8.

16 Lu Q, Zhao X, Luque R, Eid K. Structure–activity relationship of tri-metallic Pt-based nanocatalysts for methanol oxidation reaction. *Coord Chem Rev.* 2023;493:215280.

17 Feaster JT, Shi C, Cave ER, Hatsukade T, Abram DN, Kuhl KP, et al. Understanding selectivity for the electrochemical reduction of carbon dioxide to formic acid and carbon monoxide on metal electrodes. *ACS Catal.* 2017;7(7):4822–7.

18 Li C, Clament Sagaya Selvam N, Fang J. Shape-controlled synthesis of platinum-based nanocrystals and their electrocatalytic applications in fuel cells. *Nano-Micro Lett.* 2023;15(1):83.

19 Khdary NH, Alayyar AS, Alsarhan LM, Alshihri S, Mokhtar M. Metal oxides as catalyst/supporter for CO_2 capture and conversion, review. *Catalysts.* 2022;12(3):300.

20 Innocent MO, Patrick OI, Job OS, Abdulsalam M, Adams M, Sikirula A, et al. Introduction to Nanotoxicology. In: Isibor PO, Devi G, Enuneku AA, editors. *Environmental Nanotoxicology: Combatting the Minute Contaminants.* Cham: Springer Nature; 2024. p. 1–22.

21 Wu J, Liu N, Li F, Jia B, Zheng J. Iron group elements (Fe, Co, Ni) in electrocatalytic applications: evaluation, characterization and prospects. *Coord Chem Rev.* 2025;525:216343.

22 Zhang S, Zhao S, Qu D, Liu X, Wu Y, Chen Y, et al. Electrochemical reduction of CO_2 toward C_2 valuables on Cu@Ag core-shell tandem catalyst with tunable shell thickness. *Small.* 2021;17(37):2102293.

23 Gong Z, Zhang W, Chen J, Li J, Tan T. Upcycling CO_2 into succinic acid via electrochemical and engineered *Escherichia coli. Bioresour Technol.* 2024;406:130956.

24 Fiévet F, Ammar-Merah S, Brayner R, Chau F, Giraud M, Mammeri F, et al. The polyol process: a unique method for easy access to metal nanoparticles with tailored sizes, shapes and compositions. *Chem Soc Rev.* 2018;47(14):5187–233.

25 Shi Y, Lyu Z, Zhao M, Chen R, Nguyen QN, Xia Y. Noble-metal nanocrystals with controlled shapes for catalytic and electrocatalytic applications. *Chem Rev.* 2020;121(2):649–735.

26 Greeley J, Mavrikakis M. Alloy catalysts designed from first principles. *Nat Mater.* 2004;3(11):810–5.

27 Hwang J, Akkiraju K, Corchado-García J, Shao-Horn Y. A perovskite electronic structure descriptor for electrochemical CO_2 reduction and the competing H_2 evolution reaction. *J Phys Chem C.* 2019;123(40):24469–76.

28 Belgami MA, Patra A, Jeong SM, Rout CS. Exploring the synergy of magnetism and electrocatalysis: a comprehensive review on mechanism, recent developments and future perspectives. *J Mater Chem A.* 2024;12:24005–40.

29 Al Harthi A, Al Abri M, Younus HA, Al Hajri R. Criteria and cutting-edge catalysts for CO_2 electrochemical reduction at the industrial scale. *J CO_2 Util.* 2024;83:102819.

30 Orlić M, Hochenauer C, Nagpal R, Subotić V. Electrochemical reduction of CO_2: a roadmap to formic and acetic acid synthesis for efficient hydrogen storage. *Energy Convers Manag*. 2024;314:118601.

31 Sun Q, Li XH, Wang KX, Ye TN, Chen JS. Inorganic non-carbon supported Pt catalysts and synergetic effects for oxygen reduction reaction. *Energy Environ Sci*. 2023;16(5):1838–69.

32 Marcandalli G, Monteiro MC, Goyal A, Koper MT. Electrolyte effects on CO_2 electrochemical reduction to CO. *Acc Chem Res*. 2022;55(14):1900–11.

33 Zhang C, Liang X, Xu R, Dai C, Wu B, Yu G, et al. H_2 in situ inducing strategy on Pt surface segregation over low Pt doped $PtNi_5$ nanoalloy with superhigh alkaline HER activity. *Adv Funct Mater*. 2021;31(14):2008298.

34 Lima VS, Almeida TS, De Andrade AR. Glycerol electro-oxidation in alkaline medium with Pt-Fe/C electrocatalysts synthesized by the polyol method: increased selectivity and activity provided by less expensive catalysts. *Nanomaterials*. 2023;13(7):1173.

35 Hunt ST, Román-Leshkov Y. Principles and methods for the rational design of core–shell nanoparticle catalysts with ultralow noble metal loadings. *Acc Chem Res*. 2018;51(5):1054–62.

36 Abbas M, Zhang J, Mansour TS, Chen J. Hierarchical porous spinel MFe_2O_4 (M = Fe, Zn, Ni and Co) nanoparticles: facile synthesis approach and their superb stability and catalytic performance in Fischer–Tropsch synthesis. *Int J Hydrogen Energy*. 2020;45(18):10754–63.

37 Inns DR, Mayer AJ, Skukauskas V, Davies TE, Callison J, Kondrat SA. Evaluating the activity and stability of perovskite $LaMO_3$-based Pt catalysts in the aqueous phase reforming of glycerol. *Top Catal*. 2021;64(17):992–1009.

38 Van den Eynden D, Pokratath R, De Roo J. Nonaqueous chemistry of group 4 oxo clusters and colloidal metal oxide nanocrystals. *Chem Rev*. 2022;122(11):10538–72.

39 Álvarez-Gómez JM, Varela AS. Review on long-term stability of electrochemical CO_2 reduction. *Energy Fuels*. 2023;37(20):15283–308.

40 Pattadar DK, Zamborini FP. Effect of size, coverage, and dispersity on the potential-controlled Ostwald ripening of metal nanoparticles. *Langmuir*. 2019;35(50):16416–26.

41 Wang J, Zhang Y, Yu L, Cui K, Fu T, Mao H. Effective separation and recovery of valuable metals from waste Ni-based batteries: a comprehensive review. *Chem Eng J*. 2022;439:135767.

42 Zhao M, Chen Z, Lyu Z, Hood ZD, Xie M, Vara M, et al. Octahedral nanocrystals with a face-centered cubic structure, {111} facets, thermal stability up to 400 °C, and enhanced catalytic activity. *J Am Chem Soc*. 2019;141(17):7028–36.

43 Hang NT, Yang Y, Nam NQ, Nogami M, Phuc LH, Long NV. Pt-based multimetal electrocatalysts and potential applications: recent advancements in the synthesis of nanoparticles by modified polyol methods. *Crystals*. 2022;12(3):375.

44 Das S, Pérez-Ramírez J, Gong J, Dewangan N, Hidajat K, Gates BC. Core–shell structured catalysts for thermocatalytic, photocatalytic, and electrocatalytic conversion of CO_2. *Chem Soc Rev*. 2020;49(10):2937–3004.

45 Hu X, Wang T, Li F, Mao X. Surface modifications of biomaterials in different applied fields. *RSC Adv*. 2023;13(30):20495–511.

46 Gazigil L, Er E, Yonar T. Determination of the optimum conditions for electrochemical regeneration of exhausted activated carbon. *Diam Relat Mater*. 2023;133:109741.

47 Ding Y, Maitra S, Wang C, Halder S, Zheng R, Barakat T. Vacancy defect engineering in semiconductors for solar light-driven environmental remediation and sustainable energy production. *Interdiscip Mater*. 2022;1(2):213–55.

48 Jiao S, Fu X, Huang H. Descriptors for the evaluation of electrocatalytic reactions: d-band theory and beyond. *Adv Funct Mater*. 2022;32(4):2107651.

49 Do VH, Lee JM. Transforming adsorbate surface dynamics in aqueous electrocatalysis: pathways to unconstrained performance. *Adv Mater*. 2025;37(10):2417516.

50 Deng Q, Huang R, Shao LH, Mumyatov AV, Troshin PA, An C. Atomic understanding of the strain-induced electrocatalysis from DFT calculation: progress and perspective. *Phys Chem Phys*. 2023;25(18):12565–86.

51 Zong Y, Chakthranont P, Suntivich J. Temperature effect of CO_2 reduction electrocatalysis on copper: potential dependency of activation energy. *J Electrochem Energy Convers Storage*. 2020;17(4):041007.

52 Gao G, Zhu G, Chen X, Sun Z, Cabot A. Optimizing Pt-based alloy electrocatalysts for improved hydrogen evolution performance in alkaline electrolytes: a comprehensive review. *ACS Nano*. 2023;17(21):20804–24.

53 Sui L, An W, Feng Y, Wang Z, Zhou J, Hur SH. Bimetallic Pd-based surface alloys promote electrochemical oxidation of formic acid: mechanism, kinetics and descriptor. *J Power Sources*. 2020;451:227830.

54 Li C, Yan S, Fang J. Construction of lattice strain in bimetallic nanostructures and its effectiveness in electrochemical applications. *Small*. 2021;17(46):2102244.

55 Luo L, Hernandez R, Zhou XD, Yan H. Heterogeneous catalysis at metal-oxide interfaces using in situ and operando spectroscopy: from nanoparticles to single-atom sites. *Appl Catal A Gen*. 2021;624:118330.

56 Nasim F, Ali H, Nadeem MA. The pronounced effect of cobalt oxide on the electrocatalytic activity of palladium nanoparticles anchored on CoOx/NC towards the ORR with increased MA and ECSA. *Mater Adv*. 2023;4(2):578–85.

57 Abdulsalam M, Mohammed A, Lawal TO. Microbiological insights in petrochemical applications: management across synthetic fuels, pollution, and renewable energy. *Egyp J Pet*. 2025;34(1):8.

58 Cypcar AD, Yang JY. Controlling hydrogen evolution and CO_2 reduction at transition metal hydrides. *Acc Chem Res*. 2024;57(24):3488–99.

59 Song JH, Ka S, Lim C, Han MH, Lee DK, Oh HS. Unveiling the anode reaction environment in a CO_2 electrolyzer to provide a guideline for anode development. *J Mater Chem A*. 2023;11(36):19312–20.

60 Ezendam S, Herran M, Nan L, Gruber C, Kang Y, Gröbmeyer F. Hybrid plasmonic nanomaterials for hydrogen generation and carbon dioxide reduction. *ACS Energy Lett*. 2022;7(2):778–815.

61 Lu Z, Zhou H, Qian B, Wang S, Zheng Y, Ge L, et al. Y and Fe co-doped $LaNiO_3$ perovskite as a novel bifunctional electrocatalyst for rechargeable zinc-air batteries. *Int J Hydrogen Energy*. 2023;48(22):8082–92.
62 Ziehl TJ, Li J, Sun S, Zhang P. New insights into the ORR catalysis on Pt alloy nanoparticles from an element-specific d-band analysis. *J Phys Chem Lett*. 2024;15(32):8306–14.
63 Ross MB, De Luna P, Li Y, Dinh CT, Kim D, Yang P, et al. Designing materials for electrochemical carbon dioxide recycling. *Nat Catal*. 2019;2(8):648–58.
64 Dogutan DK, Nocera DG. Artificial photosynthesis at efficiencies greatly exceeding that of natural photosynthesis. *Acc Chem Res*. 2019;52(11):3143–8.
65 Tang D, Liu J, Zhang X, Chen L, Ma L, Zhang Q. Sacrifice and valorization of biomass to realize energy exploitation and transformation in a photoelectrochemical way. *Green Chem*. 2023;25(20):7843–62.
66 Islam MN, Mansoor Basha AB, Kollath VO, Soleymani AP, Jankovic J, Karan K. Designing fuel cell catalyst support for superior catalytic activity and low mass-transport resistance. *Nat Commun*. 2022;13(1):6157.
67 Athanasaki G, Jayakumar A, Kannan AM. Gas diffusion layers for PEM fuel cells: materials, properties and manufacturing – a review. *Int J Hydrogen Energy*. 2023;48(6):2294–313.
68 Ren X, Lv Q, Liu L, Liu B, Wang Y, Liu A, et al. Current progress of Pt and Pt-based electrocatalysts used for fuel cells. *Sustainable Energy & Fuels*. 2020; 4(1):15–30.
69 Lovato K, Fier PS, Maloney KM. The application of modern reactions in large-scale synthesis. *Nature Reviews Chemistry*. 2021;5(8):546–563.
70 Vermaak L, Neomagus HW, Bessarabov DG. Hydrogen separation and purification from various gas mixtures by means of electrochemical membrane technology in the temperature range 100–160 C. *Membranes*. 2021;11(4):282.
71 Liu J, Shi X, Lv Z, Yu Y, He H. Ceria–tungsten–tin oxide catalysts with superior regeneration capacity after sulfur poisoning for NH3-SCR process. *Catalysis Science & Technology*. 2022;12(8):2471–2481.
72 Zhan W, Ma L, Gan M. Engineering low platinum loaded defects enriched PtxCo wrapped by carbon layers for efficient methanol electrooxidation with CO-free dominant. *Electrochim Acta*. 2023;440:141750.
73 Saha A, Mishra P, Biswas G, Bhakta S. Greening the pathways: a comprehensive review of sustainable synthesis strategies for silica nanoparticles and their diverse applications. *RSC Adv*. 2024;14(16):11197–216.

17

Metal Oxide Nanocatalysts: Eco-friendly Solutions for Catalysis in Green Chemistry

Krishnaveni Manubolu[1] and Raveesha Peeriga[2]*

[1] *Narayana Pharmacy College, Nellore, Andhra Pradesh, India*
[2] *V.V. Institute of Pharmaceutical Sciences, Gudlavalleru, Andhra Pradesh, India*

*Corresponding author: krishnaveni.manubolu@gmail.com

17.1 Introduction

17.1.1 Overview of Green Chemistry and Its Significance

Green chemistry, also known as sustainable chemistry, is an innovative scientific approach aimed at reducing or eliminating hazardous substances in chemical processes. It mainly focuses on designing environmentally benign chemical products and processes that minimize waste, reduce energy consumption, and enhance efficiency. The 12 Principles of Green Chemistry, introduced by Paul Anastas and John Warner, emphasize the importance of sustainability through atom economy, safer solvents, renewable feedstocks, and catalysis [1, 2]. By integrating green chemistry principles, industries can lower their environmental footprint, enhance safety, and contribute to global sustainability goals, such as climate action (SDG 13), clean water (SDG 6), and responsible consumption and production (SDG 12) [3].

Figure 17.1 illustrates the 17 Sustainable Development Goals (SDGs) adopted by the United Nations in 2015 as part of the 2030 Agenda for Sustainable Development. These goals provide a comprehensive framework aimed at ending poverty, promoting health and education, achieving gender equality, ensuring clean water and energy, addressing climate change, and fostering peace and justice. The SDGs are interlinked and serve as a global blueprint for a more inclusive, equitable, and sustainable future for all.

17.1.1.1 Sustainable Production of Metal Oxide Nanocatalysts (MONCs)

Green synthesis techniques for MONCs have become more popular as a result of the drive for environmentally friendly chemical processes. Concerns about the environment

Metal Oxide Nanocatalysts for Sustainable Energy Production, First Edition. Edited by Abayomi Bamisaye et al.
Published 2026 by Wiley-VCH GmbH.

Figure 17.1 The 17 United Nations SDGs.

and human health are raised by the use of hazardous chemicals, organic solvents, and large energy inputs in traditional methods [4]. In contrast, green synthesis makes use of bio-based, renewable materials, including plant extracts, biopolymers (such as starch or chitosan), and industrial or agricultural waste. In mild, environmentally benign settings, they serve as natural reducing and stabilizing agents, facilitating the creation of nanoparticles. In addition to lowering metal ions, phytochemicals such as flavonoids and polyphenols help stabilize and cap the nanoparticles, assisting in the regulation of their size and shape. Furthermore, recycling waste materials into useful precursors encourages waste reduction and circularity. Green synthesis provides scalable, affordable solutions while adhering to the fundamentals of green chemistry, which include minimizing risks, conserving energy, and employing safer materials. In a world where resources are limited, this method is perfect for energy, environmental, and catalytic applications since it promotes cleaner manufacturing.

MONCs offer sustainable solutions for various chemical transformations due to their redox activity, surface tunability, and stability. Catalysts like titanium dioxide (TiO_2), cerium oxide (CeO_2), and MnO_2 effectively promote selective oxidation of alcohols without harsh reagents, while oxide-supported systems enhance hydrogenation reactions under mild conditions. In C—C coupling reactions, such as Suzuki and Heck, these nanocatalysts improve atom economy and product yield. Moreover, oxides like zinc oxide (ZnO) and copper oxide (CuO) enable the conversion of CO_2 into methanol or formic acid, supporting carbon capture efforts. Their reusability and low environmental impact make them ideal for green, cost-effective catalytic applications [5].

MONCs' enhanced catalytic activity in environmentally friendly settings is a result of their redox behavior, surface defects, and nanoscale characteristics. While designed imperfections like oxygen vacancies work as reactive centers that improve molecule adsorption and electron transfer, a large surface area improves the exposure of active sites. Effective catalysis in moderate or solvent-free circumstances is made possible by these flaws, which lower activation energy. Continuous electron cycling is made possible by redox flexibility, such as Ce^{3+}/Ce^{4+} in cerium oxide or Fe^{2+}/Fe^{3+} in iron oxides, which enhances catalytic efficiency and selectivity. These characteristics make it possible to create strong, environmentally friendly nanocatalysts for environmentally friendly chemical reactions.

17.1.2 Role of Catalysis in Sustainable Chemical Processes

Catalysis plays a crucial role in sustainable chemical transformations by enhancing reaction efficiency while reducing energy input and waste generation. In green chemistry, catalysts enable reduction of toxic reagents and by-products, making reactions safer, lower energy consumption by enabling reactions under mild conditions, improved selectivity, reducing unwanted side reactions, and enhanced recyclability, minimizing catalyst loss and waste [6].

Among various catalysts, heterogeneous catalysts, particularly metal oxides, have gained significant attention due to their ability to facilitate reactions on solid surfaces without being consumed, making them reusable and economically viable [6].

Figure 17.2 illustrates how catalysis serves as a cornerstone in promoting sustainable chemical transformations and addressing global climate challenges. Catalysts enhance the efficiency of chemical reactions by lowering energy requirements and increasing selectivity, which directly reduces greenhouse gas emissions. They also enable critical green technologies such as carbon capture and utilization (CCU), green hydrogen production, biomass conversion, and pollution control. By integrating catalysis into industrial and energy systems, we can significantly decrease our environmental footprint; reduce dependence on fossil fuels, and move toward a low-carbon, circular economy essential for climate change mitigation [7, 8].

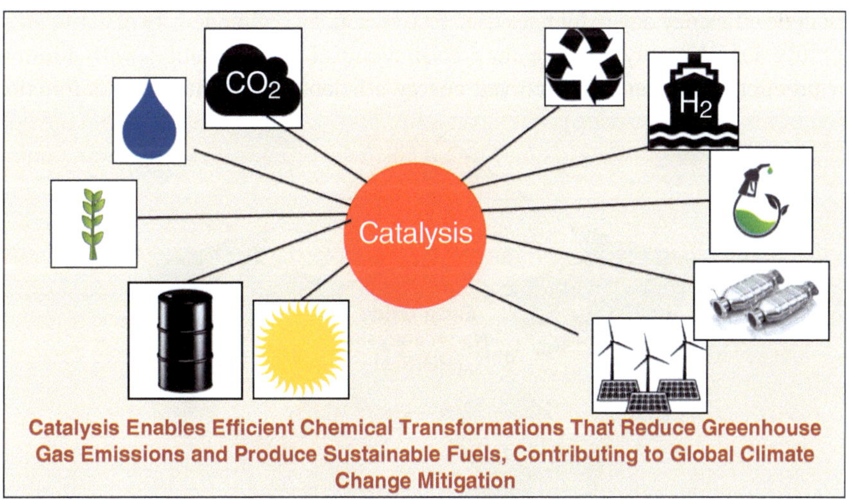

Figure 17.2 Role of catalysis in sustainable chemical processes and climate change mitigation.
Source: mipan / Adobe Stock Photo.

17.1.3 Introduction to MONCs

MONCs are a class of catalysts engineered at the nanoscale, possessing unique physicochemical properties that enhance their catalytic efficiency. These materials include TiO_2, ZnO, CeO_2, and Fe_2O_3, among others. Their nanoscale structure provides increased surface area, leading to higher catalytic activity, improved electron transport, enhanced reaction rates, tunable porosity and stability, making them ideal for multiple catalytic applications. MONCs are widely applied in environmental remediation, renewable energy production, and green industrial processes, contributing to sustainable chemistry innovations.

17.1.4 Advantages of Nanocatalysts Over Conventional Catalysts

Compared to traditional catalysts, nanocatalysts offer several advantages, including:

1. **Higher Catalytic Efficiency:** The increased surface area and active sites enhance reaction speed.
2. **Selectivity and Precision:** Nanocatalysts enable highly selective chemical transformations, reducing unwanted by-products.
3. **Energy Efficiency:** Many MONCs operate under ambient or lower-temperature conditions, reducing energy consumption.
4. **Reusability and Stability:** Metal oxides are generally stable and recyclable, making them cost-effective.
5. **Multifunctionality:** Certain MONCs can serve dual roles, such as photocatalysis and redox catalysis, broadening their applications.

Figure 17.3 radial diagram depicts the core advantages of nanocatalysts—particularly MONCs—in comparison to traditional catalytic systems. At the center lies the nanocatalyst, from which five primary benefits radiate outward: (1) enhanced catalytic efficiency due to increased surface area and a higher density of active sites; (2) superior selectivity, allowing for precise chemical transformations with minimal by-product formation; (3) improved energy efficiency, as many MONCs function

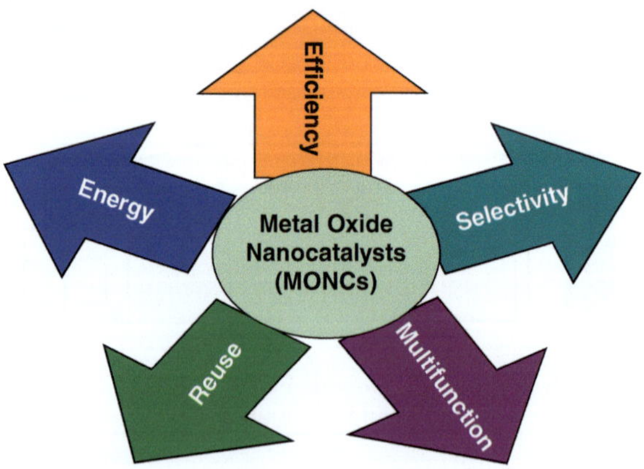

Figure 17.3 Radial illustration of the key advantages of nanocatalysts over conventional catalysts.

effectively under ambient or low-temperature conditions; (4) reusability and thermal stability, supporting multiple catalytic cycles with minimal degradation; and (5) multifunctionality, with select MONCs exhibiting dual catalytic roles, such as in photocatalysis and redox reactions. The radial format underscores the multifaceted performance of nanocatalysts in advancing sustainable and high-performance catalytic technologies.

17.1.5 Scope and Objectives of the Chapter

This chapter explores the design, mechanisms, and applications of MONCs in green chemistry. The key objectives include understanding the mechanistic role of MONCs in catalysis, exploring their environmental applications in pollution control, renewable energy, and sustainable manufacturing, highlighting recent advancements and prospects in MONCs, and discussing challenges, safety concerns, and regulatory aspects of MONC applications.

17.1.5.1 TiO_2, ZnO, and CuO: Key Players in Green Catalysis

MONCs such as "TiO_2, ZnO, and CuO" have emerged as pivotal materials in green chemistry due to their eco-friendly nature, abundance, and versatile catalytic functions. These nanomaterials are widely used in both thermal and photocatalytic applications for environmental remediation and clean energy production. These oxides offer a combination of high surface area, tunable electronic properties, and remarkable chemical stability, making them ideal for processes such as pollutant degradation, CO_2 conversion, and water splitting.

"TiO_2" is extensively studied for its photocatalytic properties under ultraviolet (UV) light, with applications ranging from wastewater treatment to air purification [9]. Its non-toxic nature and resistance to photocorrosion enhance its appeal for long-term environmental applications. ZnO, a wide-band gap semiconductor, similarly acts as a highly efficient photocatalyst, exhibiting strong activity in UV regions with notable applications in antimicrobial coatings, organic dye degradation, and solar energy harvesting. CuO, a p-type semiconductor, stands out for its narrow band gap and visible-light-driven catalytic activity, making it particularly useful for solar-driven chemical transformations and carbon monoxide oxidation. The integration of these oxides with other nanomaterials or through doping techniques can further enhance their performance by improving light absorption, charge carrier mobility, and reaction specificity [10, 11].

Together, these MONCs promote safer, energy-efficient, and environmentally benign chemical processes, aligning with the principles of green chemistry. Their adoption in catalysis not only improves reaction efficiency but also contributes significantly to climate change mitigation and sustainable industrial development.

17.2 Types of MONCs

MONCs have gained significant attention due to their unique structural, electronic, and catalytic properties. Their ability to facilitate various chemical transformations efficiently makes them valuable for environmental remediation, energy conversion, and biomedical applications. Below are some of the most extensively studied MONCs and their key functionalities [12, 13].

17.2.1 TiO_2–Photocatalytic Applications in Environmental Remediation

TiO_2 is one of the most widely used MONCs, particularly in photocatalysis [14, 15]. Its ability to generate reactive oxygen species (ROS) under UV light makes it an excellent candidate for environmental remediation. It has high stability and non-toxicity, strong oxidation capability under UV light, and high surface area and tunable band gap.

Applications:

- **Water Purification:** TiO_2-based photocatalysts break down organic pollutants and microbial contaminants in wastewater [16].
- **Air Pollution Control:** Used in self-cleaning coatings for removing nitrogen oxides (NO_x) and volatile organic compounds (VOCs) [17, 18].
- **Degradation of Persistent Organic Pollutants (POPs):** Effective in decomposing dyes, pesticides, and pharmaceutical residues [19, 20].

17.2.2 ZnO–Antibacterial and Energy Conversion Applications

ZnO is another versatile MONC with applications in antibacterial coatings, energy conversion, and photocatalysis due to its tunable electronic and optical properties.

Key Properties:

- Wide bandgap (~3.37 eV) enabling strong UV absorption.
- Intrinsic antibacterial properties against bacteria and fungi.
- Piezoelectric and optoelectronic properties suitable for energy harvesting.

Applications:

- **Antimicrobial Coatings:** ZnO nanoparticles exhibit strong antibacterial activity, making them suitable for medical devices, packaging, and textiles.
- **Solar Energy Conversion:** Used in dye-sensitized solar cells (DSSCs) and perovskite solar cells for enhanced electron transport.
- **Gas Sensors:** ZnO-based sensors detect harmful gases such as CO, NO_2, and H_2S.

Figure 17.4 is a comparative snapshot of prominent MONCs—TiO_2, ZnO, CeO_2, Fe_2O_3/Fe_3O_4, CuO, and hybrid systems—highlighting their structural features, electronic properties, and multifunctional applications across environmental, energy, and biomedical domains.

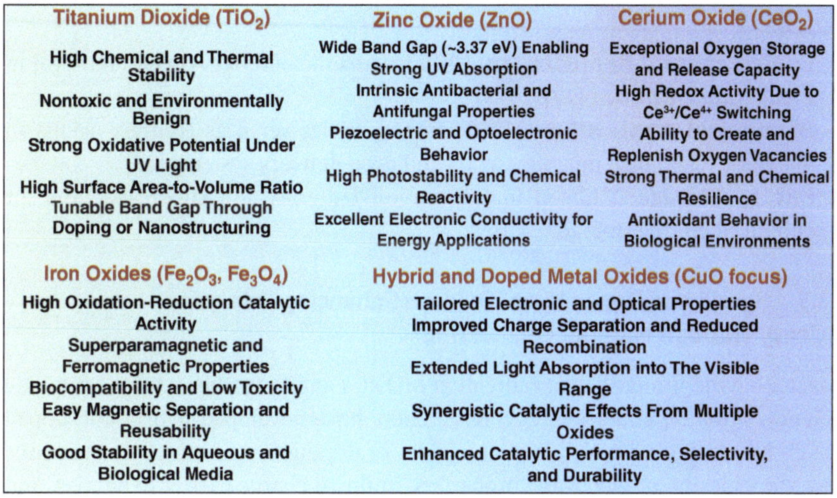

Figure 17.4 Overview of key MONCs and their distinctive characteristics.

17.2.3 CeO_2—Oxygen Storage and Catalytic Efficiency in Redox Reactions

CeO_2, also known as ceria, is a unique nanocatalyst with excellent oxygen storage capacity (OSC), making it a key material in redox catalysis and energy applications [21]. The key properties are rapid oxygen vacancy formation and replenishment, high redox activity, crucial for fuel oxidation and gas conversion, and strong thermal and chemical stability [22].

Applications:

- **Three-way Catalytic Converters (TWC):** Used in automobile exhaust systems to reduce CO, NO_x, and hydrocarbons.
- **Hydrogen Production:** Ceria-based catalysts are used in steam reforming and water-gas shift reactions [23].
- **Biomedical Applications:** CeO_2 nanoparticles exhibit antioxidant properties, reducing oxidative stress in biological systems [24].

17.2.4 Iron Oxides (Fe_2O_3, Fe_3O_4)—Environmental and Biomedical Applications

"Iron oxide nanoparticles," including Fe_2O_3 (hematite) and Fe_3O_4 (magnetite), are widely studied due to their magnetic properties and catalytic efficiency [25]. It has high catalytic activity in oxidation-reduction reactions, magnetic properties enabling easy separation and reusability, and biocompatibility, making them suitable for medical applications [26].

Applications:

- **Environmental Remediation:** Used for arsenic and heavy metal removal in wastewater treatment [27].
- **Biomedical Applications:** Fe_3O_4 nanoparticles serve as contrast agents in magnetic resonance imaging (MRI) and drug delivery systems [28].
- **Energy Storage:** Utilized in lithium-ion batteries and supercapacitors for enhanced charge storage.

17.2.5 Hybrid and Doped Metal Oxides—Enhancing Catalytic Efficiency Through Material Engineering

To overcome the limitations of individual MONCs and to further improve catalytic efficiency, stability, and selectivity, researchers have developed hybrid and doped MONCs by combining multiple metal oxides or doping them with other elements. These strategies tailor electronic properties, improve charge carrier dynamics, and extend light absorption ranges [29, 30].

Key Strategies in Hybrid and Doped MONCs:

- **Doping with Transition Metals (e.g., Fe, Cu, Mn, Ag)**: Incorporation of dopants modifies the electronic structure of metal oxides, enhancing charge separation, reducing recombination of electron-hole pairs, and increasing photocatalytic activity.
- **Composite Formation (e.g., TiO_2-ZnO, CeO_2-Fe_2O_3, CuO-TiO_2)**: Combining two or more metal oxides forms heterojunctions that facilitate efficient charge transfer and improve catalytic performance.

CuO—Key Features and Applications in Hybrid Catalysts:

"CuO" is a p-type semiconductor with a narrow band gap (~1.2–1.9 eV), making it highly effective in visible light-driven photocatalysis and catalytic applications. It is widely used both as a standalone nanocatalyst and as a dopant or component in hybrid metal oxide systems [31, 32]. It shows a narrow band gap, enabling visible light absorption and photocatalytic activity, good electrical conductivity, catalytic reactivity, and strong adsorption capacity for pollutants and gases [33].

Applications:

- **Photocatalysis:** Enhances visible light-driven degradation of organic pollutants when combined with TiO_2 or ZnO.
- **Gas Sensors:** Sensitive to gases such as CO and H_2S, useful in environmental monitoring.
- **Antimicrobial Activity:** Used in coatings and biomedical applications for its antimicrobial properties.
- **Energy Conversion:** Acts as a catalyst or co-catalyst in water splitting and fuel cells.

17.3 Mechanisms of Catalysis in Green Chemistry

Catalysis plays a fundamental role in green chemistry by enhancing reaction efficiency, reducing energy consumption, and minimizing waste production. MONCs exhibit unique catalytic properties due to their high surface area, tunable electronic structure, and nanoscale reactivity. This section explores the basic principles of catalysis, key mechanisms, and factors influencing the catalytic efficiency of MONCs.

17.3.1 Basic Principles of Catalysis

Catalysis involves the acceleration of chemical reactions by a substance (catalyst) that remains unchanged after the reaction. Catalysts provide an alternative reaction pathway with lower activation energy, increasing reaction rates and selectivity. In green chemistry, catalysts are designed to enhance sustainability by reducing hazardous byproducts and energy-intensive conditions. Types of catalysis are shown in Table 17.1.

17.3.2 Surface Area and Active Sites in Nanocatalysts

The catalytic activity of MONCs is strongly influenced by their surface area and active sites, as reactions primarily occur on the catalyst's surface. The key considerations are "high surface area-to-volume ratio" as nano-sized catalysts provide more exposed reactive sites, enhancing catalytic efficiency, defects and vacancies as structural defects (e.g., oxygen vacancies in CeO_2) create highly active catalytic sites, porosity and morphology as nanoporous and hierarchical structures facilitate mass transfer and increase reaction rates.

17.3.3 Electron Transfer Mechanisms

Electron transfer is a critical process in many catalytic reactions involving MONCs shown in Table 17.2. These reactions often proceed via redox mechanisms, where electron flow between reactants and catalysts dictates reaction kinetics.

Table 17.1 Types of catalysis.

Type of catalysis	Description	Examples
Homogeneous catalysis	Catalyst and reactants are in the same phase (usually liquid or gas). Offers high selectivity, but difficult catalyst recovery	Acid/base catalysis in solution; organometallic catalysts in hydroformylation
Heterogeneous catalysis	Catalyst is in a different phase (typically solid) than the reactants. Allows for easy separation, reusability, and industrial scalability	Metal oxides in oxidation/reduction; Ni or Pt on supports for hydrogenation
Photocatalysis	Catalysis initiated by light energy, where photoactive materials absorb photons to drive reactions. Often used for environmental remediation	TiO_2 or ZnO in water purification, solar fuel production

Table 17.2 Key mechanisms in MONCs.

Mechanism	Description	Examples/applications
1. Redox reactions	Involves oxidation-reduction cycles using variable valence states (e.g., Fe^{2+}/Fe^{3+}, Ce^{3+}/Ce^{4+}) to catalyze chemical transformations.	Fuel cells, VOC degradation, and NO_x reduction
2. Charge carrier dynamics	Upon light absorption, metal oxides generate electron-hole pairs that initiate redox reactions. Charge separation and migration are crucial here.	Photocatalytic degradation of dyes and CO_2 reduction
3. Surface adsorption and activation	Reactants are adsorbed on the catalyst's surface, where they are activated via electron transfer or bond polarization.	Catalytic cracking, hydrogenation, and CO oxidation

Table 17.3 Factors affecting catalytic efficiency of MONCs.

Factor	Description	Impact on catalysis
1. Particle size and morphology	Smaller particles offer greater surface area; specific shapes (nanorods, nanotubes) improve interaction and selectivity.	Increases active site availability and enhances reactivity.
2. Stability and aggregation	Structural stability ensures long-term activity; aggregation limits surface exposure.	Stable catalysts remain efficient; aggregation lowers activity.
3. Electronic structure and band gap	Band gap controls photon absorption; it can be tuned by doping or forming composites.	Enhances light absorption and catalytic performance.
4. Surface functionalization and defects	Functional groups improve adsorption; defects like oxygen vacancies enhance reactivity.	Boosts reaction rates and provides active catalytic sites.
5. Environmental conditions	Reaction medium parameters, such as pH, temperature, and solvent, influence kinetics and catalyst lifespan.	Tailored conditions optimize selectivity and durability.

17.3.4 Photocatalysis vs. Heterogeneous Catalysis

Both photocatalysis and heterogeneous catalysis play essential roles in green chemistry but operate via different principles [33]. Photocatalysis involves light-driven activation of catalysts such as TiO_2 and ZnO. It generates reactive species (e.g., hydroxyl radicals, superoxide anions) for pollutant degradation and energy conversion. It is used in water purification, solar fuel production, and antimicrobial coatings whereas heterogeneous catalysis utilizes solid catalysts that interact with gaseous or liquid-phase reactants. Reaction occurs at active sites, with the catalyst facilitating bond breaking and formation. Used in fuel refining, industrial chemical synthesis, and emission control systems.

17.3.5 Factors Affecting Catalytic Efficiency

Several structural and environmental factors influence the performance of MONCs tabulated in Table 17.3.

Understanding the catalytic mechanisms of MONCs provides insights into designing more efficient, selective, and environmentally friendly catalysts. By optimizing structural properties and reaction conditions, MONCs can revolutionize green chemistry applications, enabling sustainable industrial processes and pollution mitigation [34–36].

17.4 Sustainable Manufacturing Processes

MONCs contribute to green manufacturing by minimizing waste, reducing hazardous by-products, and improving process efficiency.

17.4.1 Green Synthesis of Fine Chemicals and Pharmaceuticals

"TiO_2 and ZnO nanocatalysts" facilitate eco-friendly organic transformations, reducing the need for toxic solvents and harsh reagents [37–42]. Metal oxide-supported catalysts (e.g., Fe_2O_3-based catalysts) enable selective oxidation and hydrogenation reactions in pharmaceutical synthesis. CeO_2 and Fe_3O_4 nanocatalysts promote biocompatible and recyclable catalytic processes.

17.4.2 Reduction of Hazardous By-products in Industrial Catalysis

Hybrid metal oxides with engineered surface properties play a crucial role in enhancing selectivity by suppressing unwanted side reactions, thereby improving the overall efficiency of catalytic processes. For instance, CeO_2-based redox catalysts are particularly effective in industrial oxidation reactions, where they minimize the formation of toxic by-products through controlled redox cycling. Additionally, the development of nanocatalysts compatible with green solvents further advances sustainable practices by maintaining high catalytic performance while significantly reducing environmental impact.

17.4.3 Recycling and Reusability of Nanocatalysts

To reduce waste and enhance sustainability, MONCs are designed for high reusability and resistance to leaching or deactivation. Stable catalysts that retain activity over multiple cycles minimize waste generation and environmental impact. Strategies like surface modification, doping, and immobilization improve durability, while magnetic or easily separable catalysts allow for simple recovery and reuse, supporting green and efficient chemical processes.

1. Magnetic Fe_3O_4 nanocatalysts allow easy recovery and reuse in multiple catalytic cycles [43].
2. CeO_2 and TiO_2-based catalysts exhibit high stability and recyclability in industrial processes [44].
3. Encapsulated metal oxides in mesoporous materials enhance durability, maintaining catalytic performance over prolonged use.

The application of MONCs in green chemistry is a promising approach toward achieving sustainability in environmental remediation, energy generation, and industrial manufacturing. By leveraging their superior catalytic properties, these nanomaterials contribute to cleaner, more efficient, and eco-friendly chemical processes. Future advancements in MONC design, surface modification, and hybridization will further enhance their catalytic efficiency while ensuring minimal environmental impact [45].

17.5 Challenges and Limitations

Despite the promising potential of MONCs in green chemistry, several challenges must be addressed to ensure their widespread adoption in industrial and environmental applications [46]. These challenges include stability and aggregation issues, environmental and health concerns, scalability and cost-effectiveness, and regulatory and safety considerations.

17.5.1 Stability and Aggregation Issues

One of the key challenges in the practical application of MONCs is their stability and tendency to aggregate in solution, which can significantly reduce their catalytic efficiency.

Nanoparticle Aggregation: Due to their high surface energy, MONCs tend to aggregate, leading to decreased surface area and active sites. This negatively affects their catalytic performance [46].

Structural Degradation: Some metal oxides undergo phase transformations or dissolution under harsh reaction conditions, reducing their long-term stability.

Surface Passivation: Over time, MONCs may undergo surface oxidation or poisoning, particularly in reactions involving sulfur- or carbon-based compounds, limiting their reusability [47].

Mitigation Strategies: Strategies such as surface functionalization, encapsulation in porous materials, and doping with stabilizing agents can improve the stability and dispersibility of MONCs [48].

17.5.2 Environmental and Health Concerns of Nanoparticles

The increasing application of MONCs in green chemistry raises valid concerns about their potential environmental and human health impacts. Despite their catalytic benefits, certain MONCs like ZnO and TiO_2 are known to generate ROS, which can induce oxidative stress and damage in biological systems. Additionally, their persistence and tendency to bioaccumulate may disrupt aquatic ecosystems and soil microbiota, posing ecological risks. In occupational settings, the inhalation of fine MONC particles can lead to respiratory and systemic health issues, especially where exposure controls are inadequate. To address these challenges, it is essential to develop MONCs that are biodegradable, non-toxic, and environmentally benign [49]. Incorporating comprehensive safety evaluations and life-cycle assessments during their design and application will further help in mitigating potential hazards and ensuring their responsible use in sustainable technologies.

17.5.3 Scalability and Cost-Effectiveness in Industrial Applications

While MONCs offer significant advantages, their commercial viability is often hindered by challenges in large-scale production and economic feasibility.

High Synthesis Costs: The production of high-purity and uniformly sized MONCs involves expensive precursors and complex fabrication techniques [50–51].

Energy-Intensive Processes: Some synthesis methods require high temperatures and vacuum conditions, increasing energy consumption and operational costs.

Reusability and Recycling: Ensuring the efficient recovery and reuse of MONCs in industrial processes remains a challenge, as nanoparticles may lose their activity after multiple cycles.

Mitigation Strategies: Advances in scalable green synthesis methods, such as bio-mediated and sol-gel techniques, can improve cost-effectiveness and sustainability.

17.5.4 Regulatory and Safety Concerns

The widespread adoption of MONCs in industrial and environmental applications is subject to regulatory frameworks that address their safety and environmental impact.

- **Lack of Standardized Guidelines:** Regulations governing nanomaterial safety and toxicity testing are still evolving, leading to uncertainty in industrial adoption [52].
- **Compliance with Environmental Laws:** MONCs must adhere to global environmental policies, such as "REACH (Registration, Evaluation, Authorization, and Restriction of Chemicals)" and "EPA (Environmental Protection Agency)" guidelines [53].
- **Public Perception and Ethical Concerns:** The lack of public awareness about the risks and benefits of MONCs may hinder their acceptance [54].
- **Mitigation Strategies:** Stronger collaboration between researchers, policymakers, and industry stakeholders is essential to developing clear safety protocols and ensuring responsible nanotechnology development [55].

While MONCs hold great promise in sustainable chemistry, addressing these challenges is crucial for their long-term success. Research efforts should focus on enhancing stability, reducing toxicity, improving cost-efficiency, and ensuring regulatory compliance. By overcoming these limitations, MONCs can achieve their full potential in green chemistry applications, leading to safer and more environmentally friendly industrial processes.

17.6 Future Perspectives and Emerging Trends

The field of MONCs is continuously evolving, with emerging innovations aimed at enhancing their efficiency, selectivity, and sustainability in green chemistry. Future advancements focus on surface modification, bio-inspired approaches, AI-driven optimization, and the development of multifunctional hybrid catalysts [56, 57]. These cutting-edge strategies have the potential to revolutionize catalysis, making MONCs more effective and environmentally friendly.

17.6.1 Innovations in Surface Modification and Functionalization

One of the key strategies to improve MONCs involves surface engineering, which enhances their catalytic activity, stability, and recyclability [58].

- **Doping and Defect Engineering:** Introducing dopants (e.g., transition metals, rare earth elements) and creating oxygen vacancies can significantly improve catalytic efficiency and reaction selectivity.
- **Surface Coatings and Ligand Functionalization:** Functionalizing MONCs with organic ligands, polymers, or biomolecules can increase dispersion, prevent aggregation, and enhance reusability.
- **Heterojunction Formation:** Designing heterostructures (e.g., TiO_2-ZnO, CeO_2-Fe_2O_3) can facilitate synergistic effects that enhance photocatalytic and redox activity.
- **Plasmonic Enhancements:** Incorporating noble metal nanoparticles (e.g., Au, Ag, Pt) can improve light absorption and charge separation, boosting photocatalytic efficiency.

Figure 17.5 gives a comparison of pristine and doped TiO_2 band structures. Doping introduces intermediate energy levels, reducing the band gap from ~3.2 eV to ~2.2 eV and enabling visible light absorption for enhanced photocatalytic activity.

17.6.2 Bio-Inspired and Biomimetic Nanocatalysts

Nature provides efficient catalytic systems, inspiring the design of biomimetic and bio-inspired MONCs for sustainable applications.

- **Enzyme-Mimicking Nanocatalysts:** Certain MONCs exhibit peroxidase-, oxidase-, and catalase-like activity, enabling applications in environmental detoxification and biomedical sensing [59].
- **Bio-Templated Synthesis:** Using biopolymers, bacteria, or plant extracts for nanoparticle synthesis can yield eco-friendly MONCs with controlled morphology and high catalytic efficiency [60].
- **Nanocellulose and Biopolymer Composites:** Hybridizing MONCs with natural biopolymers (e.g., chitosan, cellulose) enhances their sustainability and dispersibility in aqueous environments [61].

Figure 17.6 funnel diagram illustrates the hierarchical stages involved in designing sustainable MONCs inspired by natural systems. Starting from the fundamental inspiration

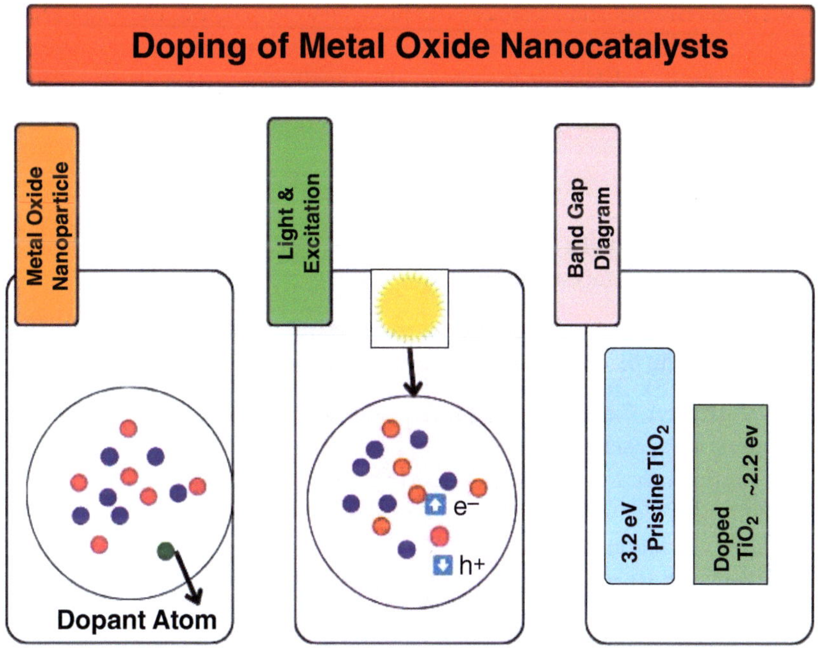

Figure 17.5 Band gap narrowing in TiO_2 via doping.

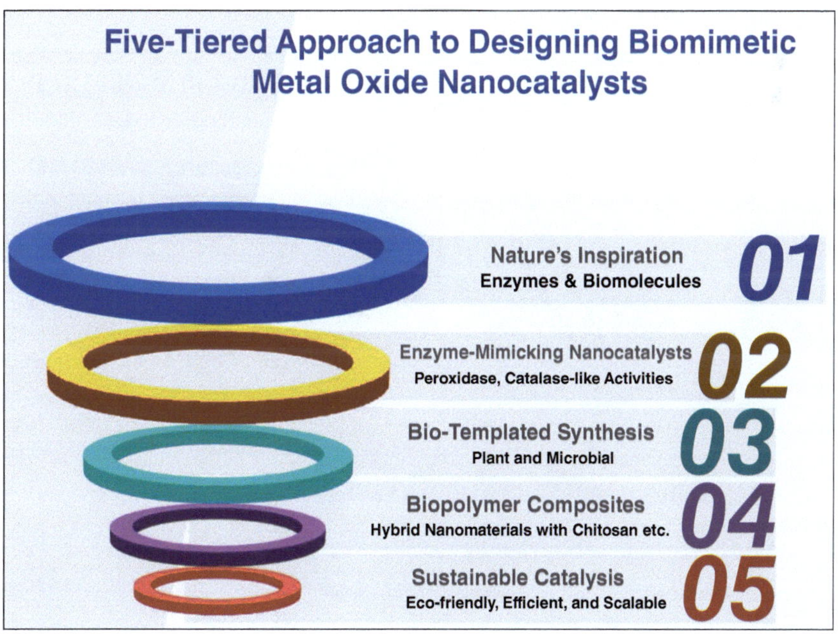

Figure 17.6 Five-tiered framework for bio-inspired MONC development.

derived from enzymes and biomolecules, the framework progresses through enzyme-mimicking activities, bio-templated synthesis methods, integration with biopolymers, and culminates in scalable, eco-friendly catalytic applications. Each tier highlights key strategies that collectively contribute to enhanced catalytic efficiency and environmental sustainability.

17.6.3 Integration with Artificial Intelligence (AI) and Machine Learning (ML) for Catalyst Optimization

"AI and ML" are transforming nanocatalyst research by accelerating material discovery, optimizing reaction conditions, and predicting catalyst performance [62].

- **Data-Driven Catalyst Design:** AI models analyze vast datasets to predict optimal compositions, structures, and reaction conditions for MONCs [63].
- **Computational Screening:** Density functional theory (DFT) and ML algorithms help identify promising catalysts with superior properties before experimental validation [64].
- **Automated Reaction Optimization:** AI-driven high-throughput experimentation and robotic synthesis enable the rapid screening of catalysts for industrial applications [57].

Figure 17.7 illustrates the key roles of AI and ML in advancing MONC research. It highlights the data-driven design of catalyst compositions, computational screening using DFT and ML algorithms, and automated reaction optimization through robotic synthesis and high-throughput experimentation. Together, these approaches

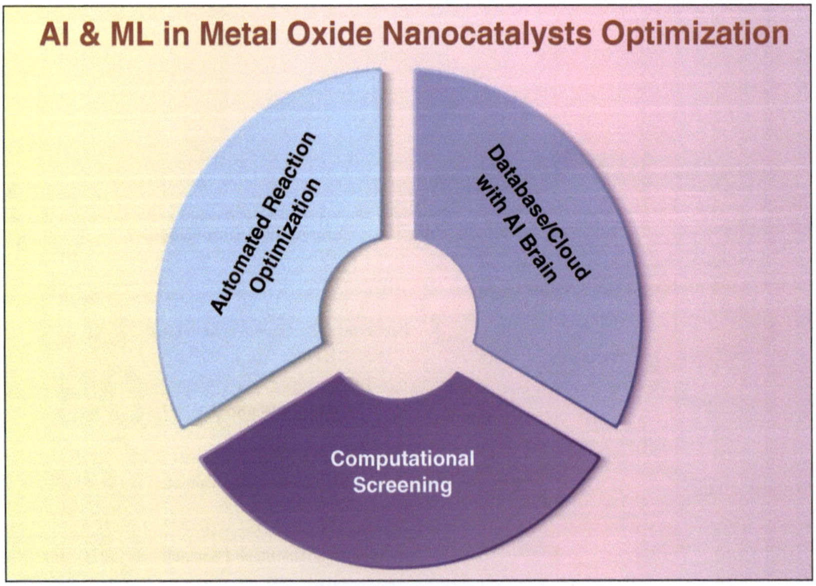

Figure 17.7 Integration of AI and ML in MONC optimization.

accelerate catalyst discovery, improve performance predictions, and enable sustainable industrial applications.

17.6.4 Hybrid and Multifunctional Nanocatalysts for Enhanced Performance

The development of hybrid and multifunctional MONCs is a key trend in achieving enhanced catalytic performance and broader applicability.

- **Dual-Function and Multicomponent Nanocatalysts:** Combining metal oxides with co-catalysts (e.g., noble metals, carbon nanomaterials, or MOFs) enhances catalytic activity, stability, and selectivity [41].
- **Self-Healing and Adaptive Catalysts:** Smart MONCs that self-regenerate active sites under reaction conditions can extend catalyst lifespan and efficiency [65].
- **Magnetic and Light-Responsive Catalysts:** Magnetic MONCs (e.g., Fe_3O_4-based) allow for easy separation and recycling, while plasmonic MONCs enhance solar-driven catalysis [66].

Future research in MONCs is expected to focus on sustainability, efficiency, and intelligent design. By leveraging surface modifications, biomimetic principles, AI-driven discovery, and hybrid materials, MONCs will continue to play a pivotal role in green chemistry and environmental sustainability. These innovations will drive the development of cost-effective, scalable, and environmentally friendly catalytic solutions, paving the way for a cleaner and more energy-efficient future.

17.6.4.1 Emerging Innovations in MONCs for Climate Change Mitigation

17.6.4.1.1 Advanced Synthesis Techniques for Tailored Nanocatalysts

Innovative synthesis methods have significantly improved the structural control and catalytic efficiency of MONCs. Techniques such as hydrothermal synthesis and sol-gel processes enable precise tuning of particle size, shape, and crystallinity, which directly affect catalytic activity. Green synthesis methods using plant extracts or biopolymers have gained attention for their eco-friendly approach, reducing hazardous chemicals and energy consumption during catalyst preparation [67]. For example, TiO_2 nanoparticles synthesized via green methods have shown enhanced photocatalytic degradation of organic pollutants compared to those prepared by conventional routes. Such approaches align well with sustainable chemistry principles, ensuring catalysts are effective and environmentally benign.

17.6.4.1.2 Composite and Hybrid Nanostructures for Enhanced Performance

Coupling metal oxides with carbon-based materials like graphene and carbon nanotubes (CNTs) creates hybrid catalysts with improved electron mobility and reduced charge recombination, essential for photocatalysis under visible light [68]. For instance, TiO_2-graphene composites have demonstrated superior photocatalytic CO_2 reduction rates compared to pristine TiO_2 due to enhanced charge separation and expanded light absorption. Similarly, heterojunctions between ZnO and CuO facilitate efficient charge transfer and broaden the absorption spectrum, improving catalytic efficiency for methane oxidation and greenhouse gas conversion [69]. Such composites are promising for large-scale applications in air purification and renewable fuel production.

17.6.4.1.3 Doping Strategies to Tune Electronic Properties

Elemental doping is a powerful tool to tailor the bandgap and surface properties of MONCs. Transition metal doping (Fe, Mn, Co) and non-metal doping (N, S, P) can narrow the band gap, enabling visible light absorption and enhancing photocatalytic activity [70]. For example, nitrogen-doped TiO_2 nanoparticles exhibit higher activity for CO_2 photoreduction due to improved electron-hole separation and increased surface oxygen vacancies, which serve as active sites [35]. These modifications also enhance catalyst stability, critical for long-term applications in harsh environmental conditions.

17.6.4.1.4 Integration into Multifunctional Systems for Renewable Energy and Environmental Remediation

MONCs have been incorporated into systems for simultaneous pollutant degradation and renewable energy generation. In photocatalytic water splitting, doped TiO_2 and ZnO nanocatalysts have been employed to efficiently generate hydrogen fuel under solar irradiation, presenting a sustainable alternative to fossil fuels [71]. Additionally, MONCs facilitate carbon capture and conversion processes, transforming CO_2 into hydrocarbons or alcohols—clean energy carriers—via photocatalytic or electrocatalytic routes [72]. These multifunctional applications contribute to circular carbon economy models by reducing greenhouse gases while producing valuable fuels.

Figure 17.8 illustrates the dual-role functionality of MONCs in tackling global environmental and energy challenges. By integrating catalytic properties for pollutant

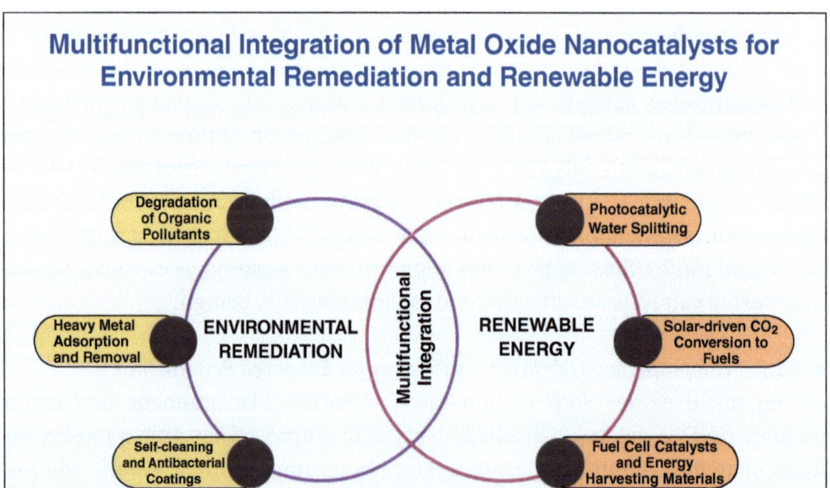

Figure 17.8 Multifunctional applications of MONCs in environmental remediation and renewable energy systems.

degradation with energy conversion capabilities, MONCs enable sustainable solutions like water splitting for hydrogen production and CO_2 reduction into clean fuels. Their multifunctionality enhances efficiency and contributes to circular carbon economy models.

17.7 Scalability and Environmental Compatibility

For practical deployment, MONCs must be cost-effective, scalable, and environmentally friendly. Research focuses on utilizing abundant, non-toxic materials and developing synthesis routes that reduce waste and energy input. Recyclability and catalyst regeneration strategies are also critical for sustainability [73]. For example, ZnO nanocatalysts synthesized from agricultural waste precursors have shown promising scalability and performance in pollutant degradation, demonstrating potential for commercial application. Such approaches ensure MONCs can be integrated into industrial processes without generating secondary pollution, reinforcing their role in sustainable climate solutions [74].

17.8 Key Findings

MONCs, such as TiO_2, ZnO, CeO_2, and Fe_2O_3 have proved to be effective in a broad spectrum of catalytic purposes. Their one-of-a-kind physicochemical characteristics allow them to be used most effectively to support vital processes like treating wastewater, reducing CO_2, cleaning up sweat water, and creating a hydrogen source through a water split, and transforming biomass products into biofuels. Such as, TiO_2 and ZnO, with their high photocatalytic performance in UV or visible light can remove long-sustained organic pollutants in the water, and the super OSC of CeO_2 improves redox reactions that are essential to environmental remediation and power generation. Such a variety of its applications highlights the potential of MONCs in resolving world challenges in clean energy and environmental sustainability.

Exquisite catalytic activity of MONCs has been ascribed to many factors, foremost is the unique combination of the nanoscale size, customizable morphologies, and surface chemistry. The smaller particles can have better surface area-to-volume ratio, and they provide more active sites per unit of catalytic volume. The light absorption, mass diffusion of reactants, and dynamics of charge carriers are further optimized by morphological engineering, that is, synthesizing nanorods, nanotubes, or hollow spheres. Further, surface-functionalization with organic ligands or dopants refines the electronic structure and reactivity and provides sophisticated controls on selectivity and efficiency on complex transformations. These mechanistic insights raise the question as to why, in many catalytic processes, MONCs can be better than their bulk analogues.

Although there are indications that point towards the use of MONCs, there are various issues that restrain its overall usage. Under reaction conditions, stability is continually a question, because with time nanostructures may experience aggregation, leaching, or degradation of structure, and functional durability becomes limited. Moreover, possible environmental toxicity of metal oxide nanoparticles—because of their stability and reactivity, the possible toxicity of metal oxide nanoparticles may challenge the safe use of nanoparticles, particularly in open ecosystems. Lastly, although synthesis and applications conducted at laboratory scale have been encouraging, there are challenges of scaling to industrial level (regarding cost-effective production, uniformity, and possibility of incorporating the projects within the existing infrastructure). The potential of nanocatalysts must be achieved by improving their designs to make them less sensitive, less environmentally hazardous, and create sustainable manufacturing conditions and processes in future studies.

17.9 Potential of MONCs in Green Chemistry

As industries shift towards more sustainable and eco-friendly technologies, MONCs are positioned to play a key role in advancing catalytic efficiency while reducing energy consumption and hazardous waste production. The integration of MONCs with hybrid materials, bio-based catalysts, and AI-driven optimization will further enhance their performance and industrial applicability.

17.10 Future Research Directions for Sustainable Nanotechnology

To maximize the potential of MONCs in green chemistry, future research should focus on:

- Developing more stable and selective nanocatalysts with enhanced recyclability and longevity.
- Exploring bio-inspired and biomimetic approaches for eco-friendly synthesis and functionalization.
- Integrating AI and ML to optimize catalyst design and reaction conditions.
- Assessing environmental and health impacts to ensure the safe and responsible use of MONCs in industrial applications.

With continuous advancements in material engineering, computational modeling, and sustainable synthesis, MONCs are poised to revolutionize the field of green chemistry. By addressing current limitations and exploring emerging trends, these nanocatalysts can significantly contribute to a cleaner, more energy-efficient, and environmentally sustainable future.

17.11 Conclusion

MONCs have emerged as powerful tools in green chemistry, offering high catalytic efficiency, recyclability, and eco-friendly applications. Their role in environmental remediation, renewable energy generation, and sustainable manufacturing highlights their potential to drive cleaner, more efficient chemical processes. By leveraging unique properties such as high surface area, tunable porosity, and enhanced electron transfer capabilities, MONCs contribute to addressing pressing global challenges, including pollution reduction, carbon capture, and sustainable energy production.

References

1 George J, Sreekala MS, Thomas SA. A review on interface modification and characterization of natural fiber reinforced plastic composites. *Polym Eng Sci.* 2001;41:1471.

2 Gholami A, Mohammadi A, Samadi A. Sustainable approaches in natural fiber composites: a review. *Compos Part B Eng.* 2020;182:107674.

3 Faruk O, Bledzki AK, Fink HP, Sain M. Biocomposites reinforced with natural fibers: 2000–2010. *Prog Polym Sci.* 2012;37(11):1552–96.

4 Mohanty AK, Misra M, Drzal LT. Sustainable bio-composites from renewable resources: opportunities and challenges in the green materials world. *J Polym Environ.* 2002;10(1):19–26.

5 Pickering KL, Efendy MGA, Le TM. A review of recent developments in natural fibre composites and their mechanical performance. *Compos Part A Appl Sci Manuf.* 2016;83:98–112.

6 John MJ, Thomas S. Biofibres and biocomposites. *Carbohydr Polym.* 2008;71(3):343–64.

7 Puglia D, Biagiotti J, Kenny JM. A review on natural fibre-based composites – part II: application of natural reinforcements in composite materials for automotive industry. *J Nat Fibers.* 2005;2(3):23–65.

8 Herrera-Franco PJ, Valadez-Gonzalez A. A study of the mechanical properties of short natural-fiber reinforced composites. *Compos Part B Eng.* 2005;36(8):597–608.

9 Sanjay MR, Arpitha GR, Naik LL, Gopalakrishna K, Yogesha B. Applications of natural fibers and its composites: an overview. *Compos Part B Eng.* 2016;91:549–65.

10 Li X, Tabil LG, Panigrahi S. Chemical treatments of natural fiber for use in natural fiber-reinforced composites: a review. *J Polym Environ.* 2007;15(1):25–33.

11 Bledzki AK, Gassan J. Composites reinforced with cellulose based fibres. *Prog Polym Sci.* 1999;24(2):221–74.

12 Thomas S, Paul SA, Pothan LA, Deepa B. Natural fibres: structure, properties and applications. In: Kalia S, Kaith BS, Kaur I, editors. *Cellulose Fibers: Bio- and Nano-Polymer Composites: Green Chemistry and Technology*. Berlin; Heidelberg: Springer; 2011. p. 3–42.

13 Akil HM, Omar MF, Mazuki AAM, Safiee SZA, Ishak ZAM, Abu Bakar A. Kenaf fiber reinforced composites: a review. *Mater Des.* 2011;32 (8–9):4107–21.

14 Alamri H, Low IM. Mechanical properties and water absorption behaviour of recycled cellulose fibre reinforced epoxy composites. *Polym Test.* 2012;31(5):620–628.

15 Kabir MM, Wang H, Lau KT, Cardona F, Chemical treatments on plant-based natural fibre reinforced polymer composites: an overview. *Compos Part B Eng.* 2012;43(7):2883–92.

16 Xie Y, Hill CAS, Xiao Z, Militz H, Mai C. Silane coupling agents used for natural fiber/polymer composites: a review. *Compos Part A Appl Sci Manuf.* 2010;41(7):806–19.

17 Ray D, Sarkar BK, Rana AK, Bose NR. The mechanical properties of vinylester resin matrix composites reinforced with alkali-treated jute fibres. *Compos Part A Appl Sci Manuf.* 2001;32(1):119–27.

18 Mishra S, Mohanty AK, Drzal LT, Misra M, Parija S, Nayak SK, et al. Studies on mechanical performance of biofibre/glass reinforced polyester hybrid composites. *Compos Sci Technol.* 2003;63(10):1377–85.

19 Ramesh M, Palanikumar K, Reddy KH. Plant fibre based bio-composites: sustainable and renewable green materials. *Renew Sustain Energy Rev.* 2017;79:558–84.

20 Dicker MPM, Duckworth PF, Baker AB, Francois G, Hazzard MK, Weaver PM. Green composites: a review of material attributes and complementary applications. *Compos Part A Appl Sci Manuf.* 2014;56:280–89.

21 Zhang J, Xu Q, Feng Z, Li M, Li C. Importance of the relationship between surface phases and photocatalytic activity of TiO_2. *Angew Chem Int Ed.* 2008;47(9):1766–69.

22 Henderson MA. A surface science perspective on TiO_2 photocatalysis. *Surf Sci Rep.* 2011;66 (6–7):185–297.

23 Diebold, U. The surface science of titanium dioxide. *Surf Sci Rep.* 2003;48 (5–8):53–229.

24 Hashimoto K, Irie H, Fujishima A. TiO_2 photocatalysis: a historical overview and future prospects. *Jpn J Appl Phys.* 2005; 44(12R):8269.

25 Chen X, Selloni A. Introduction: titanium dioxide (TiO_2) nanomaterials. *Chem Rev.* 2014;114(19):9281–82.

26 Zhang Z, Wang CC, Zakaria R, Ying JY. Role of particle size in nanocrystalline TiO_2-based photocatalysts. *J Phys Chem B.* 1998;102(52):10871–78.

27 Carp O, Huisman CL, Reller A. Photoinduced reactivity of titanium dioxide. *Prog Solid State Chem.* 2004;32 (1–2):33–177.

28 Ohtani B, Prieto-Mahaney OO, Li D, Abe R. What is Degussa (Evonik) P25? Crystalline composition analysis, reconstruction from isolated pure particles and photocatalytic activity test. *J Photochem Photobiol A Chem.* 2010;216 (2–3):179–82.

29 Hurum DC, Gray KA, Rajh T, Thurnauer MC. Recombination pathways in the Degussa P25 formulation of TiO_2: surface versus lattice mechanisms. *J Phys Chem B.* 2005;109(2):977–80.

30 Luttrell T, Halpegamage S, Tao J, Kramer A, Sutter E, Batzill M. Why is anatase a better photocatalyst than rutile? – model studies on epitaxial TiO_2 films. *Sci Rep.* 2014;4:4043.

31 Scanlon DO, Dunnill CW, Buckeridge J, Shevlin SA, Logsdail AJ, Woodley SM, et al. Band alignment of rutile and anatase TiO_2. *Nat Mater.* 2013;12(9):798–801.

32 Tang H, Prasad K, Sanjinès R, Schmid PE, Lévy F. Electrical and optical properties of TiO_2 anatase thin films. *J Appl Phys.* 1994;75(4):2042–47.

33 Kavan L, Grätzel M, Gilbert SE, Klemenz C, Scheel HJ. Electrochemical and photoelectrochemical investigation of single-crystal anatase. *J Am Chem Soc.* 1996;118(28):6716–23.

34 Hanaor DAH, Sorrell CC. Review of the anatase to rutile phase transformation. *J Mater Sci*. 2011;46:855–74.

35 Li G, Gray KA. The solid-solid interface: explaining the high and unique photocatalytic reactivity of TiO_2-based nanocomposite materials. *Chem Phys*. 2007;339 (1–3):173–87.

36 Zhang J, Xu Q, Zhao X, Zhu Y. Controllable synthesis of anatase TiO_2 nanocrystals with dominant {001} facets and enhanced photocatalytic activity. *Chem Mater*. 2009;21(17):4336–41.

37 Yang HG, Sun CH, Qiao SZ, Zou J, Liu G, Smith SC, et al. Anatase TiO_2 single crystals with a large percentage of reactive facets. *Nature*. 2008;453:638–41.

38 Megarajan S, Shanmugam S. Facile synthesis of anatase TiO_2 nanosheets with exposed (001) facets for enhanced photocatalytic activity. *J Mater Sci Mater Electron*. 2018;29(5):3588–3598.

39 Liu G, Wang L, Yang HG, Cheng HM, Lu GQ. Titania-based photocatalysts – crystal growth, doping and heterostructuring. *J Mater Chem*. 2010;20(5):831–43.

40 Chen H, Nan J, Zhang Q, Li X, Zhao Y. Facile synthesis of TiO_2 hollow spheres with exposed {001} facets and enhanced photocatalytic activity. *CrystEngComm*. 2012;14(19):6201–07.

41 Yu J, Low J, Xiao W, Zhou P, Jaroniec M. Enhanced photocatalytic CO_2-reduction activity of anatase TiO_2 by coexposed {001} and {101} facets. *J Am Chem Soc*. 2014;136(25):8839–42.

42 Prabhu MA, Sheela GE, Alzahrani H, Siddiq H, Okmi A, Alharbi N, Florence SS, Lekshmy MS, Velayuthan Pillai S. Green synthesis of ZnO, Fe_3O_4, and TiO_2 nanoparticles: exploring enhanced bacterial inhibition, catalysis, and photocatalysis for sustainable environmental applications. *Int J Environ Anal Chem*. 2024; 1–25.

43 Li R, Weng Y, Zhou X, Wang X, Mi Y, Chen J. Preparation of anatase TiO_2 hollow spheres with exposed {001} facets and enhanced photocatalytic activity. *ACS Appl Mater Interfaces*. 2014;6(5):3406–12.

44 Etacheri V, Di Valentin C, Schneider J, Bahnemann D, Pillai SC. Visible-light activation of TiO_2 photocatalysts: advances in theory and experiments. *J Photochem Photobiol C Photochem Rev*. 2015;25:1–29.

45 Asahi R, Morikawa T, Ohwaki T, Aoki K, Taga Y. Visible-light photocatalysis in nitrogen-doped titanium oxides. *Science*. 2001;293(5528):269–71.

46 Khan SUM, Al-Shahry M, Ingler WB. Efficient photochemical water splitting by a chemically modified n-TiO_2. *Science*. 2002;297(5590):2243–45.

47 Diker H, Varlikli C, Mizrak K, Dana A. Characterizations and photocatalytic activity comparisons of N-doped nc-TiO2 depending on synthetic conditions and structural differences of amine sources. *Energy*. 2011;36(2):1243–54.

48 Meng A, Zhang J, Xu D, Cheng B, Yu J. Enhanced photocatalytic H2-production activity of anatase TiO2 nanosheet by selectively depositing dual-cocatalysts on {101} and {001} facets. *Appl Catal B Environ*. 2016;198:286–94.

49 Sakthivel S, Kisch H. Daylight photocatalysis by carbon-modified titanium dioxide. *Angew Chem Int Ed*. 2003;42(40):4908–11.

50 Khan SB, Rahman MM, Akhtar K, Asiri AM. Recent development in photocatalytic application of hybrid TiO_2/graphene nanocomposites. *J Ind Eng Chem*. 2015;28:1–15.

51 Yu J, Dai G, Huang B. Fabrication and characterization of visible-light-driven plasmonic photocatalyst Ag/AgCl/TiO_2 nanotube arrays. *J Phys Chem C*. 2009;113(40):16394–401.

52 Primo A, Corma A, García H. Titania supported gold nanoparticles as photocatalyst. *Phys Chem Chem Phys*. 2011;13(3):886–910.

53 Awazu K, Fujimaki M, Rockstuhl C, Tominaga J, Murakami H, Ohki Y, et al. A plasmonic photocatalyst consisting of silver nanoparticles embedded in titanium dioxide. *J Am Chem Soc*. 2008;130(5):1676–80.

54 Chen X, Liu L, Huang F. Black titanium dioxide (TiO_2) nanomaterials. *Chem Soc Rev*. 2015;44(7):1861–85.

55 Li H, Liu L, Cui W, Yang G, Fu X. Construction of black TiO_2 via surface hydrogenation and its excellent photocatalytic performance. *Appl Catal B Environ*. 2014; 150–151:288–93.

56 Liu N, Ha D, Lee J. Black TiO_2 nanotube arrays for visible light photoelectrochemical water splitting. *J Mater Chem A*. 2014;2(21):6790–95.

57 Taheri-Ledari R, Maleki A. Photocatalytic reduction of CO_2 by advanced nanostructured materials: Fundamentals, mechanisms, and recent advances. *Journal of Nanostructure in Chemistry*. 2020;10:1–24.

58 Low J, Yu J, Jaroniec M, Wageh S, Al-Ghamdi AA. Heterojunction photocatalysts. *Adv Mater*. 2017;29(20):1601694.

59 Sharma K, Dutta V, Sharma S, Raizada P, Hosseini-Bandegharaei A, Thakur VK. Fabrication of ZnO/Fe_2O_3 heterojunction nanocomposite for enhanced photocatalytic CO_2 reduction under visible light. *Journal of Materials Science: Materials in Electronics*. 2019;30(12):11320–11332.

60 Li X, Yu J, Wageh S, Al-Ghamdi AA, Xie J. Graphene in photocatalysis: a review. *Small*. 2016;12(48):6640–96.

61 Ge M, Cao C, Huang J, Li S, Chen Z, Zhang KQ, et al. A review of one-dimensional TiO_2 nanostructured materials for environmental and energy applications. *J Mater Chem A*. 2016;4(18):6772–801.

62 Xie X, Li Y, Liu ZQ, Haruta M, Shen W. Low-temperature oxidation of CO catalysed by Co_3O_4 nanorods. *Nature*. 2009;458:746–49.

63 Wu H, Zhang L, Gao Y, Diao P. Preparation of Co_3O_4 nanorods with different aspect ratios and their electrochemical properties. *J Mater Chem*. 2011;21(18):6805–10.

64 Yu J, Yang B, Cheng B. Noble metal-modified Co_3O_4 photocatalysts for photocatalytic CO_2 reduction. *Dalton Trans*. 2018;47(27):8954–64.

65 Zhang N, Xu YJ. Aggregation- and leaching-resistant noble metal nanoparticles loaded on TiO_2: one-step synthesis and high photocatalytic activity. *J Phys Chem C*. 2013;117(18):9194–201.

66 Zhang G, Lan ZA, Wang X. Conjugated polymers: catalysts for photocatalytic hydrogen evolution. *Angew Chem Int Ed*. 2016;55(51):15712–27.

67 Wang Q, Hisatomi T, Jia Q, Tokudome H, Zhong M, Wang C, et al. Scalable water splitting on particulate photocatalyst sheets with a solar-to-hydrogen energy conversion efficiency exceeding 1%. *Nat Mater*. 2016;15:611–15.
68 Luo W, Wang Z, Wan C, Li X, Zhang L, Zhang J. Preparation of nanoporous black TiO_2 by Mg reduction and its photocatalytic performance. *J Mater Chem A*. 2013;1(20):6735–41.
69 Yang C, Wang D, Liu W. Synthesis of black TiO_2 nanoparticles by hydrogen plasma-assisted method and their photocatalytic performance. *ACS Appl Mater Interfaces*. 2013;5(19):9691–96.
70 Yu H, Irie H, Hashimoto K. Conduction band energy level control of titanium dioxide for efficient hydrogen generation. *J Am Chem Soc*. 2010;132(19):6898–99.
71 Kumar SG, Devi LG. Review on modified TiO_2 photocatalysis under UV/visible light: selected results and related mechanisms on interfacial charge carrier transfer dynamics. *J Phys Chem A*. 2011;115(46):13211–41.
72 Wang Z, Li C, Domen K. Recent developments in heterogeneous photocatalysts for solar-driven overall water splitting. *Chem Soc Rev*. 2019;48(7):2109–25.
73 Low J, Jiang C, Cheng B, Wageh S, Al-Ghamdi AA, Yu J, A review of direct Z-scheme photocatalysts. *Small Methods*. 2017;1(5):1700080.
74 Li X, Yu J, Jaroniec M. Review on Z-scheme photocatalysts. *Chem Rev*. 2016;116(19):12256–318.

Index

Note: Page numbers in *italics* and **bold** refers to figures and tables respectively.

Metal Oxide Nanocatalysts for Sustainable Energy Production, First Edition. Edited by Abayomi Bamisaye et al.
Published 2026 by Wiley-VCH GmbH.

q

r

s